Jiří Jaromír Klemeš (†), Petar Sabev Varbanov, Sharifah Rafidah Wan Alwi,
Zainuddin Abdul Manan, Yee Van Fan, Hon Huin Chin
Sustainable Process Integration and Intensification

Also of interest

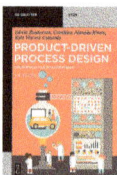

Jiří Jaromír Klemeš (†), Petar Sabev Varbanov,
Sharifah Rafidah Wan Alwi, Zainuddin Abdul
Manan, Yee Van Fan and Hon Huin Chin

Sustainable Process Integration and Intensification

Saving Energy, Water and Resources

3rd, Completey Revised and Extended Edition

DE GRUYTER

Authors

Jiří Jaromír Klemeš (†)
Formerly:
Sustainable Process Integration Laboratory (SPIL)
NETME Centre, Faculty of Mechanical Engineering
Brno University of Technology
Technická 2896/2, 616 00 Brno
Czech Republic

Petar Sabev Varbanov
Sustainable Process Integration Laboratory (SPIL)
NETME Centre, Faculty of Mechanical Engineering
Brno University of Technology
Technická 2896/2, 616 00 Brno
Czech Republic

Sharifah Rafidah Wan Alwi
Faculty of Chemical and Energy Engineering
Universiti Teknologi Malaysia
81310 UTM Johor Bahru
Johor
Malaysia

Zainuddin Abdul Manan
Faculty of Chemical and Energy Engineering
Universiti Teknologi Malaysia
81310 UTM Johor Bahru
Johor
Malaysia

Yee Van Fan
NETME Centre, Faculty of Mechanical Engineering
Brno University of Technology
Technická 2896/2, 616 00 Brno
Czech Republic

Hon Huin Chin
NETME Centre, Faculty of Mechanical Engineering
Brno University of Technology
Technická 2896/2, 616 00 Brno
Czech Republic

ISBN 978-3-11-078283-7
e-ISBN (PDF) 978-3-11-078298-1
e-ISBN (EPUB) 978-3-11-078300-1

Library of Congress Control Number: 2023935782

Bibliographic information published by the Deutsche Nationalbibliothek
The Deutsche Nationalbibliothek lists this publication in the Deutsche Nationalbibliografie;
detailed bibliographic data are available on the internet at http://dnb.dnb.de.

© 2023 Walter de Gruyter GmbH, Berlin/Boston
Cover image: Sustainable Process Integration Laboratory (SPIL), NETME Centre, Faculty of Mechanical
Engineering, Brno University of Technology
Typesetting: Integra Software Services Pvt. Ltd.
Printing and binding: CPI books GmbH, Leck

www.degruyter.com

Preface to the third edition

This is the third edition of *Sustainable Process Integration and Intensification: Saving Energy, Water and Resources*, which brings further extensions of the Process Integration methodology. They concern key problems such as multi-contaminant water integration, as well as industrial and urban symbiosis. The first two editions have been widely used by researchers, practitioners, as well as postgraduate and undergraduate students.

The resulting body of material helps the readers to advance in the fields of resource conservation and improving the sustainability performance of Process Systems. The mentioned extensions come to complement the fundament built by the previous two editions. The topics start with the founding Process Integration discipline – Heat Integration, moving to the fundamentals of Industrial Utility Systems, the choice of optimal steam pressure levels, Targeting for Total Site Heat Recovery with Total Site Heat Integration and Cogeneration, Data Extraction of Heat Sources and Demands, Total Site-Problem Table Algorithm (TS-PTA), Combined Heat and Power Generation (CHP) targeting, extended treatment of Power Integration and Carbon Emission Pinch Analysis, as well as a new chapter on Low Carbon Industrial Site Planning.

Further development of the book has considered feedback from the users, especially teachers of Process Integration courses who went through the book in very close detail. The authors would like to express their deepest thanks and appreciation to Dr Xia Liu (刘霞), presently with SINOPEC Shanghai Research Institute of Petrochemical Technology, Shanghai, China. During her stay in Europe, she has provided comprehensive feedback, and she should be very much praised for her contribution. Special and sincere thanks are also expressed to Ms Ting Pan (潘婷), a PhD Student at Brno University of Technology, Brno, the Czech Republic, for the thorough check of all numerical examples in the book.

In the years after the publication of the second edition, the field of Process Integration and Intensification kept growing even further, acquiring new methods for performance targeting and network design in fields previously thought as impossible for treatment by Pinch Analysis, especially Water/Mass Integration of multiple contaminant systems. The authors believe that the Process Integration methodology growth will pave the way for a community-driven effort of scientists and engineers of achieving substantial improvement in the sustainability of current societies in all its dimensions – environmental, economic, and social. The contribution of this book is to keep being an indispensable source of know-how to enable users to apply Process Integration techniques. Similarly, it is our hope that this book will be a useful source of reference for the users on where and how we should direct the further development.

https://doi.org/10.1515/9783110782981-202

The authors would therefore be most grateful to the readers and users for their feedbacks, comments, and suggestions.

Thank you for acquiring and using this book.

Jiří Jaromír Klemeš

Petar Sabev Varbanov
varbanov@fme.vutbr.cz

Yee Van Fan
fan@fme.vutbr.cz

Sharifah Rafidah Wan Alwi
syarifah@utm.my

Zainuddin Abdul Manan
dr.zain@utm.my

Hon Huin Chin
chin@fme.vutbr.cz

Preface to the second edition

The first edition of *Process Integration and Intensification: Saving Energy, Water and Resources* was published in 2014.

The book has been widely used by researchers, practitioners, as well as postgraduate and undergraduate students. As the methodology of Process Integration and Intensification is continuously developing, the authors, with the encouragement of the publisher, have decided to provide an improved and extended edition of the book.

Further development of the book has considered feedbacks from the users, especially teachers of Process Integration courses who went through the book in very close details. The authors would like to express their deepest thanks and appreciation to Dr Xia Liu (刘霞), presently with SINOPEC Shanghai Research Institute of Petrochemical Technology, Shanghai, China. During her stay in Europe, she has provided very comprehensive feedback, and she should be very much praised for her contribution.

The current edition includes some extensions and very novel developments of Process Integration methodology. The added topics include the choice of optimal steam pressure levels, Targeting for Total Site Heat Recovery with Total Site Heat Integration and Cogeneration, Data Extraction of Heat Sources and Demands, Total Site-Problem Table Algorithm (TS-PTA), Combined Heat and Power Generation (CHP) targeting, extended treatment of Power Integration and Carbon Emission Pinch Analysis, as well as a new chapter on Low Carbon Industrial Site Planning.

Process Integration and Intensification has been developing rapidly, covering novel features and extending implementations into new frontiers of resource utilisation. We believe that, together, we could continue to develop the methodology as well as this book into a helpful tool and source of know-how to enable users to apply Process Integration techniques. Similarly, it is our hope that this book will be a useful source of reference for the users on where and how we should direct the further development. The authors would therefore be most grateful to the readers and users for their feedbacks, comments, and suggestions.

Thank you for acquiring and using this book.

Jiří Jaromír Klemeš
klemes@fme.vutbr.cz

Petar Sabev Varbanov
varbanov@fme.vutbr.cz

Sharifah Rafidah Wan Alwi
syarifah@utm.my

Zainuddin Abdul Manan
dr.zain@utm.my

https://doi.org/10.1515/9783110782981-203

Preface

Process Integration and Pinch Technology came to eminence in the late 1970s and early 1980s and provided a phase change in the design of chemical processes, more especially in relation to energy use and energy efficiency. The central concept of this new approach was to provide targets for energy use in a chemical or related process, prior to the design required to achieve those targets. This new approach was not only elegant, but was rigorous in its thermodynamic basis, and provided a quick and systematic design of heat recovery in energy using processes.

The Centre for Process Integration in the University of Manchester Institute of Science and Technology (UMIST, now part of the University of Manchester) became the focus of this new technology, which expanded to include efficient use of raw materials, emissions reduction, and operability. The research and applied technology developed at Manchester was transferred quickly to industry and academia worldwide by the use of professional development, graduate, and undergraduate programmes.

Professor Jiří J. Klemeš has been at the leading edge of this technology since its early pioneering days in Manchester. His efforts in promoting the Process Integration were later boosted by many researchers across the globe including Prof Petar S. Varbanov, Prof Sharifah Wan Alwi, and Prof Zainuddin Abdul Manan who were themselves graduates of the Centre for Process Integration at UMIST, UK.

Their contributions to the technology and its application are brought together in this exclusive book, which not only clearly examines and evaluates the many elements of the technology, but provides clear and concise examples of their application. He and his co-authors have provided all the elements required for complete understanding of the basic concepts in heat recovery and water minimisation in chemical and related processes, and followed these with carefully selected and developed problems and solutions in order to ensure that the concepts delivered can be applied. This book then extends this targeting to the design of heat recovery and water recovery networks, and again provides carefully selected problems and solutions.

Process Integration and Process Intensification are continuing to expand and evolve. The authors have included a final chapter which provides comprehensive and extensive insights into many of these new developments. Also included are sources of materials which can be easily accessed for those who wish to further enhance their knowledge of this important area of technology.

April 2014

Simon Perry
MSc Programme Director, Centre for Process Integration,
The University of Manchester, UK

https://doi.org/10.1515/9783110782981-204

Contents

1 Process Integration and Intensification: an introduction

Numerous studies have been performed on improving the efficiencies of the supply and utilisation of energy, water, and other resources while simultaneously reducing pollutant emissions. The research and development in the area have become even stronger in the last five years since the Second Edition. This vital task is the focus of this textbook. It has been estimated that the majority of industrial plants throughout the world use up to 50% more energy than necessary (Alfa Laval, 2011). Despite the achievements in the last decades in improving energy and resource efficiency, world energy use still features an upward trend with no sign of slowing down (Ritchie et al., 2022) – even accounting for the temporary fluctuations caused by the COVID-19 pandemic (Jiang et al., 2021).

Usually, reducing resource consumption is achieved by increasing the internal recycling and reuse of energy and material streams. Projects for improving resource efficiencies can be very beneficial and also potentially improve the public perceptions of companies. Motivating, launching, and carrying out such projects, however, involve proper optimisation based on adequate process models.

As a response to these industrial and societal requirements, considerable research effort has been targeted towards Process Integration (PI) and Process Intensification (PIs). After a short assessment of these advanced engineering approaches, this handbook makes an attempt to describe the methodology that can lead a potential user through the introductory steps. This introduction provides a short overview of the historical development, achievements, and future challenges. After the introduction, the text focuses on Process Integration as an important engineering tool that can be exploited to achieve the goals of Process Intensification.

1.1 Process Intensification

There have been several initiatives supporting development in this area. One of them is the Process Intensification Network – PIN (PINET, 2013). This network declared that Process Intensification (PIs) was originally conceived in ICI (at that time, the Imperial Chemical Industries plc) as "the reduction of process plant volumes by two to three orders of magnitude". PIs targeted at that time the reduction of capital cost, primarily by minimising equipment installation factors, which involve piping, support structures, etc. It has since become apparent that a rigorously pursued strategy of PIs has far wider benefits than mere CAPEX (Capital Expenditure funds used by a company to acquire or upgrade physical assets such as property, industrial buildings, or equipment) reduction. Its definition has accordingly been softened to include very significant plant

https://doi.org/10.1515/9783110782981-001

size reductions based upon revolutionary or "step-out" new technology. PIs is not a mere evolutionary "apple-polishing" exercise of incremental development.

The benefits of PIs have extended far beyond the CAPEX reductions envisaged years ago. The iron grip, which can now be imposed upon the fluid dynamic environment within a reactor, means that improved selectivities and conversions can usually be achieved.

However, the former ICI head of manufacturing technology, Roger Benson, said that after years of false starts, the time has come for Process Intensification (PIs) technologies to make a major breakthrough in the chemical and process industries (IChemE, 2011).

The PIN website (PINET, 2013) summarises the advantages as:
1. Better product quality;
2. Just-in-time manufacture becomes feasible with ultra-short residence times;
3. Distributed (rather than centralised) manufacture may become economical;
4. Lower waste levels reduce downstream purification cost and are conducive to "Green" manufacture;
5. Smaller inventories lead to improved intrinsic safety;
6. Better control of process irreversibilities can lead to lower energy use.

PIs can help companies and others meet all of these demands in the process industries and in other sectors. The PIN network helps companies to compete and helps researchers target successful research goals. HEXAG is an international association of organisations involved in the manufacture and use of heat exchangers. PIN and HEXAG (HEXAG and PIN, 2017) help students gain an awareness of PIs for future use in their employment.

A kind of bible of PIs has been published by Reay et al. (2008) that has a very recently updated version (Reay et al., 2013). It covers main issues such as PIs as compact and micro-heat exchangers; reactors; intensification of separation processes; mixing; application areas in petrochemicals and fine chemicals; off-shore processing; miscellaneous process industries; the built environment, electronics, and the home; specifying, manufacturing and operating PI plants.

Reay et al. (2013) correctly stated that the heat transfer engineer notes that "intensification" is analogous to "enhancement", and intensification is based to a substantial degree on active and, to a lesser extent, passive enhancement methods that are used widely in heat and mass transfer.

There has also been a Working Party (WP) on Process Intensification (EFCE, 2022b). Their mission statement declares:

> Process Intensification presents one of the most significant developments in chemical and process engineering of the past decennia. It attracts more and more attention of the chemical engineering community. Several international conferences, smaller symposia/workshops every year, books and a number of dedicated issues of professional journals are a clear proof of it. Process Intensification with its ambition and ability to make chemical processing plants substantially

smaller, simpler, more controllable, more selective and more energy-efficient, addresses the fundamental sustainability issues in process industry and presents the core element of Green Chemical Engineering. In many research centres throughout Europe and the world numerous PIs-oriented research programs are carried out. Process Intensification is taught at various courses and gradually enters the regular university curricula. In the UK and in the Netherlands national PIs networks have been operated for a number of years. A similar network is being formed in Germany (DECHEMA, 2013). Process Intensification plays an important role in the CEFIC's Technology Platform on Sustainable Chemistry (CEFIC, 2013). Process Intensification has now established its organisational position within the European Federation of Chemical Engineering.

The PIs WP of the EFCE (2022b) has been organising annual conferences under the title European Process Intensification Conference (EPIC, 2013), and the success story continues. The previous EPIC conferences were held in Copenhagen in 2007, Venice in 2009, and Manchester in 2011. The WP keeps organising events, as can be seen on their website, with the most recent to take place on 2 December 2022 (EFCE, 2022a).

1.2 Process Systems Engineering and Process Integration

PIs have been very much targeted at processing units. On the other hand, Process Integration (PI) has developed a methodology by which PIs can be very beneficial at the system level. PI has been an important part of Process Systems Engineering, which is handled by the Working Party of Computer-Aided Process Engineering of the European Federation of Chemical Engineering (CAPE-WP, 2022). It has been one of the longest-serving working parties. Its definition is as follows:

> Computer Aided Process Engineering (CAPE) concerns the management of complexity in systems involving physical and chemical change. These systems usually involve many time and scale lengths and their characteristics, and the influences on them are uncertain. CAPE involves the study of approaches to the analysis, synthesis, and design of complex and uncertain process engineering systems and the development tools and techniques required for this. The tools enable process engineers to systematically develop products and processes across a wide range of domains involving chemical and physical change: from molecular, thermodynamic, and genetic phenomena to manufacturing processes and to related business processes. The Working Party promotes the development, study, and use of CAPE tools and techniques.
>
> The main characteristics are: Multiple scales, uncertainty, multidisciplinary, complexity.
>
> Core competencies: Modelling, synthesis, design, control, optimisation, problem-solving. Domains: manufacturing products and processes involving molecular change, sustainability, business processes, biological systems, energy, water.

The CAPE Working Party (CAPE-WP, 2022) main venue has been ESCAPE – European Symposium of Computer-Aided Process Engineering, whose first venue was already organised in 1992 in Helsingør, Denmark. The 2014 ESCAPE conferences were in Budapest (ESCAPE 24, 2014), Copenhagen in 2015, Portoroz in Slovenia in 2016, Barcelona in 2017 (CACE, 2017), and Graz in 2018 (CACE, 2018). The last venue was ESCAPE 32, organised in Toulouse – France, to be followed by ESCAPE 33 in Athens – Greece in 2023. The CAPE

WP has also collaborated with EURECHA to organise the CAPE Forum (2022) – with the last venue just organised by the University of Twente in The Netherlands.

1.3 Contributions to PIs and PI to energy and water saving

To save energy on a large scale and on a global basis, companies taking ownership and responsibility for new plants clearly need to question the energy efficiency of the equipment and layout recommended to them by the proposing contractors. Historically, designers and builders of process plants have not been asked or been paid to critically review the energy-efficiency options available to their clients, preferring to offer low-risk, easy-to-replicate, and therefore, easy-to-guarantee generic designs. For existing plants, reducing energy consumption can be more challenging than is the case for grassroots developments.

PI supporting Process Design, Integration, and Optimisation has been around for more than 40 years (Klemeš and Kravanja, 2013), and at present, the history has been reaching half of the century (Klemeš et al., 2022). Its ongoing development has been closely related to the development of Chemical, Power, and Environmental Engineering, the implementation of mathematical modelling, and the application of information technology. Its development has accelerated over the years, as its methodology has been able to provide answers and support on important issues regarding economic development – better utilisation and savings regarding energy, water, and other resources.

PIs field is further developing as a discipline under Process Engineering. A comprehensive review of the topic was presented by Keil (2018). The review compares the available definitions, areas of application, as well as a classification of its constitutive elements. Further aspects are discussed as energy-saving potential, GHG reduction potential and cost-saving.

A dedicated optimisation of a chemical process design has been investigated by Demirel et al. (2020). As an example process, ethylene glycol production was taken. The authors used two criteria in their optimisation – the minimisation of GHG Footprint and the rate of return on investment. The obtained results show that there is a limit to the effect of Process Intensification alone, and the best results (minimal GHG Footprint combined with a high return on investment) were achieved for a process design combining partial intensification with Heat Integration. While at first sight, this can be a counter-intuitive result, it should be noted that the optimisation involved the process of economic performance – accounting simultaneously for costs (operating and investment) and emissions. The major non-linearity and limitation to the effect of Process Intensification are reported to be caused by the investment costs. The considered process configurations can be summarised as follows (most to least beneficial, including emissions and Return on Investment – ROI):

(1) Partial Process Intensification of Reactive Distillation with Heat Integration (up to 80%/y ROI, ≈ 44–46 kt CO_2-eq/y);
(2) Partial Process Intensification of Reactive Distillation (up to 75%/y ROI, ≈ 52–58 kt CO_2-eq/y);
(3) Other configurations, with or without intensification, having performance close to the traditional non-intensified, non-reactive distillation design (up to 30–50%/y ROI, ≈ 51–53 kt CO_2-eq/y).

This brief analysis outlines the need to combine both PIs and PI, to achieve tangible results in energy saving and pollution reduction. The main lesson is that PIs on its own not always has the goals that lead to those savings. This is where the synergy with PI paves the way to achieving the benefits. PI is based on an analysis of the key performance indicators of the considered systems and sets appropriate performance targets.

1.4 What is Process Integration?

Process Integration (PI) is a family of methodologies for combining several parts of processes or whole processes to reduce the consumption of resources and prevent the release of harmful emissions to the environment. PI began mainly as Heat Integration (HI) stimulated by the energy crises of the 1970s. HI has been extensively used in the processing and power-generating industries for half a century. PI examines the potential for improving and optimising the Heat Exchange between heat sources and sinks, aimed at reducing the amount of external heating and cooling, together with the related cost and emissions. It provides a systematic design procedure for energy recovery networks.

HI (using Pinch Technology) has several definitions, almost invariably referring to the thermal combination of steady-state process streams or batch operations for achieving Heat Recovery via Heat Exchange. More broadly, the definition of PI, as adopted by the International Energy Agency (Gundersen, 2000), reads as follows:

> Systematic and General Methods for Designing Integrated Production Systems ranging from Individual Processes to Total Sites, with special emphasis on the Efficient Use of Energy and reducing Environmental Effects.

Reducing an external heating utility is usually accompanied by an equivalent reduction in the cooling utility demand (Linnhoff and Flower, 1978). This also tends to reduce the CO_2 emissions from the corresponding sites. Reduction of wastewater effluents based on Water (Mass) Integration can also lead to reduced freshwater intake (Wang and Smith, 1994), as demonstrated in industrial implementations elsewhere (Thevendiraraj et al., 2003).

1.5 A short history of the development of Process Integration

Process Integration (PI) roots go back as deep as 1972. The methodology has been pioneered by several research centres, originally in the UK (UMIST Manchester), Japan, and the US. Over the years, the methodology and research spread out over the world. There have been numerous publications devoted to PI for more than 40 years of development covering methodology – starting with the pivotal work on Heat Exchanger Networks (Linnhoff and Flower, 1978), the Pinch Design Method (Linnhoff and Hindmarsh, 1983), the Total Site Integration development (Klemeš et al., 1997) and its elaborations to Locally Integrated Energy Sectors (Perry et al., 2008) and the Total Site Heat Cascade with Time Slices (Varbanov and Klemeš, 2011). Attention has also been paid to industrial implementations, providing valuable feedback to researchers (Chew et al., 2013).

Several methodologies emerged in the late 1970s in response to these industrial and societal challenges. One of them was Process System Engineering (PSE) (Sargent, 1979), later extended again by Sargent (1983). Another methodology that received world prominence was Process Integration (PI) (Linnhoff and Flower, 1978). PI was first formulated in the first PI book by Linnhoff et al. (1982). Its development was further contributed to by a number of works from UMIST, Manchester, UK and other research groups.

It is remarkable that researchers have never lost interest in PI during the last nearly 50 years, and it has even been flourishing recently. HI has proved itself to have considerable potential for reducing the overall energy demand and emissions across a site, leading to a more effective and efficient site utility system. One of the first related works was by Hohmann (1971) in his PhD thesis at the University of Southern California. This work was the first to introduce systematic thermodynamics-based reasoning for evaluating the minimum energy requirements of a Heat Exchanger Network (HEN) synthesis problem. Various approaches dealing with the optimum HEN synthesis have since been published. Some of them have become very popular, such as the work by Ponton and Donaldson (1974). The comprehensive overview of HEN synthesis presented by Gundersen and Naess (1990) and the overview of process synthesis presented earlier by Nishida et al. (1977) provided considerable impetus for further research and development in this field – as can be witnessed in the more recent overview by Furman and Sahinidis (2002).

In some literature, such as (Gundersen, 2000), it is declared that the concept of HI based on Recovery Pinch was independently discovered by Hohmann (1971) and by two research groups: (i) the two-part paper Linnhoff and Flower (1978) and Flower and Linnhoff (1978), followed up by the Linnhoff PhD thesis (1979) and (ii) by the group around Umeda et al. (1978).

Gundersen (2000) also stated that Hohmann (1971) was the first to provide a systematic way of obtaining energy targets by using his Feasibility Table. Hohmann (1971) wrote his PhD, in which some basic principles were included. However, with the exception of a conference presentation (Hohmann and Lockhart, 1976), he has never extensively published the results in a way that would attract a wider audience. Moreover, a lesser-known important part leading to the Problem Table Algorithm was published at that time by MSc student, Bodo Linnhoff at ETH, Zürich (Linnhoff, 1972). During those times before the development of information technology, interactions among researchers were slower and more difficult. It was only possible to discover what other researchers were working on after the printed publications were made available.

During the remaining part of the 1970s, Bodo Linnhoff, who was at that time a PhD student at the University in Leeds, again perused and realised the potential of HI. The beginning was difficult, as his first paper (Linnhoff and Flower, 1978), which later became very highly cited, was nearly rejected by a leading journal of that time. Bodo's strong will and persistence prevailed in getting the idea published and off the ground. After the first paper's initial difficult birth, others followed: Flower and Linnhoff (1978) and Flower and Linnhoff (1979) followed smoothly.

The other group that produced an interesting contribution was from Japan at the Chiyoda Chemical Engineering & Construction Co., Ltd., Tsurumi, Yokohama. They published a series of works on HEN synthesis – Umeda et al. (1978, 1979) –, optimum water reallocation in a refinery – Takama et al. (1980) – and applications of the T-Q diagram for heat-integrated system synthesis – Itoh et al. (1986).

The publication of the first "red" book by Linnhoff et al. (1982) played a key role in the dissemination of HI. More recently, this book received a new Foreword (Linnhoff et al., 1994) and content update. This User Guide through Pinch Analysis provided insight into the more common process network design problems, including Heat-Exchanger Network synthesis, Heat Recovery Targeting, and selecting multiple utilities. As a spin-off from the Leeds centre, work originating in Central Europe has been published. Firstly, Klemeš and Ptáčník (1985) presented an attempt to computerise HI (Ptáčník and Klemeš, 1987), followed by HEN synthesis development and mathematical methods for HENs (Ptáčník and Klemeš, 1988).

The full-scale development and application of these methodologies have been pioneered by the Department of Process Integration, UMIST (now The University of Manchester) in the 1980s and 1990s. Among other earlier key publications was Linnhoff and Hindmarsh (1983), with presently more than 1,233 citations in SCOPUS, as of 24/10/2022, followed by a number of works dealing with extensions, e.g. the summary by Smith et al. (1995), a first updated Russian version (Smith et al., 2000) and more updated version Smith (2005). A specific food industry overview of HI was presented by Klemeš and Perry (2008) in a book edited by Klemeš et al. (2008), later by Klemeš et al. (2010), and more recently by Klemeš et al. (2013). Tan and Foo (2007) successfully applied the Pinch Analysis approach to carbon-constrained energy sector planning. More recently, Foo et al. (2009) applied the Cascade Analysis technique to carbon and footprint-constrained energy planning, while Wan Alwi and Manan (2008) provided a holistic framework for designing cost-effective minimum water utilisation networks.

Setting targets for HI has been widely publicised by Linnhoff et al. (1982), followed by two books by Smith (1995) and updated versions (Smith, 2005, 2016). Smith et al. (2000) presented a Russian version extended by a number of examples, and readers who can read Russian are encouraged to look at this source. The second edition of Linnhoff et al. (1994) was elaborated by Kemp (2007).

A very good analysis has been developed by Gundersen (2013) in his chapter of the PI Handbook edited by Klemeš (2013). Gundersen (2013) summarised the important elements of basic Pinch Analysis as:

1. Performance Targets ahead of design;
2. The Composite Curves representation can be used whenever an "amount" (such as heat) has a "quality" (such as temperature);
3. The fundamental Pinch Decomposition into a heat deficit region and a heat surplus region.

Heat Integration has been extended in various directions, such as Total Site Heat Integration (Dhole and Linnhoff, 1993), further extended by Klemeš et al. (1997), and to Locally Integrated Energy Systems (Perry et al. 2008). Mohammad Rozali et al. (2012) extended the Heat Integration on Total Site by including the power (electricity) in the game as well.

Direct implementation of Process Intensification by heat transfer intensification pioneered by Zhu et al. (2000) was extended by Kapil et al. (2012). It opened another exciting avenue for PI. Pan et al. (2011) implemented this methodology to retrofit industrial plants.

However, very important for all design and optimisation problems are the quality of the input data and their verification. Manenti et al. (2011) contributed to the data reconciliation, which is one of the key issues for accurate mining data representing the real plant. Klemeš and Varbanov (2010) provided one of the surprisingly few works published, so far, advising about the potential pitfalls of Process Integration.

The most obvious analogy to heat transfer is provided by mass transfer. In heat transfer, heat is transferred with temperature difference as the driving force. Similarly, in mass transfer, mass (or certain components) is transferred using concentration difference as the driving force. The corresponding Mass Pinch, developed by El-Halwagi and Manousiouthakis (1989), has a number of industrial applications whenever process streams are exchanging mass in a number of mass transfer units, such as absorbers, extractors, etc.

Mass Integration (MI) is a branch of PI providing a systematic methodology for optimising the global flow of mass within a process based on setting performance targets and for optimising the allocation, separation, and generation of streams and species. Within the context of wastewater minimisation, an MI problem involves transferring mass from rich process streams to lean process streams in order to achieve their target outlet concentrations, while simultaneously minimising the waste generation and the consumption of the utilities, including freshwater and external mass separating agents.

On this topic, El-Halwagi has produced three valuable books on PI; the first focuses on pollution prevention through PI (El-Halwagi, 1997); the second is a more general discussion on PI and includes MI (El-Halwagi, 2006); and the most recent book investigates sustainable design through PI (El-Halwagi, 2012).

The process network synthesis associated with these chemical properties cannot be addressed by conventional MI techniques, so another generic approach has been developed to deal with this problem, firstly by Shelley and El-Halwagi (2000), and further extended by El-Halwagi et al. (2003). For systems that are characterised by one key property, Kazantzi and El-Halwagi (2005) introduced a Pinch-based graphical targeting technique that establishes rigorous targets for minimum usage of fresh materials, maximum recycling, and minimum waste discharge.

One specific application of the Mass Pinch is in the area of Wastewater Minimisation, where the optimal use of water and wastewater is achieved through reuse, regeneration, and possibly recycling. The corresponding Water Pinch, developed by Wang and Smith (1994), can also be applied to the design of Distributed Effluent Treatment processes. The Universiti Teknologi Malaysia (UTM) group has delivered substantial work – as Wan Alwi and Manan (2006) developed SHARPS – a New Cost-Screening Technique To Attain a Cost-Effective Minimum Water Utilisation Network; Wan Alwi and Manan (2010), STEP – A new graphical tool for simultaneous targeting and design of a Heat Exchanger Network, and Wan Alwi et al. (2011) – a new graphical approach for simultaneous mass and energy minimisation. An interesting extension has been Process Integration for Resource Conservation (Foo, 2012).

In recent years, substantial Process Integration developments have been published. A visualisation development aiding in HEN retrofit has been introduced by Yong et al. (2015) called Shifted Retrofit Thermodynamic Grid Diagram, which assumed a single heat exchanger type over the network. It properly visualises heat exchanger networks and their key parameters, such as heat capacity flowrate, temperature, and temperature differences. A follow-up improvement – the Shifted Retrofit Thermodynamic Grid Diagram with the Shifted Temperature Range of Heat Exchangers was proposed by Wang et al. (2020), which enables the visualisation of identifying the potential retrofit plan of HEN with heat-exchanger type selection. Similarly, the visualisation of thermodynamic targets of HEN retrofit – the Energy Transfer Diagram, was developed by Bonhivers et al. (2014) and evolved in a paper presented at the PRES'17 conference (Walmsley et al., 2017). Further developments followed – such as:
-   Multi-contaminant material recycle/reuse (Chin et al., 2021c) – for treating site or regional symbiosis networks
-   Total Site Materials Integration – network design and headers targeting (Chin et al., 2021b)
-   Pinch-based Water Network Design with Water Mains (Chin et al., 2021a)
-   Water Scarcity Pinch Analysis (Jia et al., 2020) and regional water recyclability (Chin et al., 2022a) – for regional water management

– Integrated regional waste management (Fan et al., 2021a), Total EcoSite Integration (Fan et al., 2021b), Pinch Analysis for the valorisation of organic waste (Chee et al., 2022) – for maximising the benefits of regional waste management
– Plastic Pinch Analysis (Chin et al., 2022b) for targeting the performance of plastics management systems
– Energy Quality Pinch Analysis (Varbanov et al., 2022) for energy performance targeting of symbiosis networks

A number of excellent reviews have also been presented – by Klemeš and Kravanja (2013) – outlining the historical development of Heat Integration until 2014 and the interactions of Pinch Analysis with Mathematical Programming to achieve symbiotic methods. That was followed by a wider overview of the Process Integration developments (Klemeš et al., 2013) for the same period for wider applications – besides Heat Integration, also including Water and Mass Integration. The most recent comprehensive and updated review was published by a team from the Sustainable Process Integration Laboratory at the Brno University of Technology (Klemeš et al., 2018). Some of the most important updates in that review include the targeting of Total Sites with process-specific temperature differences, HEN heat transfer enhancement, Energy Transfer Diagram extension for HEN retrofits, new improved variants of the Grid Diagram, Exergy Pinch, regional planning of renewable energy supply, Power Pinch Analysis, and GHG emissions targeting. Reviews and information can also be found in the Handbook of Process Integration (Klemeš, 2013) as well as in its updated second edition (Klemeš, 2022). The vast majority of leading PI researchers contributed to that handbook with their unique expertise.

PI has been discussed at dedicated conferences for a quarter of a century – this is the series PRES – Process Integration, Modelling and Optimisation for Energy Saving and Pollution Reduction. The 20-year Jubilee Conference was at PRES'17 (2017), organised by Tianjin University and Hebei University of Technology, in Tianjin, China, 21–24 August 2017 (Klemeš et al, 2017). During the pandemic, two PRES conferences were organised. The 23rd Conference of Process Integration, Modelling and Optimisation for Energy Saving and Pollution Reduction was organised by Xi'an Jiaotong University from China as a fully virtual event during 17–21 August 2021. The conference featured 549 presentations and nearly 1,000 discussion threads. The hybrid 24th Conference of Process Integration, Modelling and Optimisation for Energy Saving and Pollution Reduction (PRES'21) was held in Brno, Czech Republic (NETME, 2021). Pandemic restrictions were still in force, and the conference needed to develop a novel hybrid and environmentally friendly style (Tao et al., 2021). Submitted were 509 abstracts, 397 manuscripts and 1,008 authors from 67 countries. Three Plenary Lectures, six Guests of Honour, 257 Oral (210 online and 47 on-site) presentations and 94 (82 and 12) posters were presented. The virtual conference received 1,685 Q & A in the COMET conference system, which had a significant positive impact on the spreading of the most recent research works and knowledge, setting up a good example for virtual conferences under the COVID-19 circumstance. Twenty-one session topics were involved in the conference. In addition to that, several special sessions were organised, including the SPIL Symposium, Transmission Enhancement and Energy Optimised Integration of Heat Exchangers in Petrochemical Industry Waste Heat Utilisation, Jubilee Session (Prof Urbaniec and Prof Taler), iNETME – International Cooperation and Projects in Engineering, Energy Saving and Pollution Reduction, UMIST – Sustainable Process Integration Excellence Roots and Horizon 2020: RESHeat: Renewable Energy heating of buildings (Klemeš et al., 2021). Overcoming the consequences of the COVID-19 pandemic, the last edition – the 25th PRES'22 was held in Bol – Croatia during 5–8 September 2022 (Klemeš et al., 2022), attracting over 400 delegates, of which more than 200 were on-site.

1.6 The aim and scope of this textbook, acknowledgements

This textbook is more focused on the detailed description of the selected principles to the extent that the reader should be able to solve both the illustration problems and also real-life industrial tasks.

It has been supplemented by a number of hands-on working sessions to allow readers to practise and deepen their understanding of the problem and to avoid potential pitfalls during the solution. The working sessions have been based on the authors' long-term academic teaching experience as well as on delivering further development courses for the industry worldwide.

The authors would like to express deep gratitude to all their colleagues at UMIST, The University of Manchester, Universiti Teknologi Malaysia, the University of Pannonia, Hungary, the University of Nottingham Malaysian Campus, De La Salle University, Manila, Philippines, the University of Maribor, Slovenia, the Hebei University of Technology in Tianjin, Fudan University Shanghai, the University of Waikato in Hamilton, Xi'an Jiaotong University; the South China University of Technology, Guangzhou, Xi'an Jiaotong-Liverpool University Suzhou, JiangSu, and Tianjin University in China. Also, collaboration with the Cracow University of Technology, Poland; Kharkiv National University "Kharkiv Polytechnic Institute", Ukraine; The University in Tomsk, Russian Federation, University POLITEHNICA Bucharest, Romania, Széchenyi István University Györ, Hungary has been acknowledged.

The third edition was created at SPIL, NETME Centre at the Brno University of Technology – VUT Brno – and many others – fellow staff members, postgraduate and even undergraduate students who, by using this material for their education, provided the most valuable feedback and comments, which substantially contributed to the testing, verification, and development of the handbook material. A special acknowledgement should be given to the delegates from the industry whose comments have been most valuable. The authors would like to mention at least two of those: MOL Hungarian Oil Company and PETRONAS Malaysia.

This book, besides reflecting the overall state-of-the-art in the area of Process Integration, is based on extensive research by the authors. Gratitude and appreciation are expressed for the following projects:

– **2022–2025** Bilateral Slovenian-Czech Project Sustainable Plastic Value Chain to Support a Circular Economy Transition Bilateral Lead Agency Project, Reg # 21-45726L, via Brno University of Technology (CZ)
– **2020–2024** Renewable Energy System for Residential Building Heating and Electricity Production (RESHeat), Horizon 2020 Grant Agreement 956255, started December 2020, via Brno University of Technology (CZ)
– **2019–2021** Project LTACH19033 "Transmission Enhancement and Energy Optimised Integration of Heat Exchangers in Petrochemical Industry Waste Heat Utilisation", under the bilateral collaboration of the Czech Republic and the People's Republic of China (partners Xi'an Jiaotong University and Sinopec Research Institute Shanghai; SPIL VUT, Brno University of Technology and EVECO sro, Brno), programme INTER-EXCELLENCE, INTER-ACTION of the Czech Ministry of Education, Youth and Sports; and the National Key Research and Development Program of China (2018YFE0108900).
– **2017–2022** Sustainable Process Integration Laboratory (SPIL), NETME Centre, Vysoké učení technické v Brně, Fakulta strojního inženýrství, project No. CZ.02.1.01/0.0/0.0/15_003/0000456 funded by EU "CZ Operational Programme Research, Development and Education", Priority 1: Strengthening capacity for quality research, via Brno University of Technology (CZ)
– **2013–2015** The Hungarian State and the European Union under the project TAMOP-4.2.2.A-11/1/KONV-2012-0072 TÁMOP-4.2.2.A-11/1/KONV-2012-0072 "Energia ellátó és hasznosító rendszerek korszerűsítésének és hatékonyabb üzemeltetésének tervezése és optimalizálása megújuló energiaforrások és infokommunikációs technológiák felhasználásával", via University of Pannonia (HU)
– **2013–2015** The Hungarian State and the European Union under the project TÁMOP-4.1.1.C-12/1/KONV-2012-0017 "Green Energy" – Cooperation of the higher education sector for the development of green economy in the area of energetics, via University of Pannonia (HU)
– **2012–2015** Efficient Energy Integrated Solutions for Manufacturing Industries (EFENIS), via University of Pannonia (HU)
– **2012–2016** Distributed Knowledge-Based Energy Saving Networks (DISKNET), via University of Pannonia (HU)

References

Bonhivers J-C, Korbel M, Sorin M, Savulescu L, Stuart P R. (2014). Energy Transfer Diagram for Improving Integration of Industrial Systems, Applied Thermal Engineering, 63, 468–479. doi: 10.1016/j.applthermaleng.2013.10.046.

CEFIC. (2013). The European Chemical Industry Council. www.cefic.org, accessed 16/04/2013.

CACE. (2017). 27 European Symposium on Computer Aided Process Engineering, Barcelona, Spain, Computer Aided Chemical Engineering, Computer Aided Chemical Engineering, Vol. 40. ISBN: 978-0-444-63965-3.

CACE. (2018). 28 European Symposium on Computer Aided Process Engineering, Graz, Austria., Computer Aided Chemical Engineering, Vol. 43. ISBN: 978-0-444-64172.

CAPE Forum. (2022). CAPE Forum – Computer Aided Process Engineering (CAPE). <https://www.wp-cape.eu/index.php/cape-forum/>, accessed 25.10.2022.

CAPE-WP. (2017). Working Party of Computer Aided Process Engineering of European Federation of Chemical Engineering, 2022. <https://www.wp-cape.eu/>, accessed 25/10/2022.

Chee W C, Ho W S, Mah A X Y, Klemeš J J, Fan Y V, Bong C P C, Wong K Y, Hashim H, Wan Alwi S R, Muis Z. (2022). Maximising the Valorisation of Organic Waste Locally available via Carbon-to-nitrogen ratio Supply Composite Curve Shifting, Journ of Cleaner Production, 362, 132389. doi: 10.1016/j.jclepro.2022.132389.

Chew K H, Klemeš J, Wan Alwi S R, Manan Z A. (2013). Industrial Implementation Issues of Total Site Heat Integration, Applied Thermal Engineering, 61, 17–25.

Chin H H, Jia X, Varbanov P S, Klemeš J J, Liu Z-Y. (2021). Internal and Total Site Water Network Design with Water Mains using Pinch-Based And Optimization Approaches, ACS Sustainable Chemistry and Engineering, 9, 6639–6658. doi: 10.1021/acssuschemeng.1c00183.

Chin H H, Varbanov P S, Klemeš J J, Wan Alwi S R. (2021b). Total Site Material Recycling Network Design and Headers Targeting Framework with Minimal Cross-Plant Source Transfer, Computers & Chemical Engineering, 151, 107364. doi: 10.1016/j.compchemeng.2021.107364.

Chin H H, Varbanov P S, Klemeš J J, Tan R R. (2022). Accounting for Regional Water Recyclability or Scarcity using Machine Learning and Pinch Analysis, Journal of Cleaner Production, 368, 133260. doi: 10.1016/j.jclepro.2022.133260.

Chin H H, Varbanov P S, You F, Sher F, Klemeš J J. (2022). Plastic Circular Economy Framework using Hybrid Machine Learning and Pinch Analysis, Resources, Conservation & Recycling, 184, 106387. doi: 10.1016/j.resconrec.2022.106387.

Chin H H, Varbanov P S, Liew P Y, Klemeš J J. (2021c). Pinch-based Targeting Methodology for Multi-contaminant Material Recycle/Reuse, Chemical Engineering Science, 230, 116129. doi: 10.1016/j.ces.2020.116129.

DECHEMA. (2013). Gesellschaft für Chemische Technik und Biotechnologie e.V. (Society for Chemical Engineering and Biotechnology). www.dechema.de/en/start_en.html, accessed 12/04/2013

Demirel S E, Li J, El-Halwagi M, Hasan M M F. (2020). Sustainable Process Intensification Using Building Blocks, ACS Sustainable Chemistry & Engineering, 8, 17664–17679. doi: 10.1021/acssuschemeng.0c04590.

Dhole V R, Linnhoff B. (1993). Total Site Targets for Fuel, Co-generation, Emissions and Cooling, Computers & Chemical Engineering, 17(Suppl.), 101–109.

EFCE. (2022a). EFCE Spotlight Talks – WP on Process Intensification. Lifelong Learning of Process Intensification for an Innovative industry. <https://efce.info/Spotlight_Talks/_/EFCE%20Spotlight%20Talks%20-%20WP%20Process%20Intensification.pdf>, accessed 25.10.2022.

EFCE. (2022b). European Federation of Chemical Engineering – Process Intensification. <https://efce.info/WP_PI.html>, accessed 25.10.2022.

El-Halwagi M M. (1997). Pollution Prevention through Process Integration: Systematic Design Tools. Academic Press, San Diego, USA.

El-Halwagi M M. (2012). Sustainable Design through Process Integration – Fundamentals and Applications to Industrial Pollution Prevention, Resource Conservation, and Profitability Enhancement. Elsevier, Amsterdam, The Netherlands, www.knovel.com/web/portal/browse/display?_EXT_KNOVEL_DISPLAY_ bookid$=$5170&VerticalID$=$0, accessed 01/07/2013.

El-Halwagi M M, Manousiouthakis V. (1989). Synthesis of Mass-Exchange Networks, American Institute of Chemical Engineering Journal, 35, 1233–1244. doi: 10.1002/aic.690350802.

El-Halwagi M M. (2006). Process Integration. Academic Press, Amsterdam, Netherlands.

El-Halwagi M M, Gabriel F, Harell D. (2003). Rigorous Graphical Targeting for Resource Conservation via Material Recycle/Reuse Networks, Industrial & Engineering Chemistry Research, 42(19), 4319–4328.

EPIC. (2013). European Process Intensification Conference. www.chemistryviews.org/details/event/ 2502531/EPIC_2013_-_European_ProcessIntensification_Conference_EPIC_2013.html, accessed 06/07/ 2013.

ESCAPE 24. (2013). European Symposium on Computer Aided Process Engineering, Budapest, Hungary. www.escape24.mke.org.hu/home.html, accessed 17/04/2014.

Fan Y V, Jiang P, Klemeš J J, Liew P Y, Lee C T. (2021a). Integrated Regional Waste Management to Minimise the Environmental Footprints in Circular Economy Transition, Resources, Conservation & Recycling, 168, 105292. doi: 10.1016/j.resconrec.2020.105292.

Fan Y V, Varbanov P S, Klemeš J J, Romanenko S V. (2021b). Urban and Industrial Symbiosis for Circular Economy: Total EcoSite Integration, Journal of Environmental Management, 279, 111829. doi: 10.1016/ j.jenvman.2020.111829.

Flower J R, Linnhoff B. (1978). Synthesis of Heat Exchanger Network. 2. Evolutionary Generation of Networks with Various Criteria of Optimality, AIChE Journal, 24(4), 642–654.

Flower J R, Linnhoff B. (1979). Thermodynamic Analysis in the Design of Process Networks, Computers & Chemical Engineering, 3(1–4), 283–291.

Foo D C Y. (2012). Process Integration for Resource Conservation. CRC Press, Boca Raton, Florida, USA.

Foo D C Y. (2009). State-of-the-art Review of Pinch Analysis Techniques for Water Network Synthesis, Industrial & Engineering Chemistry Research, 48(11), 5125–5159.

Furman K C, Sahinidis N V. (2002). A Critical Review and Annotated Bibliography for Heat Exchanger Network Synthesis in the 20th Century, Industrial & Engineering Chemistry Research, 41(10), 2335–2370.

Gundersen T. (2000). A Process Integration Primer – Implementing Agreement on Process Integration. International Energy Agency, SINTEF Energy Research, Trondheim, Norway.

Gundersen T. (2013). Heat Integration: Targets and Heat Exchanger Network Design, Chapter 4. In: Klemeš J J (ed.), Handbook of Process Integration (PI): Minimisation of Energy and Water Use, Waste and Emissions. Woodhead Publishing/Elsevier, Cambridge, United Kingdom, 129–167.

Gundersen T, Naess L. (1990). The Synthesis of Cost Optimal Heat Exchanger Networks: An Industrial Review of the State of the Art, Heat Recovery Systems and CHP, 10(4), 301–328.

HEXAG and PIN. (2017). www.hexag.org/index.html, accessed 14/06/2017.

Hohmann E, Lockhart F. (1976). Optimum Heat Exchangers Network Synthesis. In: Proceedings of the American Institute of Chemical Engineers. American Institute of Chemical Engineers, Atlantic City, NJ, USA.

Hohmann E C (1971). Optimum Networks for Heat Exchange. PhD thesis, University of Southern California, Los Angeles, USA.

IChemE. (2011). Advanced Chemical Engineering Worldwide, News archive. 21 June 2011, www.icheme.org/ media_centre/news/2011/process%20intensification%20the%20time%\penalty\z@20is%20now.aspx#. UXBG-kpChuM, accessed 24/06/2013.

Itoh J, Shiroko K, Umeda T. (1986). Extensive Applications of the T-Q Diagram to Heat Integrated System Synthesis, Computers & Chemical Engineering, 10(1), 59–66.

Jia X, Klemeš J J, Alwi S R W, Varbanov P S. (2020). Regional Water Resources Assessment using Water Scarcity Pinch Analysis, Resources, Conservation & Recycling, 157, 104749. doi: 10.1016/j.resconrec.2020.104749.

Jiang P, Fan Y V, Klemeš J J. (2021). Impacts of COVID-19 on Energy Demand and Consumption: Challenges, Lessons and Emerging Opportunities, Applied Energy, 285, 116441.

Kapil A, Bulatov I, Smith R, Kim J K. (2012). Process Integration of Low Grade Heat in Process Industry with District Heating Networks, Energy, 44(1), 11–19.

Kazantzi V, El-Halwagi M M. (2005). Targeting Material Reuse via Property Integration, Chemical Engineering Progress, 101(8), 28–37.

Keil F J. (2018). Process Intensification, Reviews in Chemical Engineering, 34, 135–200. https://doi.org/10.1515/revce-2017-0085.

Kemp I C. (2007). Pinch Analysis and Process Integration. In: Linnhoff B, Townsend D W, Boland D, Hewitt G F, Thomas B E A, Guy A R, Marsland R. (authors of the first edition), A User Guide on Process Integration for Efficient Use of Energy. Elsevier, Amsterdam, 416 ps.

Klemeš J. (ed.). (2013). Handbook of Process Integration (PI), Woodhead Publishing Series in Energy No. 61. Woodhead Publishing Limited/Elsevier, Cambridge, UK & Philadelphia, USA.

Klemeš J J. (2022). Handbook of Process Integration (PI). In Handbook of Process Integration (PI) Minimisation of Energy and Water Use, Waste and Emissions, 2nd ed. Woodhead Publishing / Elsevier, Cambridge, United Kingdom. ISBN 978-0-12-823850-9, 1190 ps.

Klemeš J, Dhole V R, Raissi K, Perry S J, Puigjaner L. (1997). Targeting and Design Methodology for Reduction of Fuel, Power and CO_2 on Total Sites, Applied Thermal Engineering, 17(8–10), 993–1003.

Klemeš J J, Kravanja Z. (2013). Forty Years of Heat Integration: Pinch Analysis (PA) and Mathematical Programming (MP), Current Opinion in Chemical Engineering, Biotechnology and Bioprocess Engineering / Process Systems Engineering, 2, 461–474. https://doi.org/10.1016/j.coche.2013.10.003.

Klemeš J, Perry S J. (2008). Methods to Minimise Energy Use in Food Processing. In: Klemeš J, Smith R, Kim J K (eds.), Handbook of Water and Energy Management in Food Processing. Woodhead Publishing Limited, Cambridge, UK, pp 136–199.

Klemeš J, Ptáčník R. (1985). Computer-Aided Synthesis of Heat Exchange Network, Journal of Heat Recovery Systems, 5(5), 425–435.

Klemeš J, Smith R, Kim J K. (eds.). (2008). Handbook of Water and Energy Management in Food Processing, Woodhead Publishing Limited, Cambridge, UK.

Klemeš J, Varbanov P. (2010). Process Integration – Successful Implementation and Possible Pitfalls, Chemical Engineering Transaction, 21, 1369–1374.

Klemeš J J, Varbanov P S, Fan Y V, Seferlis P, Wang X-C, Jia X. (2021). Twenty-Four Years of PRES Conferences: Recent Past, Present and Future-Process Integration Towards Sustainability, Chemical Engineering Transaction, 88, 1–12. doi: 10.3303/CET2188001.

Klemeš J J, Varbanov P S, Fan Y V, Seferlis P, Wang X-C, Wang B-H. (2022). Silver Jubilee of PRES Conferences: Contributions to Process Integration Towards Sustainability, Chemical Engineering Transaction, 94, 1–12. doi: 10.3303/CET2294001.

Klemeš J J, Varbanov P S, Kravanja Z. (2013). Recent Developments in Process Integration, Chemical Engineering Research and Design, the 60th Anniversary of the European Federation of Chemical Engineering (EFCE), 91, 2037–2053. https://doi.org/10.1016/j.cherd.2013.08.019.

Klemeš J J, Varbanov P S, Fan Y V, Seferlis P, Wang X-C, Wang B-H. (2022). Silver Jubilee of PRES Conferences: Contributions to Process Integration Towards Sustainability, Chemical Engineering Transaction, 94, 1–12.

Klemeš J J, Varbanov P S, Lam H L. (2017). Twenty Years of PRES: Past, Present and Future – Process Integration Towards Sustainability, Chemical Engineering Transaction, 61, 1–24. doi: 10.3303/CET1761001.

Klemeš J J, Varbanov P S, Walmsley T G, Jia X. (2018). New Directions in the Implementation of Pinch Methodology (PM), Renewable and Sustainable, Energy Reviews, 98, 439–468. https://doi.org/10.1016/j.rser.2018.09.030

Klemeš J J, Friedler F, Bulatov I, Varbanov P S. (2010). Sustainability in the Process Industry – Integration and Optimization. McGraw-Hill, New York, USA.

Laval A. (2011). Don't think of it as waste – it's energy in waiting. https://local.alfalaval.com/en-gb/aboutus/news/Pages/WasteHeatRecovery.aspx, accessed 02/07/2013.

Linnhoff B (1972). Thermodynamic Analysis of the Cement Burning Process (Thermodynamische Analyse des Zementbrennprozesses), Diploma thesis, Abteilung IIIa, ETH Zurich (1972), in German.

Linnhoff B (1979). Thermodynamic Analysis in the Design of Process Networks, PhD Thesis, The University of Leeds, Leeds, United Kingdom.

Linnhoff B, Flower J R. (1978). Synthesis of Heat Exchanger Networks: I. Systematic Generation of Energy Optimal Networks, AIChE Journal, 24(4), 633–642.

Linnhoff B, Hindmarsh E. (1983). The Pinch Design Method for Heat Exchanger Networks, Chemical Engineering Science, 38, 745–763.

Linnhoff B, Townsend D W, Boland D, Hewitt G F, Thomas B E A, Guy A R, Marsland R H. (1982). A User Guide on Process Integration for the Efficient Use of Energy. IChemE (revised edition published in 1994), Rugby, U.K.

Manenti F, Grottoli M G, Pierucci S. (2011). Online Data Reconciliation with Poor-Redundancy Systems, Industrial & Engineering Chemistry Research, 50(24), 14105–14114.

Mohammad Rozali N E, Wan Alwi S R, Manan Z A, Klemeš J J, Hassan M Y. (2012). Process Integration Techniques for Optimal Design of Hybrid Power Systems, Applied Thermal Engineering. doi: 10.1016/j.applthermaleng.2012.12.038.

Nishida N, Liu Y A, Lapidus L. (1977). Studies in Chemical Process Design and Synthesis: III. A Simple and Practical Approach to the Optimal Synthesis of Heat Exchanger Networks, AIChE Journal, 23(1), 77–93.

Pan M, Bulatov I, Smith R, Kim J K. (2011). Novel Optimization Method for Retrofitting Heat Exchanger Networks with Intensified Heat Transfer, Computer Aided Chemical Engineering, 29, 1864–1868.

Perry S, Klemeš J, Bulatov I. (2008). Integrating Waste and Renewable Energy to Reduce the Carbon Footprint of Locally Integrated Energy Sectors, Energy, 33(10), 1489–1497.

PINET. (2013). Process Intensification Network. www.pinetwork.org, accessed 13/06/2013.

Ponton J W, Donaldson R A B. (1974). A Fast Method for the Synthesis of Optimal Heat Exchanger Networks, Chemical Engineering Science, 29, 2375–2377.

PRES – Process Integration, Modelling and Optimisation for Energy Saving and Pollution reduction (2013). www.confrence.pres.com, accessed 08/06/2013.

PRES '17 (2017). Conference Process Integration, Modelling and Optimisation for Energy Saving and Pollution Reduction. www.conferencepres.com, accessed 28/8/2017.

Ptáčník R, Klemeš J. (1987). Synthesis of Optimum Structure of Heat Exchange Networks, Theoretical Foundations of Chemical Engineering, 21(4), 488–499.

Ptáčník R, Klemeš J. (1988). An Application of Mathematical Methods in Heat-Exchange Network Synthesis, Computers & Chemical Engineering, 12(2/3), 231–235.

Reay D, Ramshaw C, Harvey A. (2008). Process Intensification: Engineering for Efficiency, Sustainability and Flexibility, Butterworth-Heinemann, UK.

Reay D, Ramshaw C, Harvey A. (2013). Process Intensification, Engineering for Efficiency, Sustainability and Flexibility, 2nd ed. Butterworth Heinemann, UK.

Ritchie H, Roser M, Rosado P. (2022). Energy Production and Consumption. Our World in Data, October. < https://ourworldindata.org/energy-production-consumption>, accessed 24/10/2022.

Sargent R W H. (1979). Flowsheeting, Computers & Chemical Engineering, 3(1–4), 17–20.

Sargent R W H. (1983). Computers in Chemical Engineering – Challenges and Constraints, AIChE Symposium Series, 79(235), 57–64.

Shelley M D, El-Halwagi M M. (2000). Component-less Design of Recovery and Allocation Systems: A functionality-based clustering approach, Computers & Chemical Engineering, 24(9–10), 2081–2091.

Smith S. (2005, 2016). Chemical Process: Design and Integration (1st edn. 2005, 2nd edn. 2016). John Wiley & Sons Ltd., Chichester, UK.

Smith R. (1995). Chemical Process Design. McGraw-Hill, New York, USA.

Smith R, Klemeš J, Tovazhnyansky L L, Kapustenko P A, Uliev L M. (2000). Foundations of Heat Processes Integration. NTU KhPI, Kharkiv, Ukraine, in Russian.

Takama N, Kuriyama T, Shiroko K, Umeda T. (1980). Optimal Water Allocation in a Petroleum Refinery, Computers & Chemical Engineering, 4(4), 251–258.

Tan R, Foo D C Y. (2007). Pinch Analysis Approach to Carbon-Constrained Energy Sector Planning, Energy, 32(8), 1422–1429.

Tao Y, Steckel D, Klemeš J J, You F. (2021). Trend Towards Virtual and Hybrid Conferences may be an Effective Climate Change Mitigation Strategy, Nature Communications, 12, 1, 7324. doi: 10.1038/s41467-021-27251-2.

Thevendiraraj S, Klemeš J, Paz D, Aso G, Cardenas G J. (2003). Water and Wastewater Minimisation on a Citrus Plant, Resources, Conservation and Recycling, International Journal of Sustainable Resource Management and Environmental Efficiency., 37(3), 227–250.

Umeda T, Harada T, Shiroko K A. (1979). Thermodynamic Approach to the Synthesis of Heat Integration Systems in Chemical Processes, Computers & Chemical Engineering, 3(1–4), 273–282.

Umeda T, Itoh J, Shiroko K. (1978). Heat Exchanger Systems Synthesis, Chemical Engineering Progress, 74 (7), 70–76.

Varbanov P S, Chin H H, Klemeš J J, Cucek L. (2022). Sustainability of a Plastic Recycling/Symbiosis Network via Energy Quality Pinch Analysis, Chemical Engineering Transaction, 94, 97–102. doi: 10.3303/CET2294016.

Varbanov P S, Klemeš J J. (2011). Integration and Management of Renewables into Total Sites with Variable Supply and Demand, Computers & Chemical Engineering, Energy Systems Engineering, 35, 1815–1826. doi: https://doi.org/10.1016/j.compchemeng.2011.02.009.

Walmsley M R W, Lal N S, Walmsley T G, Atkins M J. (2017). A Modified Energy Transfer Diagram for Heat Exchanger Network Retrofit Bridge Analysis, Chemical Engineering Transaction, 61, 907–912. doi: 10.3303/CET1761149.

Wan Alwi S R, Manan Z A. (2008). A Holistic Framework for Design of Cost-effective Minimum Water Utilisation Network, Journal of Environmental Management, 88, 219–252. doi: 10.1016/j.jenvman.2007.02.011.

Wan Alwi S R, Manan Z A. (2006). SHARPS – A New Cost-Screening Technique to Attain Cost-Effective Minimum Water Utilisation Network, American Institute of Chemical Engineering Journal, 11(52), 3981–3988.

Wan Alwi S R, Ismail A, Manan Z A, Handani Z B. (2011). A New Graphical Approach for Simultaneous Mass and Energy Minimisation, Applied Thermal Engineering, 31(6–7), 1021–1030.

Wan Alwi S R, Manan Z A. (2010). Step – A New Graphical Tool for Simultaneous Targeting and Design of a Heat Exchanger Network, Chemical Engineering Journal, 162(1), 106–121.

Wang B, Klemeš J J, Varbanov P S, Zeng M. (2020). An Extended Grid Diagram for Heat Exchanger Network Retrofit Considering Heat Exchanger Types, Energies, 13, 2656. doi: 10.3390/en13102656.

Wang Y P, Smith R. (1994). Wastewater Minimisation, Chemical Engineering Science, 49(7), 981–1006.

Yong J Y, Varbanov P S, Klemeš J J. (2015). Heat Exchanger Network Retrofit Supported by Extended Grid Diagram and Heat Path Development, Applied Thermal Engineering, 89, 1033–1045. doi: https://doi.org/10.1016/j.applthermaleng.2015.04.025.

Zhu X X, Zanfir M, Klemeš J. (2000). Heat Transfer Enhancement for Heat Exchanger Network Retrofit, Heat Transfer Engineering, 21(2), 7–18.

2 Setting energy targets and Heat Integration

Heat Recovery can provide heating and cooling to various processes optimally – minimising the use of hot and cold utilities by replacing them with recovered heat. This is widely practised in industry and has an extensive historical record. Systematic methods for Heat Recovery have emerged in the last decades, inspired by the 1970s oil crises. Heat Recovery may take various forms. Some examples include transferring heat between process streams, generating steam from higher temperature process waste heat, or preheating a service stream (air for a furnace, as well as air or feed water for a boiler) by using excess process heat.

This chapter describes a systematic methodology for analysing the options for Heat Recovery in processes using limiting values of utility demands called targets, based on the fundamental work by Linnhoff and Flower (1978). It also provides insights on how to combine one multistream process with energy-intensive process units or how to improve the processes internally using as a criterion the reduction of the utility targets. The main text is supplemented by working sessions for consolidating the knowledge and deepening the readers' understanding.

2.1 Introduction

Reducing the consumption of resources is typically achieved by increasing internal recycling and the reuse of energy and material streams to replace the intake of fresh resources and utilities. Projects for improving process resource efficiency can offer economic benefits and improve public perceptions of the company undertaking them. However, motivating, launching, and carrying out such projects require a combination of several elements. First, as a necessary condition, is the proper optimisation studies based on adequate models of the process plants. The sufficient condition is to provide an integrated conceptual framework where fundamentally sound concepts and efficient visualisation tools are used, combined with mathematical optimisation. These together provide the plant and company managers with the necessary assurance and confidence in the proposed engineering solutions.

Process Integration (PI) is a family of methodologies for combining several processes to reduce the consumption of resources and/or harmful emissions (Klemeš et al., 2018). A fundamental resource, which drives all processes and enables the delivery and processing of other resources, is energy. Energy is currently made available to industrial and business processes as a combination of streams – fuels, electricity, heating, and cooling.

This chapter introduces the fundamentals of Heat Integration – the founding discipline of Process Integration, which deals with the recovery and efficient reuse of heat. The thinking, the model, and the procedure are introduced – for obtaining Heat

https://doi.org/10.1515/9783110782981-002

Recovery Targets, Heat Exchange area targets, and the basics of designing Heat Recovery networks. This book should serve for the initial mastering of the methodology, and for this reason, some issues are simplified. However, they can be dealt with in detail when solving real-life industrial problems. This includes, e.g. fouling of heat exchangers (Gogenko et al., 2007).

2.1.1 Overall development of Heat Integration

The initial development and a brief history of Heat Integration – the first part of Process Integration – have been described in more detail in Chapter 1. It is now over 40 years from the first conception of the methodology until the most recent extensions and achievements. An overview can be found in Friedler (2009) and a more extended form in (Friedler, 2010). Further, deeper reviews following up on the extensive methodology development can be read in (Klemeš et al., 2013) with a wider Process Integration discussion on thermodynamic-based models and the use of Mathematical Programming. A more recent overview can be found in (Klemeš et al., 2018), which focuses on developments based on Pinch Analysis. The methodology presented in this chapter has also been extended for batch processes. Some more information can be found elsewhere, e.g. by Foo et al. (2008). A recommended case study has been published by Atkins et al. (2010). For a review on the topic, the reader can look up (Magege and Majozi, 2021).

2.1.2 Pinch Technology and targeting Heat Recovery: the thermodynamic roots

The book by Klemeš et al. (2010) includes a review of works tracing the development of the research on Heat Exchanger Networks. It is discussed in an even more extended form in the Process Integration Handbook (Klemeš et al., 2022). There was only mild interest in Heat Recovery and energy efficiency until the early 1970s, by which time just a few works in the field had appeared. But between the oil crises of 1973–1974 and 1979, there were significant advances made in Heat Integration. Although capital cost remained important, the major focus was on saving energy and reducing the related cost. It is exactly this focus that resulted in attention being paid to energy flows and to the energy quality represented by temperature. The result was the development of Pinch Technology, which is firmly based on the first and second laws of thermodynamics (Linnhoff and Flower, 1978).

In this way, Heat Exchanger Network (HEN) synthesis – one of the very important and common tasks of process design – has become the starting point for the Process Integration revolution in industrial systems design. HENs in the industry are used mainly to save on energy costs. For many years the HEN design methods relied mostly on heuristics, as necessitated by the large number of permutations in which the necessary heat exchangers could be arranged. (Masso and Rudd, 1969) was a pioneering work that defined the problem of HEN synthesis. The paper proposed an evolutionary synthesis procedure based on heuristics. A complete timeline and thorough bibliography of HEN design and optimisation works are provided by Furman and Sahinidis (2002). The paper covers many more details, including the earliest known HEN-related scientific article Ten Broeck (1944).

The discovery of the Heat Recovery Pinch concept (Linnhoff and Flower, 1978) was a critical step in the development of HEN synthesis. The main idea behind the formulated procedure was to obtain, prior to the core design steps, guidelines and targets for HEN performance. This procedure is possible, thanks to thermodynamics. The hot and cold streams for the process under consideration are combined to yield (1) a Hot Composite Curve, collectively representing the process heat sources (the hot streams) and (2) a Cold Composite Curve, representing in a similar way the process heat sinks (the cold streams). For a specified minimum allowed temperature difference ΔT_{min}, the two curves are combined in one plot (Fig. 2.1), providing a clear thermodynamic view of the Heat Recovery problem. This is a fundamental concept of problem abstraction – representing a complex process arrangement that seemingly has no orientation or gradients, with a simple diagram that has an explicit orientation with gradients that show, in a single view, all essential process characteristics related to heat exchange.

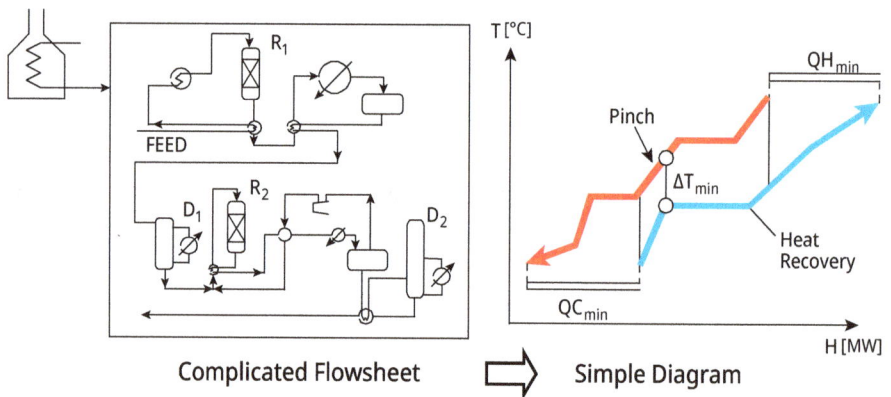

Fig. 2.1: Summary of Heat Recovery Targeting.

The overlap between the two Composite Curves on the Heat Exchange axis represents the Heat Recovery Target, i.e. the maximum amount of process heat that can be internally recovered. The vertical projection of the overlap indicates the temperature range where the maximum Heat Recovery should take place. The targets for external (utility) heating and cooling are represented by the non-overlapping segments of the Cold and Hot Composite Curves and their projections on the X-axis.

2.1.3 Supertargeting: full-fledged HEN targeting

After obtaining targets for the utility demands of a Heat Exchanger Network, the next logical step is to estimate the targets for capital and total costs. The capital cost of a Heat Exchanger Network is determined by many factors. The most significant one is the total heat transfer area and its distribution among the heat exchangers. Townsend and Linnhoff (1984) proposed a procedure for estimating HEN capital cost targets by using the Balanced Composite Curves, which are obtained by adding utilities to the Composite Curves obtained previously. Improvements to this procedure have been proposed, and they involve one or more of the following factors.

1. Obtaining more accurate surface area targets for HENs that feature non-uniform heat transfer coefficients: Colberg and Morari (1990) use NLP transshipment models and account for forbidden matches; Jegede and Polley (1992) explored the distribution of the capital cost among different heat exchanger types; Zhu et al. (1995) investigated integrated HEN targeting and synthesis within the context of the so-called "block decomposition" for HEN synthesis; non-uniform heat exchanger specifications; Serna-González et al. (2007) account also for different cost laws but do not put constraints on matches.
2. Accounting for practical implementation factors, such as construction materials, pressure ratings, and different heat exchanger types (Hall et al., 1990).
3. Accounting for additional constraints such as safety and prohibitive distance (Santos and Zemp, 2000).

In the years following those initial developments, the idea of capital cost targeting and supertargeting has been further extended, and improved methods have been provided. A few of the most notable extensions and improvements are:
- A numerical procedure with an enhanced algorithm by Sun et al. (2013). This procedure is based on the STEP method for targeting and synthesis of HENs by the same team (Wan Alwi and Manan, 2010). Using the advanced STEPS algorithm and allowing the use of different heat exchanger types within the same HEN, the authors achieved a nearly 50% reduction of the targeting imprecision. This method was further incorporated into an overall advanced procedure for HEN retrofit by Wang et al. (2021).
- The capital cost targeting and supertargeting principle have been propagated to the Total Site level – for the Heat Recovery part (Boldyryev et al., 2014) and for the steam turbines cogeneration part (Boldyryev et al., 2013).

2.1.4 Modifying the Pinch Idea for HEN retrofit

Zhu et al. (2000) proposed a heat transfer enhancement method for HEN retrofit design from which Heat Integration could benefit substantially. This approach is worthy of wider implementation, especially in the context of retrofit studies. There have been further developments in this direction: for example, Pan et al. (2013a) have developed an optimal HEN retrofit procedure for large-scale HENs, with a follow-up focusing on shell-and-tube heat exchangers (Pan et al., 2013b).

Heat Exchanger Network retrofit is a special case of optimisation. In retrofit problems, an existing network with existing heat exchangers that had been already paid for has to be accommodated as much as possible. This requirement substantially alters the economics of the problem compared with developing a new design.

2.1.5 Benefits of Process Integration

The Composite Curves plot is a visual tool that summarises the important energy-related properties of a process in a single view (Fig. 2.1). It was the recognition of the thermodynamic relationships and limitations in the underlying Heat Recovery problem that led to the development of the Pinch Design Method (Linnhoff and Hindmarsh, 1983). In addition, further applications have extended the PI approach to water minimisation as well as regional energy and emissions planning, financial planning, and batch processes.

The most important property of thermodynamically derived Heat Recovery Targets is that they cannot be improved upon by any real system. The Composite Curves play an important role in process design. HEN synthesis algorithms provide strict targets for maximum energy recovery. For process synthesis based on Mathematical Programming, the Composite Curves establish relevant lower bounds on utility requirements and capital cost, thereby narrowing the search space for the following superstructure construction and optimisation.

The preceding observation highlights an important characteristic of process optimisation problems, specifically those that involve process synthesis and design. By strategically obtaining key data about the system, it is possible to evaluate processes based on limited information before too much time or other resources are spent on more detailed analysis. This approach follows the logic of oil drilling projects: potential sites are first evaluated in terms of key preliminary indicators, and further studies or drilling commence only if the preliminary evaluations indicate that the revenues could justify further investment. The logic of this approach has been codified by Smith in his books on Process Integration (Smith, 2016). The paper by Daichendt and Grossmann (1997) integrates hierarchical decomposition and Mathematical Programming to solve process synthesis problems.

Process Integration has a direct impact on improving the sustainability of a given industrial process. All Process Integration techniques are geared toward reducing the intake of resources and minimising the release of harmful effluents, goals that are directly related to the corresponding footprints. Hence, employing Process Integration and approaching the targeted values will help minimise those footprints.

2.2 Pinch analysis for maximising energy efficiency

2.2.1 Introduction to Heat Exchange and Heat Recovery

Large amounts of thermal energy are used in the industry to perform heating. Examples of this can be found in crude oil preheating before distillation, preheating of feed flows to chemical reactors, and heat addition to carry out endothermic chemical reactions. Some processes, such as condensation, exothermal chemical reactions, and product finalisation, similarly require heat to be extracted, which results in process cooling. There are several options for utility heating; these include steam, hot mineral oils, and direct-fired heating. Steam is the most prevalent option because of its high specific heating value in the form of latent heat. Utility cooling options include water (used for moderate-temperature cooling when water is available), air (used when water is scarce or not economical to use), and refrigeration (when sub-ambient cooling is needed). Heat Recovery can be used to provide either heating or cooling to processes. Heat Recovery may take various forms: transferring heat between process streams, generating steam from higher temperature process waste heat, or preheating (air for a furnace, as well as air or feed water for a boiler).

Heat transfer takes place in heat exchangers, which can employ either direct mixing or indirect heat transfer via a wall. Direct Heat Exchange is also referred to as non-isothermal mixing because the temperatures of the streams being mixed are different. Mixing heat exchangers are efficient at transferring heat and usually have low capital costs. In most industries, the bulk of the heat exchange must occur without mixing the heat-exchanging streams. To exchange only heat while keeping fluids separate, surface heat exchangers are employed. In these devices, heat is exchanged through a dividing wall. Because of its high thermal efficiency, the counter-current stream arrangement is the most common with surface heat exchangers. Counter-current heat exchangers are assumed unless stated otherwise, to simplify the discussion. In terms of construction types, the traditional shell-and-tube heat exchanger is still the most common. However, plate-type heat exchangers are gaining increased attention (Gogenko et al., 2007). Their compactness, together with significant improvements in their resistance to leaking, has made them preferable in many cases (Klemeš et al., 2015).

2.2.1.1 Heat exchange matches

A hot process stream can supply heat to a cold one when paired with one or several consecutive heat exchangers. Each such pairing is referred to as a Heat Exchange Match. The form of the steady-state balance equations for Heat Exchange Matches that is most convenient for Heat Integration calculations is based on modelling a match as consisting of hot and cold parts, as shown in Fig. 2.2. The hot and cold part

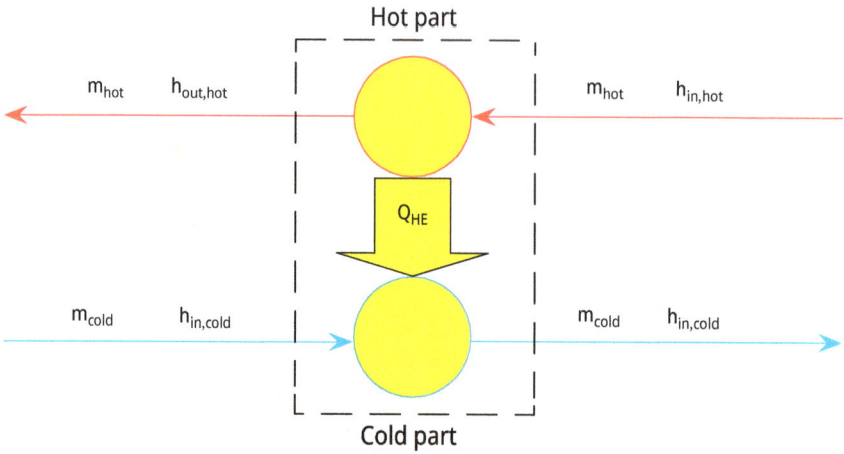

Fig. 2.2: Process flow diagram of a heat exchange match.

each have a simple, steady-state enthalpy balance that involves just one material stream and one heat transfer flow.

The main components of the model are: 1) calculations of the heat transfer flows accounted for by the enthalpy balances and 2) estimation of the necessary heat transfer area. For the latter, both the log-mean temperature difference and the overall heat transfer coefficient are used. The enthalpy balances of the hot and cold parts, and the equations for the heat transfer area estimation, may be written as follows:

$$Q_{HE} = m_{hot} \cdot (h_{in,\,hot} - h_{out,\,hot}), \tag{2.1}$$

$$Q_{HE} = m_{cold} \cdot (h_{out,\,cold} - h_{in,\,cold}), \tag{2.2}$$

$$Q_{HE} = U \cdot A \cdot \Delta T_{LM}, \tag{2.3}$$

where Q_{HE} (kW) is the heat flow across the whole heat exchanger, U (kW·m^{-2}·°C^{-1}) is the overall heat transfer coefficient, A (m^2) is the heat transfer area, and ΔT_{LM} (°C) is the logarithmic-mean temperature difference. More information can be found in Shah and Sekulić (2003) – on fundamentals of heat exchanger design, Klemeš et al. (2015) – on industrial plate heat exchangers, and Shilling et al. (2008) – containing general modelling and calculation.

2.2.1.2 Implementing heat exchange matches

Heat Exchange Matches are often viewed as identical to heat exchangers, but this is not always the case. A given Heat Exchange Match may be implemented by devices of different construction or by a combination of devices. For example, two counter-current heat exchangers in sequence may implement a single Heat Exchange Match.

The distinction between the concept of a Heat Exchange Match and its implementation via heat exchangers is important because of capital cost considerations.

2.2.2 Basic principles

2.2.2.1 Process Integration and Heat Integration

At the time of the conception of Process Integration, the attention was focused on reusing any waste heat generated on different sites. Each surface heat exchanger was described with only a few steady-state equations, and the thermal energy saved by reusing waste heat led to reductions in the expense of utility resources. This approach became popular under the names Heat Integration (HI) and Process Integration (PI). In this context, PI means integrating different processes to achieve energy savings. Engineers realised that integration could also reduce the consumption of other resources as well as the emission of pollutants. Heat and Process Integration came to be defined more widely in response to similar developments in water reuse and wastewater minimisation.

2.2.2.2 Hierarchy of process design

Process design has an inherent hierarchy that can be exploited for making design decisions more efficiently. This hierarchy may be represented by the so-called Onion Diagram (Linnhoff et al., 1982) and more extended (Linnhoff et al., 1994), as shown in Fig. 2.3(a). The design of an industrial process starts with the reactors or other key operating units (the Onion's core). This is supplemented and served by other parts of the process, such as the separation subsystem (the next layer) and the Heat Exchange Network (HEN) subsystem. The remaining heating and cooling duties, as well as the power demands, are handled by the utility system. There are further variations of the Onion Diagram within the Process Integration context. Fig. 2.3(b) shows an extended version used in the analysis of the environmental and resource implications of process systems and their energy supply.

2.2.2.3 Performance targets

The thermodynamic bounds on Heat Exchange can be used to estimate the utility usage and heat exchange area for a given Heat Recovery problem. The resulting estimates of the process performance are lower bounds on the utility demands and a lower bound on the required heat transfer area. These bounds are known as targets for the reason that those Heat Recovery estimates are achievable in practice and tend to minimise the total cost of the HEN being designed.

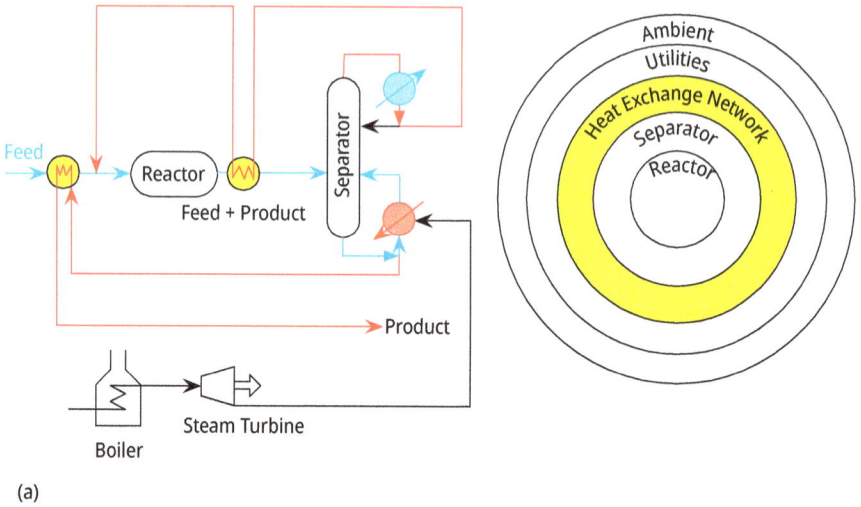

(a)

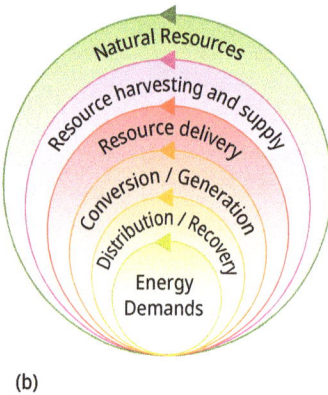

(b)

Fig. 2.3: The onion diagram (a) The version defined for Chemical Process Design. (b) An extended version for wider system boundaries (Seferlis et al., 2021).

2.2.2.4 Data extraction: Heat Recovery problem identification

For efficient Heat Recovery in industry, the relevant data must be identified and presented systematically. In the field of Heat Integration, this process is referred to as data extraction. The Heat Recovery problem data are extracted in several steps.

1. Inspect the general process flowsheet, which may contain Heat Recovery exchangers.
2. Remove the recovery heat exchangers and replace them with equivalent "virtual" heaters and coolers.
3. Lump together all consecutive heaters and coolers.
4. The resulting virtual heaters and coolers represent the net heating and cooling demands of the flowsheet streams.

5. The heating and cooling demands of the flowsheet streams are then listed in a tabular format, where each heating demand is referred to as a cold stream and, conversely, each cooling demand as a hot stream.

This procedure is best illustrated by an example. Figure 2.4 shows a process flowsheet involving two reactors and a distillation column. The process already incorporates two recovery heat exchangers. The utility heating demand of the process is $Q_H =$ 1,760 kW, and the utility cooling demand is $Q_C = 920$ kW.

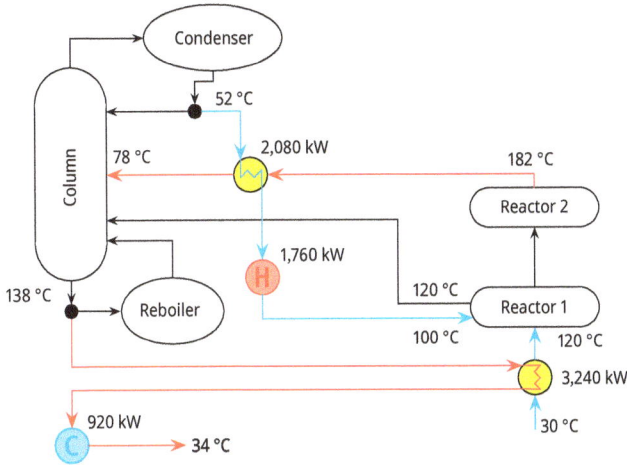

Fig. 2.4: Data extraction: an example process flowsheet.

The necessary thermal data have to be extracted from the initial flowsheet. Fig. 2.5 shows the flowsheet after applying to Fig. 2.4, steps 1 through 4 in the above list. The heating and cooling demands of the streams have been consolidated by removing the existing exchangers. Reboiler and condenser duties have been left out of the analysis for simplicity (although these duties would be retained in an actual study). It is assumed that any process cooling duty is available to match up with any heating duty.

Applying step 5 to the data in Fig. 2.5 produces the data set in Tab. 2.1. By convention, heating duties are positive, and cooling ones are negative. The subscripts S and T denote "supply" and "target" temperatures for the process streams.

The last column gives the heat capacity flowrate (CP). For streams that do not change phase (from liquid to gas or vice versa), CP is defined as the product of the specific heat capacity and the mass flow rate of the corresponding stream:

$$CP = m_{\text{stream}} \cdot C_{p,\,\text{stream}} \qquad (2.4)$$

The CP can also be calculated using the following simple equation:

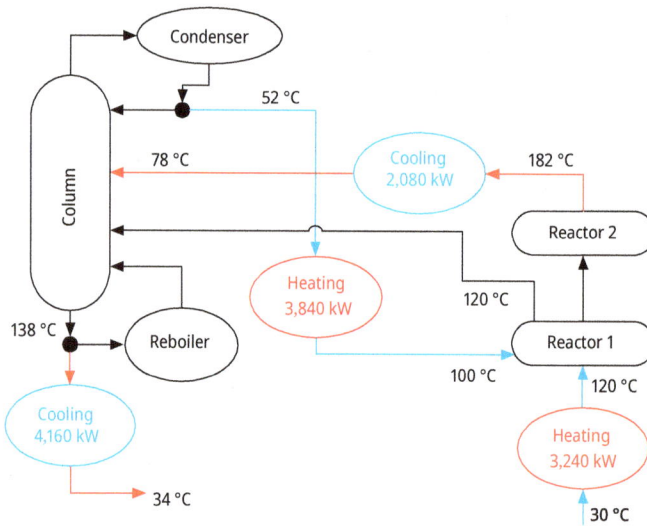

Fig. 2.5: Data extraction: heating and cooling demands.

Tab. 2.1: Data set for Heat Recovery analysis.

Stream	Type	T_S (°C)	T_T (°C)	ΔH (kW)	CP (kW/°C)
H1	Hot	182	78	−2,080	20
H2	Hot	138	34	−4,160	40
C3	Cold	52	100	3,840	80
C4	Cold	30	120	3,240	36

$$CP = \frac{\Delta H}{T_T - T_S} \tag{2.5}$$

When phase transition occurs, the latent heat is used instead of CP to calculate the stream duties.

2.2.2.5 Working session "Introduction to Heat Integration"

Assignment

Consider the flowsheet in Fig. 2.6. This is a modification of the example from the Data Extraction example in Fig. 2.5.

It features two reactors: R1 and R2, and a distillation column, C1. They are connected by piping, and the connections are characterised by temperature changes, necessitating heating and cooling. Focusing only on the heating and cooling requirements yields the

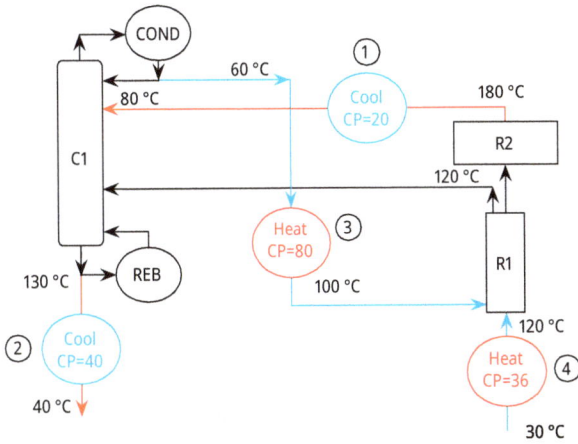

Fig. 2.6: Initial flowsheet for 2.2.2.5 Working session "Introduction to Heat Integration".

entries presented in Tab. 2.2. This Heat Recovery problem comes with a constraint of the minimum allowed temperature difference ΔT_{min} = 10 °C. Also given are two utilities:

− Steam (hot utility available) at 200 °C.
− Cooling water (CW, cold utility) is available between 25 °C and 30 °C.

Tab. 2.2: Heat Recovery problem for 2.2.2.5 working session "Introduction to Heat Integration".

Stream	T_S (°C)	T_T (°C)	CP (kW/°C)
1: Hot	180	80	20
2: Hot	130	40	40
3: Cold	60	100	80
4: Cold	30	120	36

The assignment is to design a network of steam heaters, water coolers, and exchangers. Use Heat Recovery in preference to utilities.

Solution

Evaluation of heat exchanger placement options
For an ad hoc design of a Heat Exchanger Network, the first step is to construct a Cross-Grid Diagram, as shown in Fig. 2.7, having the hot streams running from right to left and cold streams from bottom to top. The grid features points where the hot and the cold streams intersect.

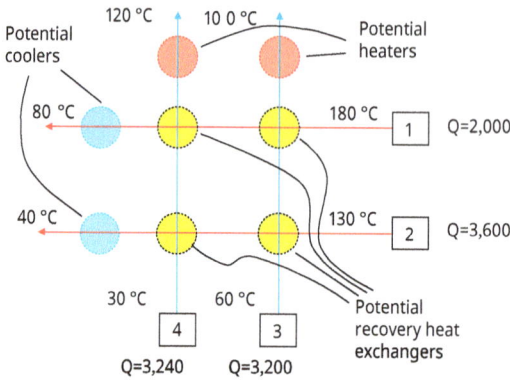

Fig. 2.7: Initial grid with placeholders for 2.2.2.5 working session "Introduction to Heat Integration".

The next step is to evaluate the overall heating and cooling demands of the process streams:

1. (Hot): $Q_1 = CP_1 \cdot (T_{S,1} - T_{T,1}) = 20 \cdot (180 - 80) = 2{,}000\,\text{kW}$ (2.6)

2. (Hot): $Q_2 = CP_2 \cdot (T_{S,2} - T_{T,2}) = 40 \cdot (130 - 40) = 3{,}600\,\text{kW}$ (2.7)

3. (Cold): $Q_3 = CP_3 \cdot (T_{T,3} - T_{S,3}) = 80 \cdot (100 - 60) = 3{,}200\,\text{kW}$ (2.8)

4. (Cold): $Q_4 = CP_4 \cdot (T_{T,4} - T_{S,4}) = 36 \cdot (120 - 30) = 3{,}240\,\text{kW}$ (2.9)

Further, one can evaluate the options for placing heat exchangers. Cooling of Stream 1 (hot) can be done by matching it with Stream 3 (cold), as shown in Fig. 2.8. It can be seen that the smaller temperature difference for this potential match is at the cold end, and it is 20 °C. In the direction of increasing the temperatures, the two T-H profiles diverge, which indicates that such a match is feasible. Stream 1 has a smaller load (2,000 kW). This is, therefore, selected as the heat exchanger duty: $Q = 2{,}000$ kW. Such a selection means that the final temperature of the hot stream of 80 °C will be reached, and its cooling requirement will be completely satisfied.

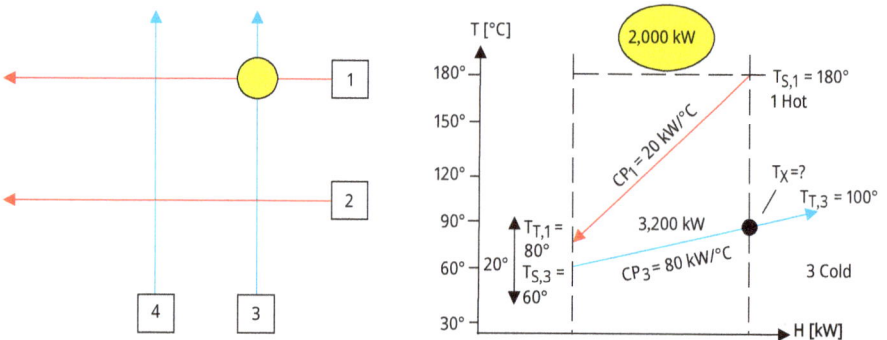

Fig. 2.8: Option 1: Matching streams 1 and 3 (2.2.2.5 working session "Introduction to Heat Integration").

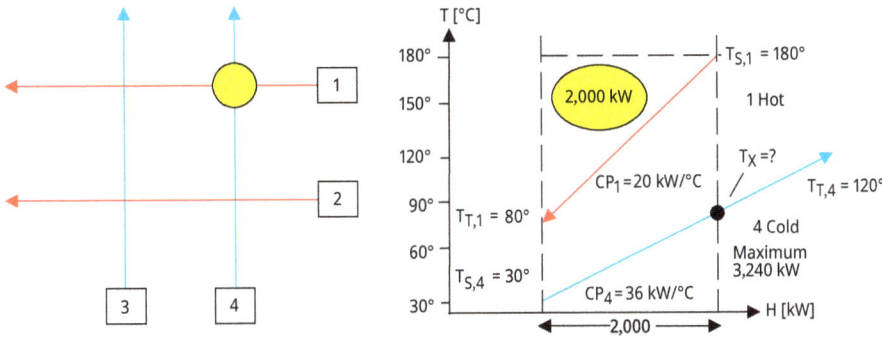

Fig. 2.9: Option 2: Matching streams 1 and 4 (2.2.2.5 working session "Introduction to Heat Integration").

The exit temperature T_X of Stream 3 will be then lower than the final one, and the heating requirement of Stream 3 will be satisfied only partially. The value of T_X (from Fig. 2.8) can be calculated as follows:

$$T_X = T_{S,3} + \frac{Q}{CP_3} = 60 + \frac{2{,}000}{80} = 60 + 25 = 85°C. \tag{2.10}$$

Option 2 involves matching Streams 1 (hot) and 4 (cold) (Fig. 2.9). In a similar analysis as for Option 1, the heat exchanger duty would be 2,000 kW, and Stream 4 will be satisfied only partially, reaching temperature 85.6 °C instead of the desired 120 °C.

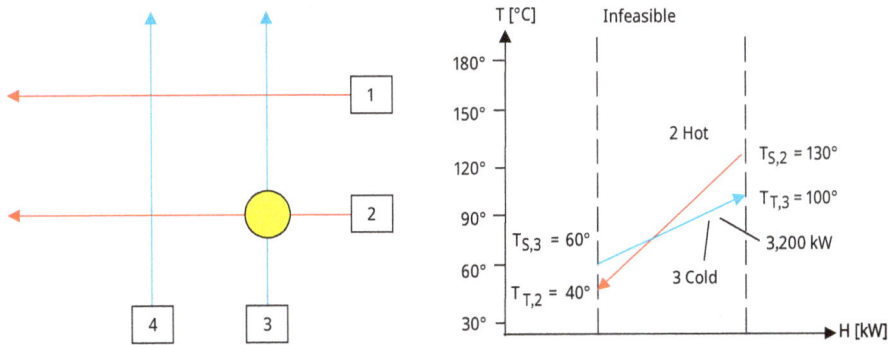

Fig. 2.10: Option 3: Matching streams 2 and 3 (2.2.2.5 working session "Introduction to Heat Integration").

The third option is to place a heat exchanger match between Streams 2 (hot) and 3 (cold), as shown in Fig. 2.10. Since Stream 2 (hot) has a smaller CP value than that of Stream 3 (cold), when starting to follow the profiles from right to left (from higher temperatures to lower), the T-H profiles of the two streams converge. This indicates that starting from a temperature difference of 30 °C, inside the heat exchanger, the temperature difference would become smaller and smaller, passing the lower bound of $\Delta T_{min} = 10$ °C and eventually zero. This is an infeasible condition, and such an option cannot be used.

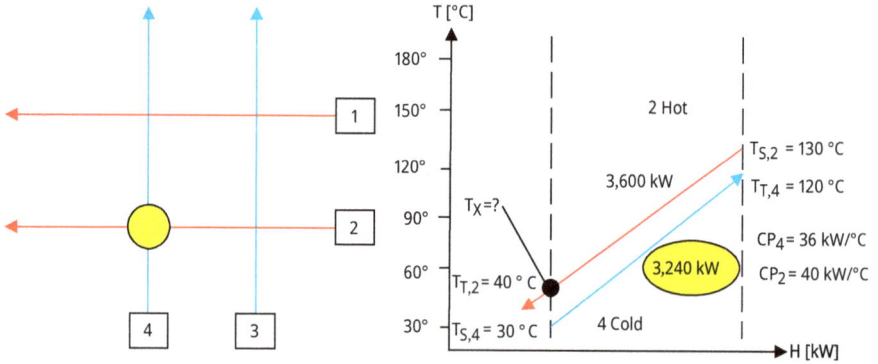

Fig. 2.11: Option 4: Matching streams 2 and 4 (2.2.2.5 working session "Introduction to Heat Integration").

Option 4 provides the final slot for a match: to match Streams 2 (hot) and 4 (cold); see Fig. 2.11. Stream 4 has the smaller load (3,240 kW), and this is selected as the heat exchanger duty: $Q = 3,240$ kW:

– The final temperature of Stream 4 (120 °C) would be reached.
– The exit temperature T_X of Stream 2 would be higher than the final one:

$$T_X = T_{S,2} - \frac{Q}{CP_2} = 130 - \frac{3,240}{40} = 130 - 81 = 49 \text{ °C} \tag{2.11}$$

Possible solutions

One possible solution is to select Options 1 and 4 for recovery heat exchange and satisfy the remaining heating requirement with a heater and the remaining cooling requirement with a cooler. The resulting network is shown in Fig. 2.12. The network in Fig. 2.12 is not the only solution. A different network allowing maximum Heat Integration is shown in Fig. 2.13.

Another possible network featuring an intermediate number of heat exchanger units can be seen in Fig. 2.14.

Another interesting solution can be obtained if the minimum allowed temperature difference is reduced to $\Delta T_{min} = 6.6$ °C. In this case, the maximum Heat Recovery eliminates the need for cooling water. The network of this solution is shown in Fig. 2.15.

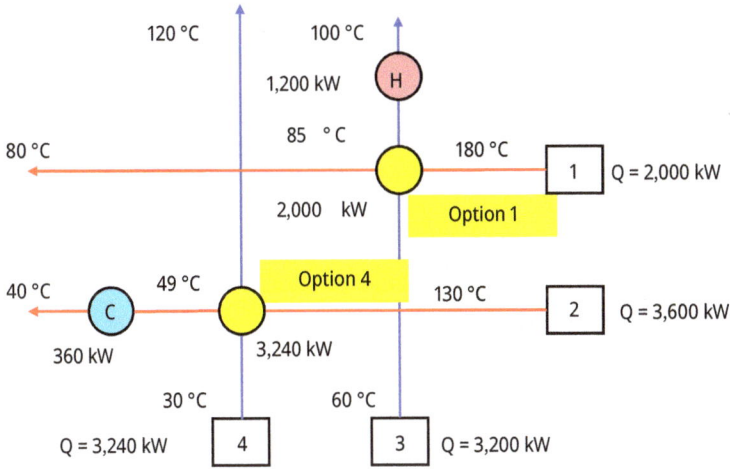

Fig. 2.12: One possible HEN (2.2.2.5 working session "Introduction to Heat Integration").

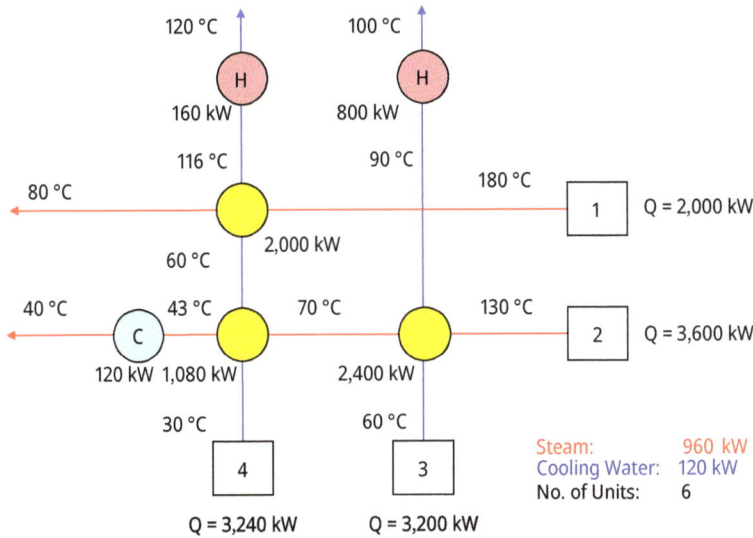

Fig. 2.13: HEN with maximum Heat Integration (2.2.2.5 working session "Introduction to Heat Integration").

2.2.3 Basic Pinch Technology

The main strategy of Pinch-based Process Integration is to identify the performance targets before starting the core process design activity. Following this strategy yields important clues and design guidelines. The most common hot utility is steam. Heating with

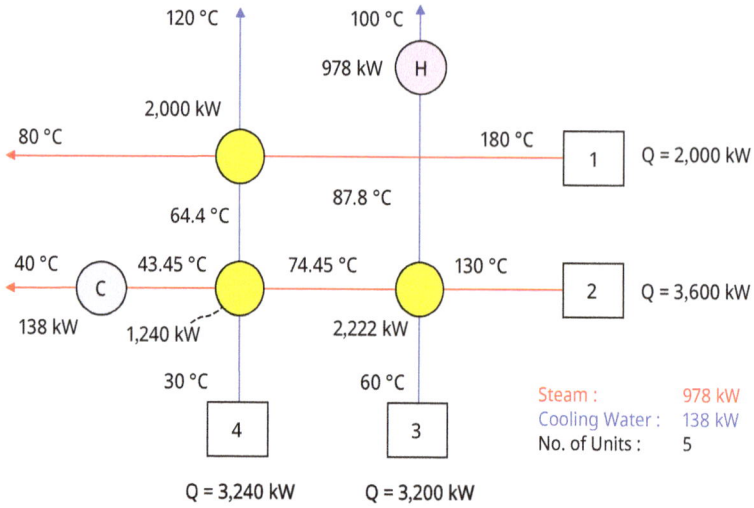

Fig. 2.14: HEN featuring fewer heat exchange matches (2.2.2.5 working session "Introduction to Heat Integration").

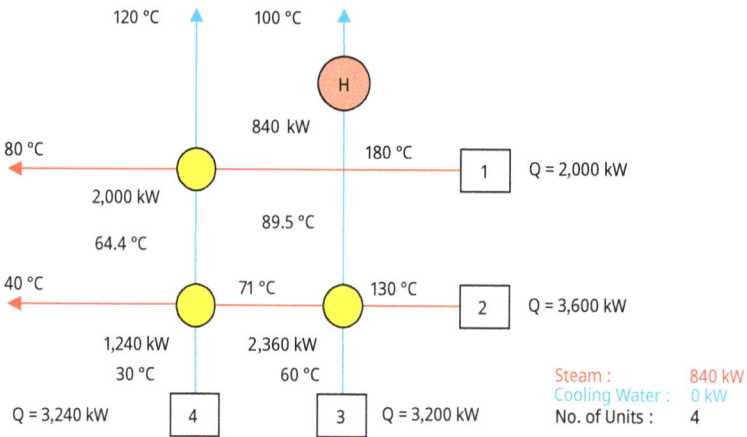

Fig. 2.15: HEN eliminating the need for cooling water (2.2.2.5 working session "Introduction to Heat Integration").

steam is usually approximated as a constant-temperature heating utility. Cooling with water is non-isothermal for that carrier because the cooling effect results from sensible heat absorption into the water stream and thus leads to increasing its temperature.

2.2.3.1 Setting energy targets

Heat Recovery between one hot and one cold stream

The Second Law of Thermodynamics implies that heat flows from higher-temperature to lower-temperature locations. As shown in eq. (2.3) (p. 18), in a heat exchanger, the required heat transfer area is proportional to the temperature difference between the streams.

In heat exchanger design, the minimum allowed temperature difference (ΔT_{min}) is the lower bound on any temperature differences to be encountered in any heat exchanger in the network. The value of ΔT_{min} is a design parameter determined by exploring the trade-offs between more Heat Recovery and the larger heat transfer area requirement. Any given pair of hot and cold process streams may exchange as much heat as allowed by their temperatures and the minimum temperature difference.

Consider the two-stream example shown in Fig. 2.16a. The amount of Heat Recovery is 10 MW, which is achieved by allowing $\Delta T_{min} = 20$ °C. If $\Delta T_{min} = 10$ °C, as in Fig. 2.16b, then it is possible to "squeeze out" one more MW of Heat Recovery. For each value of ΔT_{min}, it is possible to identify the maximum possible Heat Recovery between two process streams. This principle needs to be extended to handle multiple streams to obtain the Heat Recovery Targets for a HEN design problem.

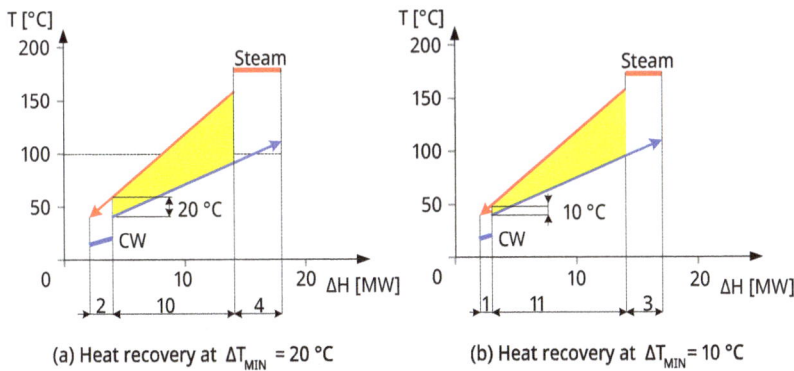

(a) Heat recovery at $\Delta T_{MIN} = 20$ °C (b) Heat recovery at $\Delta T_{MIN} = 10$ °C

Fig. 2.16: Thermodynamic limits on Heat Recovery.

Evaluation of Heat Recovery for multiple streams: the Composite Curves

The analysis starts by combining all hot streams and all cold streams into two Composite Curves (CCs) (Linnhoff et al., 1982). For each process, there are two curves: one for the hot streams (Hot Composite Curve – HCC) and another for the cold streams (Cold Composite Curve – CCC). Each CC consists of a temperature-enthalpy (T-H) profile, representing the overall heat availability in the process (the HCC) and the overall heat demands of the process (the CCC). The procedure of HCC construction is illustrated in Fig. 2.17 on the data from Tab. 2.1.

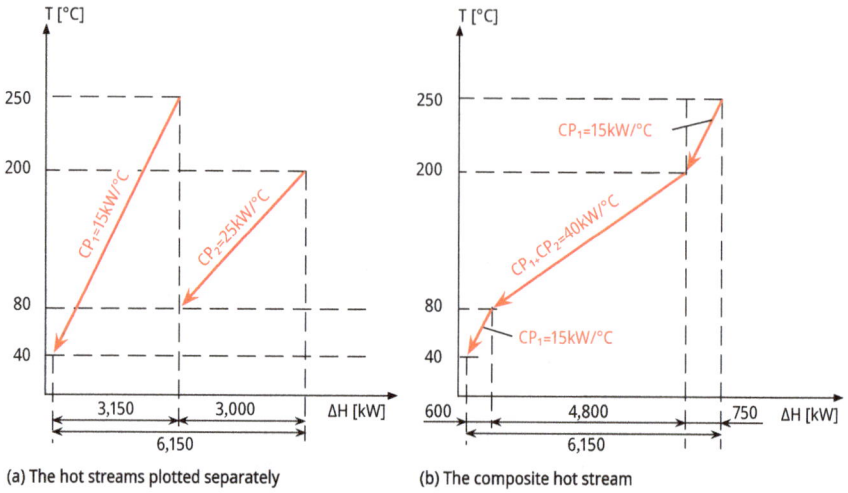

(a) The hot streams plotted separately   (b) The composite hot stream

Fig. 2.17: Constructing the Hot Composite Curve (HCC).

All temperature intervals are formed by the starting and target temperatures of the hot process streams (Fig. 2.17a). Within each temperature interval, in Fig. 2.17b, a composite segment is formed consisting of: 1) a temperature difference equal to that of the interval and 2) a total cooling requirement equal to the sum of the cooling requirements of all streams within the interval by summing up the heat capacity flow rates of the streams. The composite segments from all temperature intervals are combined to form the HCC. The construction of the Cold Composite Curve is entirely analogous.

The two Composite Curves are combined in the same plot in order to identify the maximum overlap, which represents the maximum amount of heat that could be recovered. The HCC and CCC for the example from Tab. 2.1 are shown together in Fig. 2.18.

Both CCs can be moved horizontally (i.e. along the ΔH axis), but usually, the HCC position is fixed, and the CCC is shifted. This is equivalent to varying the amount of Heat Recovery and (simultaneously) the amount of required utility heating and cooling. Where the curves overlap, heat can be recovered between the hot and cold streams. More overlap means more Heat Recovery and smaller utility requirements, and vice versa. As the overlap increases, the temperature differences between the overlapping curve segments decrease. Finally, at a certain overlap, the curves reach the minimum allowed temperature difference, ΔT_{min}. Beyond this point, no further overlap is possible. The closest approach between the curves is termed the *Pinch Point* (or simply the *Pinch*); it is also known as the *Heat Recovery Pinch*.

It is important to note that the amount of the largest overlap (and thus the maximum Heat Recovery) would be different if the minimum allowed temperature difference is changed for the same set of hot and cold streams. The larger the value of ΔT_{min} is, the smaller the possible maximum Heat Recovery. Specifying the minimum utility heating, the minimum utility cooling, or the minimum temperature difference

Figure The HCC and CCC at ΔT_{MIN} = 10 °C

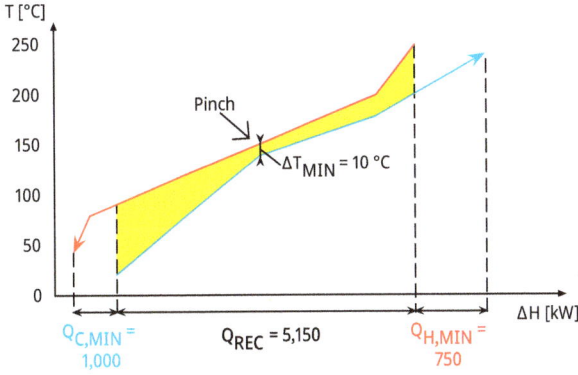

Fig. 2.18: The HCC and CCC at ΔT_{min} = 10 °C.

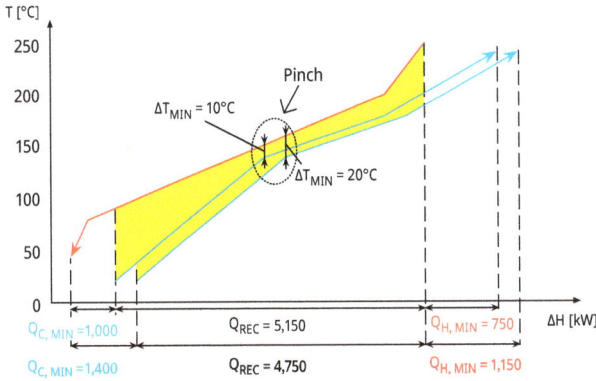

Fig. 2.19: Variation of Heat Recovery Targets with ΔT_{min}.

fixes the relative position of the two Composite Curves and the maximum possible amount of Heat Recovery. The identified Heat Recovery Targets are relative to the specified value of ΔT_{min}. If that value is increased, then the minimum utility requirements also increase and the potential for maximum recovery drops (Fig. 2.19).

The appropriate value for ΔT_{min} is determined by economic trade-offs. Increasing ΔT_{min} results in larger minimum utility demands and increased energy costs; choosing a higher value reflects the need to reduce heat transfer area and its corresponding investment cost. Conversely, if ΔT_{min} is reduced, then the utility cost goes down, but the investment costs go up. This trade-off is illustrated in Fig. 2.20.

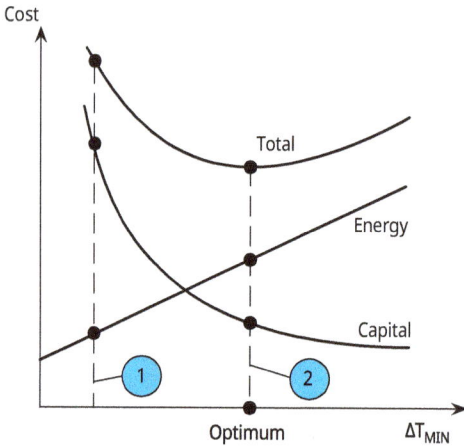

Fig. 2.20: Trade-off between investment and energy costs as a function of ΔT_{min}.

2.2.3.2 Working session "Setting energy targets"

Assignment

Consider the flowsheet again from Fig. 2.6 (p. 23). As already discussed, applying data extraction to that produces the stream data set from Tab. 2.2. To refresh, the minimum allowed temperature difference is selected as $\Delta T_{min} = 10$ °C and also given are two utilities:

– Steam (hot utility) is available at 200 °C.
– CW (cold utility) is available from 25 °C.

The assignment is to construct the Composite Curves and then identify the utility targets.

Solution

First, the Hot Composite Curve is constructed. This starts with plotting the hot streams separately, as shown in Fig. 2.21.

The starting and ending temperatures of the hot streams are listed and sorted in descending order (Fig. 2.22). The sorted list is afterwards used to form temperature intervals and calculate the enthalpy balances (Fig. 2.23).

The Hot Composite Curve is plotted using the obtained enthalpy balances by starting from enthalpy change zero and the lowest interval temperature (40 °C), then moving upward in the table from Fig. 2.23. The resulting plot is shown in Fig. 2.24.

The Cold Composite Curve is constructed in a similar way (illustrated in Fig. 2.25).

The two Composite Curves are put together, obtaining the plot in Fig. 2.26. The resulting targets are:

- Pinch location at 70/60 °C
- Minimum utility heating 960 kW
- Minimum utility cooling 120 kW

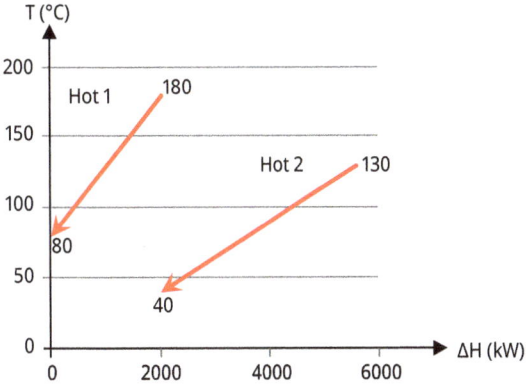

Fig. 2.21: The hot streams for working session "Setting Energy Targets" plotted separately.

Initial list

Stream	T [°C]
Hot 1	180
Hot 1	80
Hot 2	130
Hot 2	40

Sorted

Stream	T [°C]
Hot 1	180
Hot 2	130
Hot 1	80
Hot 2	40

Fig. 2.22: Starting and ending temperatures of the hot streams for working session "Setting Energy Targets".

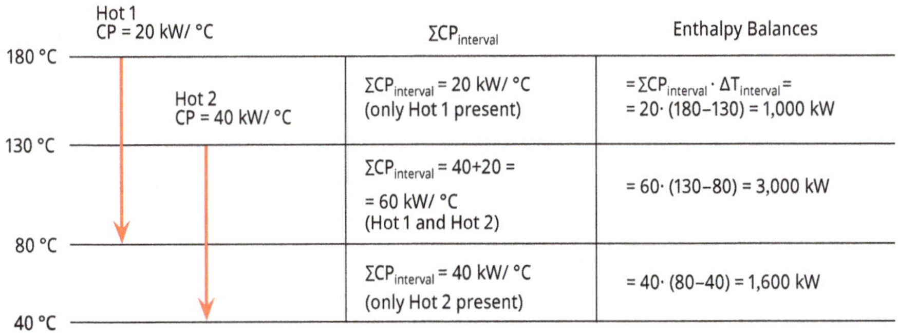

Fig. 2.23: Enthalpy balances for combining the hot streams for Working Session "Setting Energy Targets".

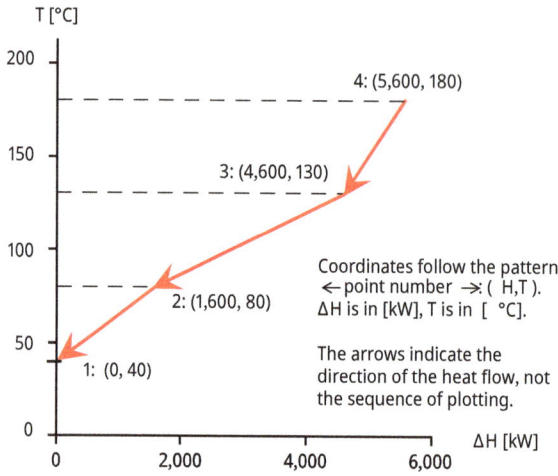

Fig. 2.24: The Hot Composite Curve for working session "Setting Energy Targets".

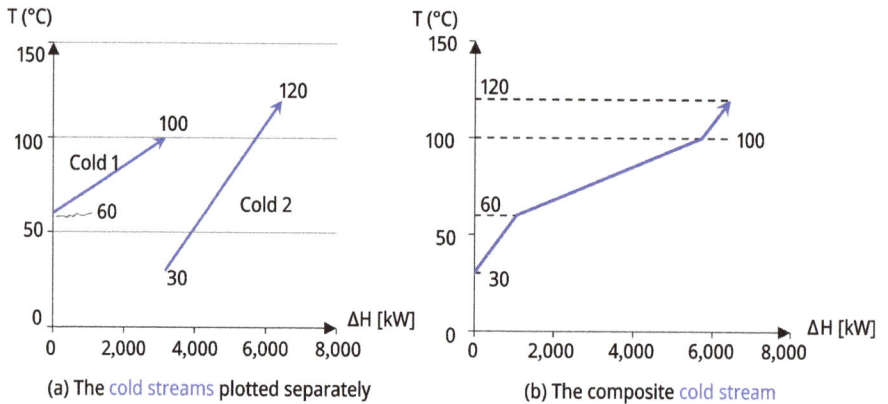

(a) The cold streams plotted separately

(b) The composite cold stream

Fig. 2.25: The Cold Composite Curve for working session "Setting Energy Targets".

A typical problem with the visual identification of the Heat Recovery Targets is what scale of the plot to use. The scale of Fig. 2.26 does not allow us to visualise the 120 kW target for minimum utility cooling. This amount was verified using calculations. This type of problem shows the obvious need to use a numerical tool for the identification of Heat Recovery Targets. Such a numerical tool is discussed later in this chapter.

2.2.3.3 The Heat Recovery Pinch

The Heat Recovery Pinch has important implications for the HEN being designed. As illustrated in Fig. 2.27, the Pinch sets the absolute limits for Heat Recovery within the process.

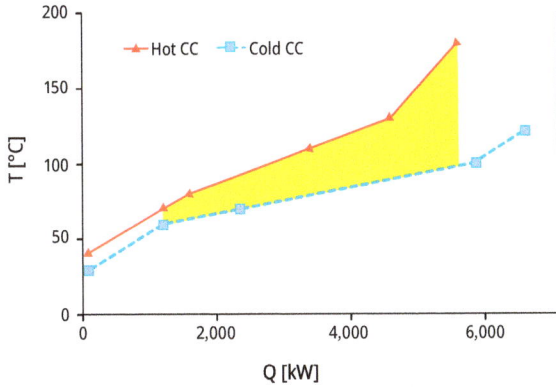

Fig. 2.26: The two Composite Curves for working session "Setting Energy Targets".

The Pinch Point divides the Heat Recovery problem into a net heat sink Above the Pinch Point and a net heat source below it (Fig. 2.28). At the Pinch Point, the temperature difference between the hot and cold streams is exactly equal to ΔT_{min}, which means that across this point, the streams are not allowed to exchange heat. As a result, the heat sink Above the Pinch is in balance with the minimum hot utility ($Q_{H,min}$), and the heat source Below the Pinch is in balance with the minimum cold utility ($Q_{C,min}$); thus, no heat is transferred Across the Pinch via utilities or via process-to-process heat transfer.

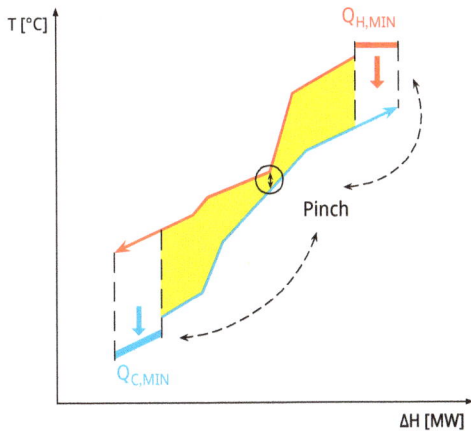

Fig. 2.27: Limits for process Heat Recovery set by the Pinch.

No heat can be transferred from Below to Above the Pinch because this is thermodynamically infeasible. However, it is feasible to transfer heat from hot streams Above the Pinch to cold streams Below the Pinch. All cold streams, even those Below the Pinch, could be heated by a hot utility. Likewise, the hot streams (even Above the Pinch) could be cooled

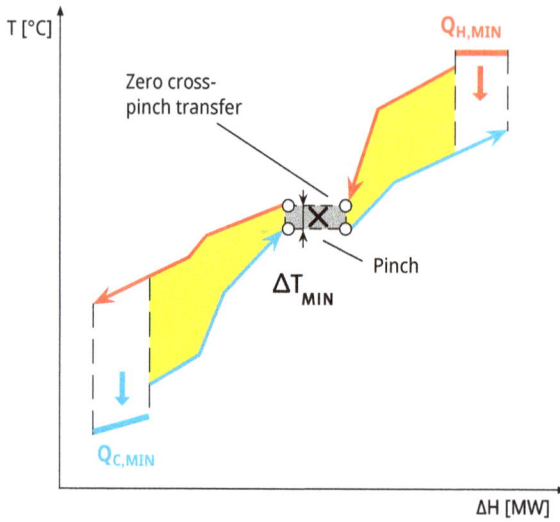

Fig. 2.28: Partitioning the Heat Recovery problem.

by a cold utility. Although these arrangements are thermodynamically feasible, applying them would cause utility use to exceed the minimum, as identified by the Pinch Analysis. This is why the Pinch is fundamental to the design of Heat Recovery systems.

It is important to discuss what happens if heat is transferred Across the Pinch. Recall that it is possible to transfer heat only from Above to Below the Pinch. If, say, **XP** units of heat are transferred Across the Pinch (Fig. 2.29), then $Q_{H,min}$ and $Q_{C,min}$ will each increase by the same amount in order to maintain the heat balances of the

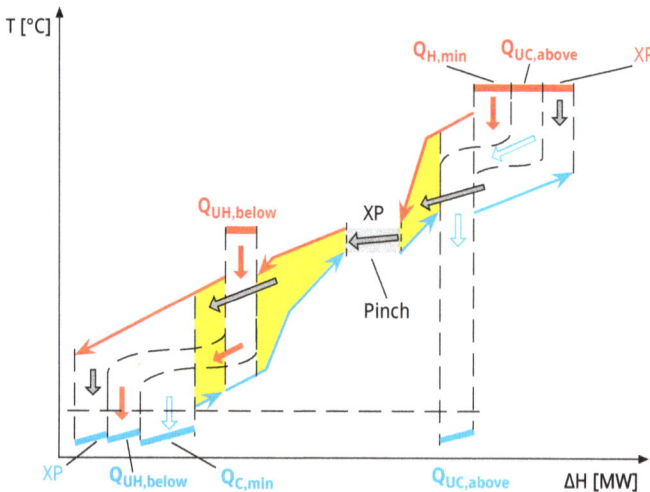

Fig. 2.29: More in – more out.

two problem parts. Any extra heat that is added to the system by the hot utility has to be taken away by the cold utility, in addition to the minimum requirement $Q_{C,min}$.

Cross-Pinch process-to-process heat transfer is not the only way by which a problem's thermodynamic Pinch partitioning can be violated. This could also happen if the external utilities are placed incorrectly. For example, any heating Below the Pinch creates a need for additional cooling in that part of the system (Fig. 2.29). Conversely, any utility cooling Above the Pinch creates a need for additional utility heating. The implications of the Pinch for Heat Recovery problems can be distilled into the following three conditions, which have to hold if the minimum energy targets for a process are to be achieved:

1. Heat must not be transferred Across the Pinch.
2. There must be no external cooling Above the Pinch.
3. There must be no external heating Below the Pinch.

Violating any of these rules will lead to an increase in energy utility demands. The rules are applied explicitly in the context of HEN synthesis by the Pinch Design Method (Linnhoff and Hindmarsh, 1983) and also before starting a HEN retrofit analysis to identify causes of excessive utility demands by a process. Other HEN synthesis methods, if they achieve the minimum utility demands, also conform to the Pinch rules (though sometimes only implicitly).

2.2.3.4 Numerical targeting: the Problem Table Algorithm

The Composite Curves are a useful tool for visualising Heat Recovery Targets. However, it can be time-consuming to draw for problems that involve many process streams. In addition, targeting that relies solely on such graphical techniques cannot be very precise, as noted in the section explaining the construction of the Composite Curves.

For that reason, the process of identifying numerical targets is usually based on an algorithm known as the Problem Table Algorithm (PTA). Some authors employ the equivalent "transshipment" model (Cerdá et al., 1990). The algorithm steps are as follows.

1. Shift the process stream temperatures.
2. Set up temperature intervals.
3. Calculate interval heat balances.
4. Assuming zero hot utility, cascade the balances as heat flows.
5. Ensure positive heat flows by increasing the hot utility as needed.

The algorithm will be illustrated using the sample data in Tab. 2.3.

Tab. 2.3: PTA example: process streams data (ΔT_{min} = 10 °C).

No.	Type	T_S (°C)	T_T (°C)	CP (kW/°C)	T_S^* (°C)	T_T^* (°C)
1	Cold	20	180	20	25	185
2	Hot	250	40	15	245	35
3	Cold	140	230	30	145	235
4	Hot	200	80	25	195	75

Step 1

Because the PTA uses temperature intervals, it is necessary to set up a unified temperature scale for the calculations. If the real stream temperatures are used, then some of the heat content would be left out of the recovery. The problem is avoided by obtaining shifted stream temperatures (T*) for PTA calculations. Thus, the hot streams are shifted to be colder by $\Delta T_{min}/2$, and the cold streams are shifted to be hotter by $\Delta T_{min}/2$. If the shifted temperatures (T*) of a cold and a hot stream (or their parts) are the same, then their real temperatures are still actually ΔT_{min} apart, which allows for feasible heat transfer. This operation is equivalent to shifting the Composite Curves toward each other vertically, as illustrated in Fig. 2.30. The last two (shaded) columns in Tab. 2.3 show the shifted process stream temperatures.

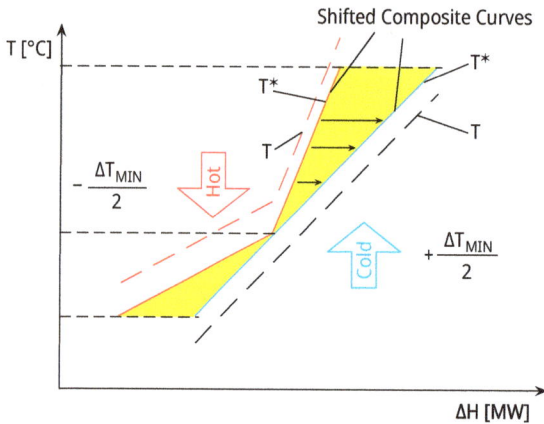

Fig. 2.30: Temperature shifting to ensure feasible heat transfer.

Step 2

Temperature intervals are formed by listing all shifted process stream temperatures in descending order (any duplicate values are considered just once). This action creates Temperature Boundaries (TBs), which form the temperature intervals for the problem. For the example from in Tab. 2.3, the TBs are 245 °C, 235 °C, 195 °C, 185 °C, 145 °C, 75 °C, 35 °C, and 25 °C.

Step 3

The heat balance is calculated in each temperature interval. First, the stream population of the process segments falling within each temperature interval (the first two columns of Tab. 2.4) is identified. Next, the sums of the segment CPs (heat capacity flow rates) in each interval are calculated; then, that sum is multiplied by the interval temperature difference (i.e. the difference between the TBs that define each interval). This calculation is also illustrated in Tab. 2.4.

Tab. 2.4: Problem Table Algorithm (PTA) for the streams in Tab. 2.3.

Interval temp (°C)	Stream population	$\Delta T_{INTREVAL}$ (°C)	$\Sigma CP_H - \Sigma CP_C$ (kW/°C)	$\Delta H_{INTREVAL}$ (kW)	Surplus/deficit
245					
		10	15	150	Surplus
235					
		40	−15	−600	Deficit
195					
		10	10	100	Surplus
185					
		40	−10	−400	Deficit
145					
		70	20	1,400	Surplus
75					
		40	−5	−200	Deficit
35					
		10	20	−200	Deficit
25					

Stream population diagram annotations: 2, 4, 3, 1; $CP = 15$, $CP = 30$, $CP = 25$, $CP = 20$.

Step 4

The Problem Heat Cascade shown in Fig. 2.31 has a box allocated to each temperature interval; each box contains the corresponding interval enthalpy balances. The boxes are connected with heat flow arrows in order of descending temperature. The top heat flow represents the total hot utility provided to the cascade, and the bottom heat flow represents the total cold utility. The hot utility flow is initially assumed to be zero, and this value is combined (summed up) with the enthalpy balance of the top cascade interval to produce the value for the next lower cascade heat flow. This operation is repeated for the lower temperature intervals and connecting heat flows until the bottom heat flow is calculated, resulting in the cascade shown in Fig. 2.31(a).

Step 5

The resulting heat flow values in the cascade are examined, and a feasible heat cascade is obtained; see Fig. 2.31(b). From the cascading heat flows, the smallest value is identified; if it is non-negative (i.e. positive or zero), then the heat cascade is thermodynamically feasible. If a negative value is obtained, then a positive utility flow of the same absolute value

Figure Heat Cascade for the example from Table 3.2

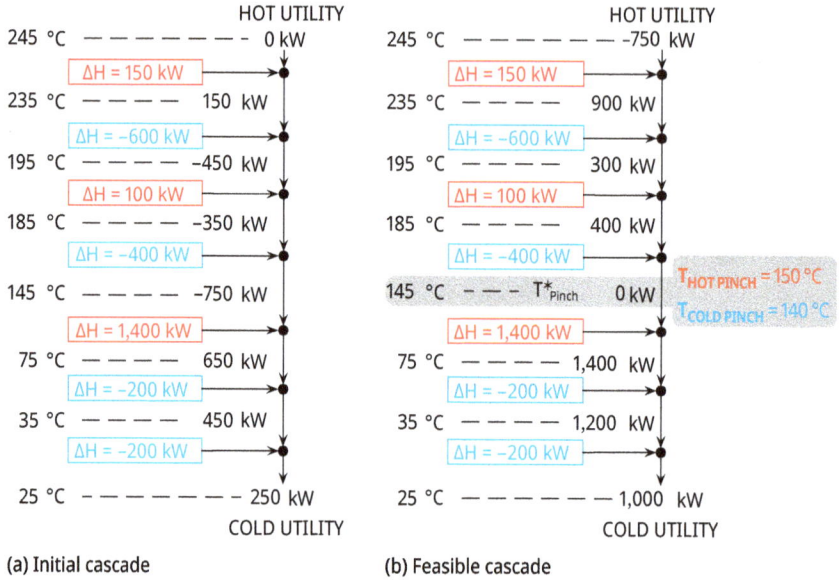

Fig. 2.31: Heat cascade for the process data in Tab. 2.3.

has to be provided at the topmost heat flow, after which the cascading described in Step 4 is repeated. The resulting heat cascade is guaranteed to be feasible and provides numerical Heat Recovery Targets for the problem. The topmost heat flow represents the minimum hot utility, the bottommost heat flow represents the minimum cold utility, and the TB with zero heat flow represents the location of the (Heat Recovery) Pinch. It is often possible to obtain more than one zero-flow temperature boundary, each representing a separate Pinch point.

2.2.3.5 Working session "The Problem Table Algorithm"

Assignment

The same flowsheet as before is considered (Fig. 2.6, p. 23). The process stream data is the same (Tab. 2.2, p. 23), and so are the remaining specifications of the Heat Recovery problem (ΔT_{min} = 10 °C, Steam at 200 °C, CW at 25 °C). The task is to calculate the minimum hot and cold utility requirements and the location of the Pinch, using the Problem Table Algorithm.

Solution

First, the stream temperatures are shifted by $\Delta T_{min}/2$. Hot streams are shifted down, and cold streams up, as shown in Tab. 2.5.

Tab. 2.5: Obtaining the shifted temperatures (Working Session "The Problem Table Algorithm").

Stream	T_S (°C)	T_T (°C)	T_S^* (°C)	T_T^* (°C)
1 (hot)	180	80	175	75
2 (hot)	130	40	125	35
3 (cold)	60	100	65	105
4 (cold)	30	120	35	125

After eliminating the duplicates and sorting them in descending order, the following list of temperature boundaries remains: 175 °C, 125 °C, 105 °C, 75 °C, 65 °C, and 35 ° C. This allows us to further draw the process stream population against the temperature boundaries, form the temperature intervals and calculate the enthalpy balances of the intervals (Tab. 2.6).

Tab. 2.6: The Problem Table for working session "The Problem Table Algorithm".

Shifted interval temperature	Stream population	$\Delta T_{interval}$	$\Sigma CP_H -$ ΣCP_C	$\Delta H_{interval}$	Surplus/ deficit
175					
		50	20	1,000	Surplus
125					
		20	24	480	Surplus
105					
		30	−56	−1,680	Deficit
75					
		10	−76	−760	Deficit
65					
		30	4	120	Surplus
35					

The enthalpy balances of the temperature intervals are used to construct the heat cascade. At the first stage, a zero hot utility is assumed and the balances propagated (Fig. 2.32(a)), and then the negative flows are eliminated to obtain the feasible heat cascade (Fig. 2.32(b)).

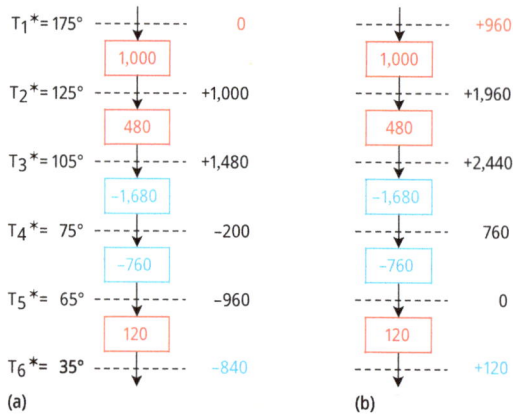

$T_1{}^* = 175°$ — 0
1,000
$T_2{}^* = 125°$ — +1,000
480
$T_3{}^* = 105°$ — +1,480
−1,680
$T_4{}^* = 75°$ — −200
−760
$T_5{}^* = 65°$ — −960
120
$T_6{}^* = 35°$ — −840

(a)

— +960
1,000
— +1,960
480
— +2,440
−1,680
— 760
−760
— 0
120
— +120

(b)

Fig. 2.32: Heat Cascade for working Session "The Problem Table Algorithm".

In the feasible cascade (Fig. 2.32(b)), the temperature boundary with zero heat flow is at 65 °C, which indicates the Pinch location as 70/60 °C. The minimum utility heating is 960 kW, identified by the heat flow at the top of the cascade and the minimum utility cooling is 120 kW, identified by the heat flow at the bottom of the cascade.

2.2.3.6 Threshold problems

Threshold problems feature only one utility type, either hot or cold. They are important mostly because they often result in no utility–capital trade-off below a certain value of ΔT_{min}, since the minimum utility demand (hot or cold) becomes invariant; see Fig. 2.33.

Typical examples of threshold Heat Integration problems involve high-temperature fuel cells, which usually have large cooling demands but no heating demands; some examples are the flowsheets of Molten Carbonate Fuel Cells (Varbanov et al., 2006) and Solid Oxide Fuel Cells (Varbanov and Klemeš, 2008). An essential feature that distinguishes threshold problems is that, as ΔT_{min} is varied, demands for only one utility type (hot or cold) are identified over the variation range; in contrast, pinched problems require both hot and cold utilities over this range. When synthesising HENs for threshold problems, one can distinguish between two subtypes (see Fig. 2.34):

1. Low-threshold ΔT_{min}: Problems of this type should be treated exactly as Pinch-type problems.
2. High-threshold ΔT_{min}: For these problems, it is first necessary to satisfy the required temperature for the no-utility end before proceeding with the remaining design.

(a) Heat Recovery, hot and cold utilities

(b) More Heat Recovery, no hot utility

(c) No increase in Heat Recovery

(d) Utility substitution

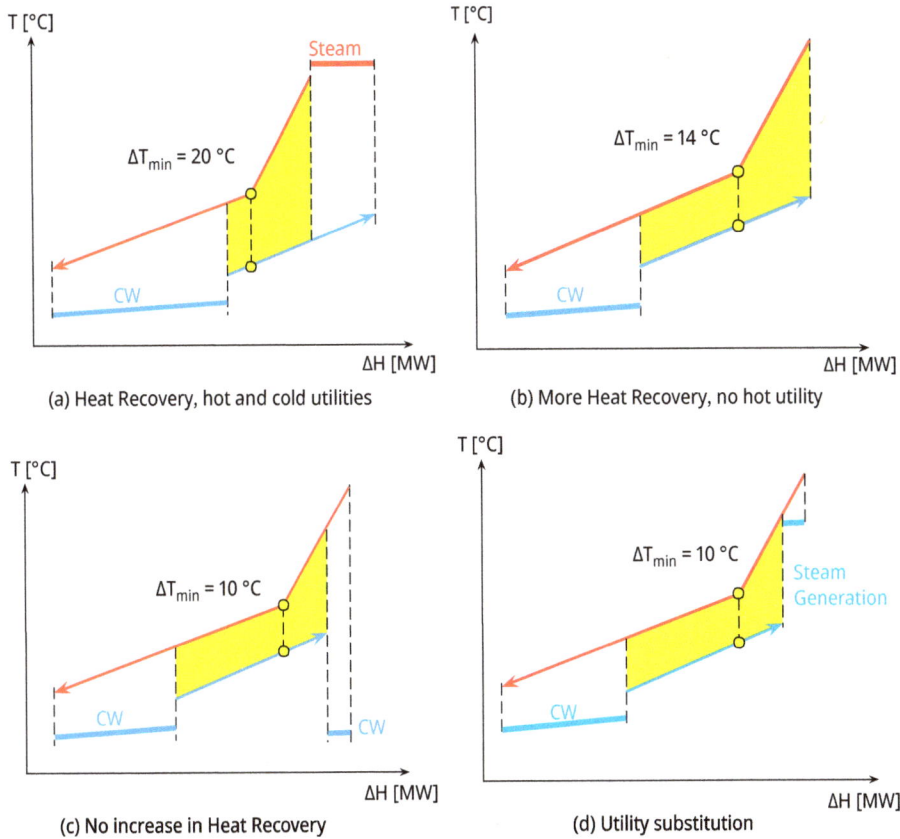

Fig. 2.33: Threshold problems.

2.2.3.7 Multiple utilities targeting

Utility placement: Grand Composite Curve (GCC)

In many cases, more than one hot and one cold utility are available for providing the external heating and cooling requirements after energy recovery. It thus becomes necessary to find the cheapest and most effective combination of the available utilities (Fig. 2.35).

To assist with this choice and to enhance the information derived from the HCC and CCC, another graphical construction has been developed: the Grand Composite Curve (GCC) (Townsend and Linnhoff, 1983). The heat cascade and the PTA (Linnhoff and Flower, 1978) offer guidelines for the optimum placement of hot and cold utilities. These allow for determining the heat loads associated with each utility. To this point, the assumption has been that only one cold and one hot utility are available, albeit with sufficiently low and

(a) Low Threshold ΔT_{min}

(b) High Threshold ΔT_{min}

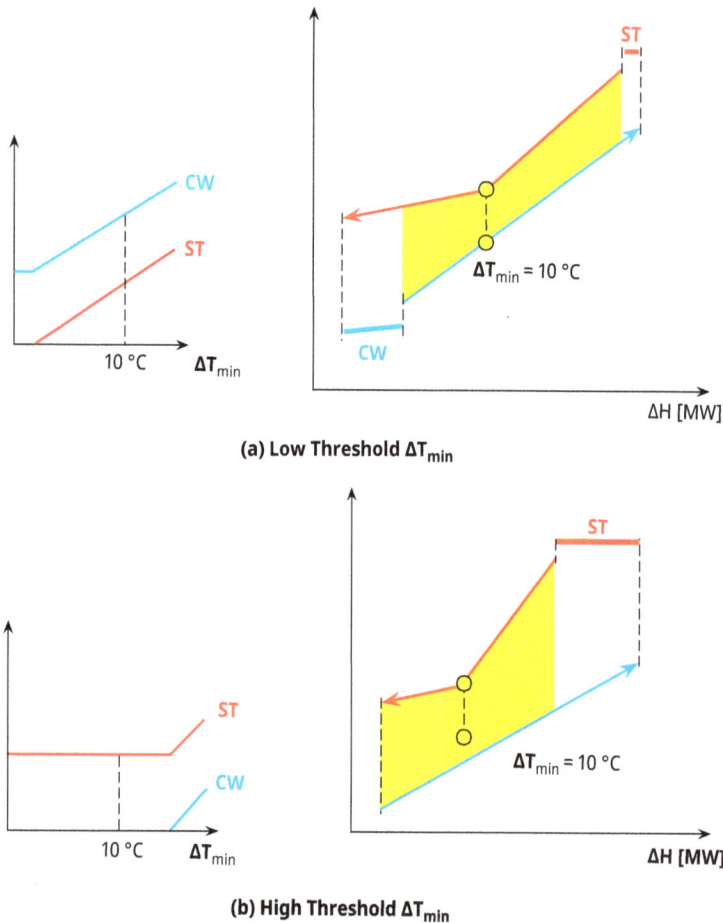

Fig. 2.34: Threshold HEN design cases.

high temperatures, to satisfy the cooling and/or heating demands of the process. However, most industrial sites feature multiple heating and cooling utilities at several different temperature levels (e.g. steam levels, refrigeration levels, hot oil circuits, furnace flue gas). Each utility has a different unit cost. Usually, the higher-temperature hot utilities and the lower-temperature cold utilities cost more than the ones with temperatures closer to the ambient. This fact underscores the need to choose a mix resulting in the lowest utility cost. The general objective is to maximise the use of cheaper utilities and to minimise the use of more expensive utilities. For example, it is usually preferable to use low-pressure (LP) instead of high-pressure (HP) steam and to use cooling water instead of refrigeration. The Composite Curves (CC) plot (Fig. 2.18) provides a convenient view for evaluating the process driving force and the general Heat Recovery Targets. However, the CC is not useful for identifying targets when multiple utility levels are available. The GCC is used for this task.

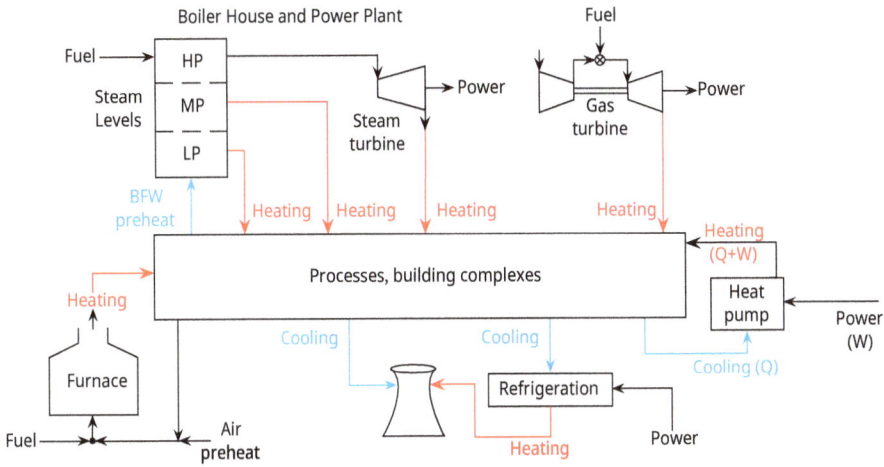

Fig. 2.35: Choices of hot and cold utilities (amended after CPI 2004).

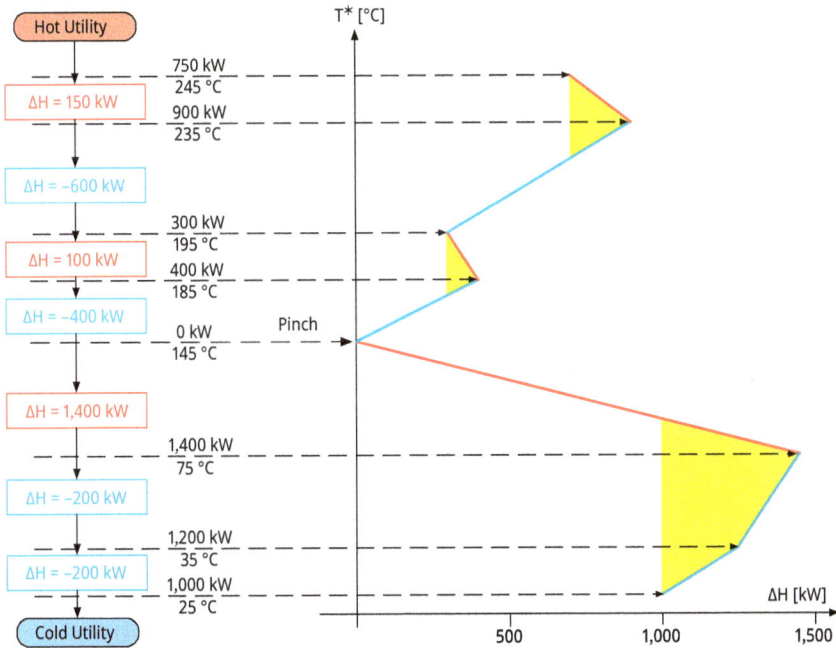

Fig. 2.36: Constructing the GCC for the streams in Tab. 2.3.

Construction of the Grand Composite Curve

The GCC is constructed using the Problem Heat Cascade (Fig. 2.31, p. 39). The heat flows are plotted in the T-ΔH space, where the heat flow at each temperature boundary corresponds to the x-coordinate and the temperature to the y-coordinate (Fig. 2.36).

The Grand Composite Curve can be directly related to the Shifted Composite Curves (SCCs), which are the result of shifting the CCs toward each other by $\Delta T_{min}/2$, so that the curves touch each other at the Pinch (see Fig. 2.37). At each temperature boundary, the heat flow in the Problem Heat Cascade and GCC corresponds to the horizontal distance between the SCCs.

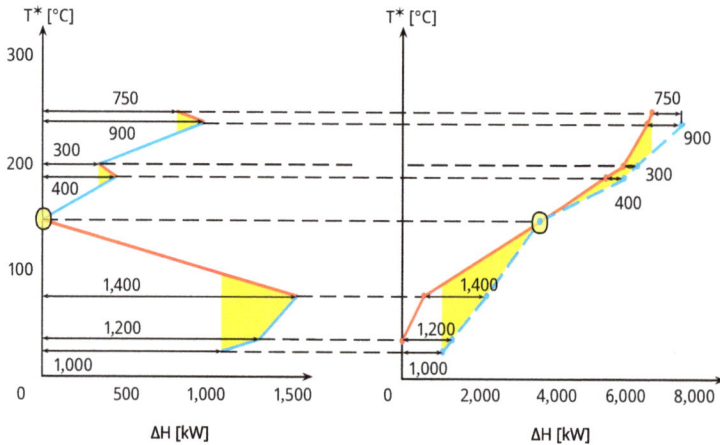

Fig. 2.37: Relation between the GCC (left) and the SCC (right) for the streams in Tab. 2.3.

The GCC has several fundamental properties that facilitate an understanding of the underlying Heat Recovery problem. The parts with positive slope (i.e. running uphill from left to right) indicate that cold streams dominate (Figs. 2.36 and 2.37). Similarly, the parts with negative slopes indicate excess hot streams. The shaded areas, which signify opportunities for process-to-process Heat Recovery, are referred to as *Heat Recovery Pockets*.

Utility placement options

The GCC shows the hot and cold utility requirements of the process in terms of both enthalpy and temperature. This allows one to distinguish between utilities at different temperature levels. There are typically utilities at several different temperature levels available on a site. For example, it can be a supply of both high-pressure (HP) and medium-pressure (MP) steam. As indicated previously, it is desirable to maximise the use of cheaper utilities and to minimise the use of more expensive ones. Utilities at higher temperatures and/or pressure are usually more expensive; see Fig. 2.38. Therefore, MP steam is used first: ranging from the y-axis until it touches the GCC, which maximises its usage. Only then is HP steam used.

– When a utility line or profile touches the GCC, a new Pinch Point is created, termed a "Utility Pinch" (Fig. 2.38). Each additional steam level creates another Utility Pinch, and so increases the complexity of the utility system. Higher complexity has several negative consequences, including increased capital costs, greater potential for leaks, reduced safety, and more maintenance expenses. Limits are typically placed on the number of steam levels.

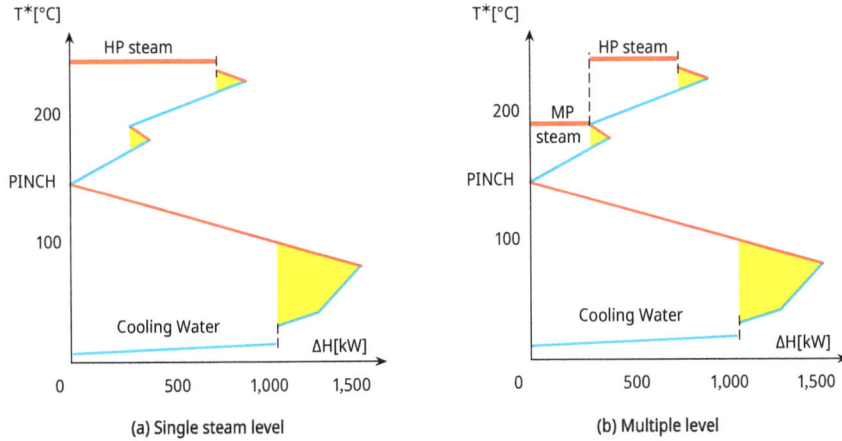

Fig. 2.38: Targeting for two steam levels using the GCC.

Higher-temperature heating demands are satisfied by non-isothermal utilities. These include hot oil and hot flue gas, both of which maintain their physical phase (liquid and gaseous, respectively) across a wide range of temperatures. The operating costs associated with such utilities are largely dependent on furnace efficiency and on the intensity and efficiency of the pumping or fan blowing. When targeting the placement of a non-isothermal hot utility, its profile is represented by a straight line, which runs from the upper right to the lower left in the graph of Fig. 2.39. The line's starting point corresponds to the utility supply temperature and also to the rightmost point for the utility's heating duty. The utility use endpoint corresponds either to the zero of the ΔH axis, in which case, all utility heating is covered by the current non-isothermal utility, or to the rightmost point on the ΔH axis for other, cheaper hot utilities.

As plotted in the figure, the non-isothermal utility's termination point corresponds to the ambient temperature. Thus, the distance from this point to the zero of the ΔH axis represents the thermal losses from using the utility. The heat capacity flow rate of the non-isothermal utility target is determined by making the utility line as steep as possible, thereby minimising its CP and the corresponding losses (Fig. 2.39). Its supply temperature is usually fixed at the maximum allowed by the furnace and the heat carrier composition;

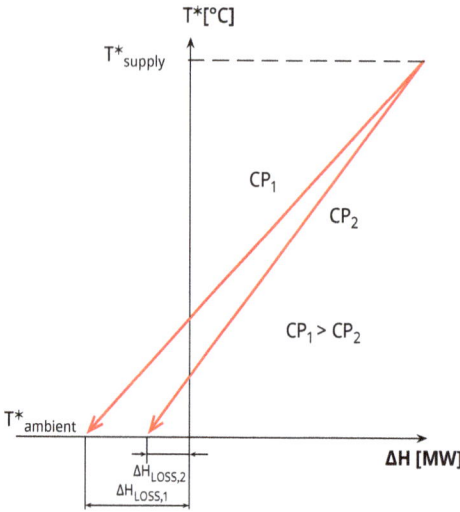

Fig. 2.39: Properties of non-isothermal hot utilities.

the remaining degree of freedom corresponds to the utility's exact CP. Smaller CP values result in steeper slopes and smaller losses. The placement for a non-isothermal utility (e.g. hot oil) may be constrained by two problem features: the process Pinch point and a "kink" in the GCC at the top end of a Heat Recovery Pocket; see Fig. 2.40, for an example.

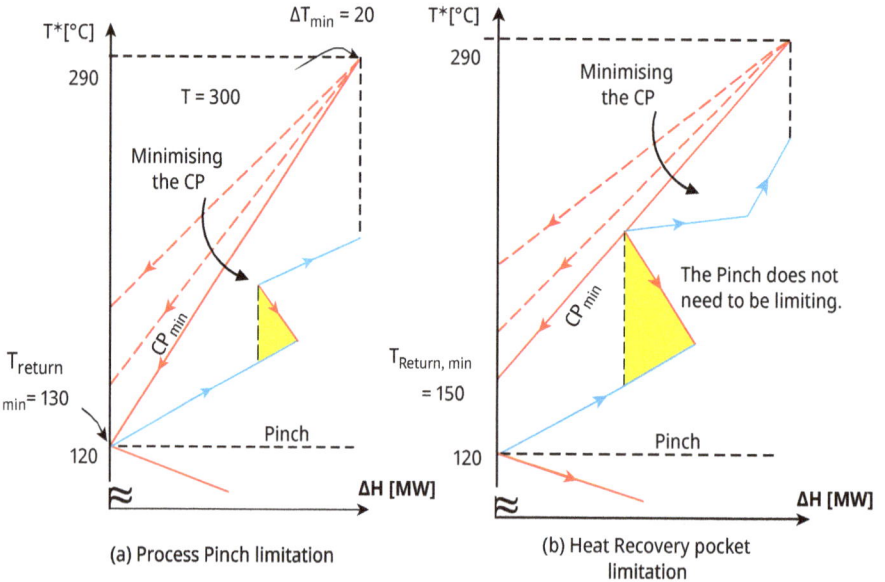

Fig. 2.40: Constraints for placing hot oil utilities.

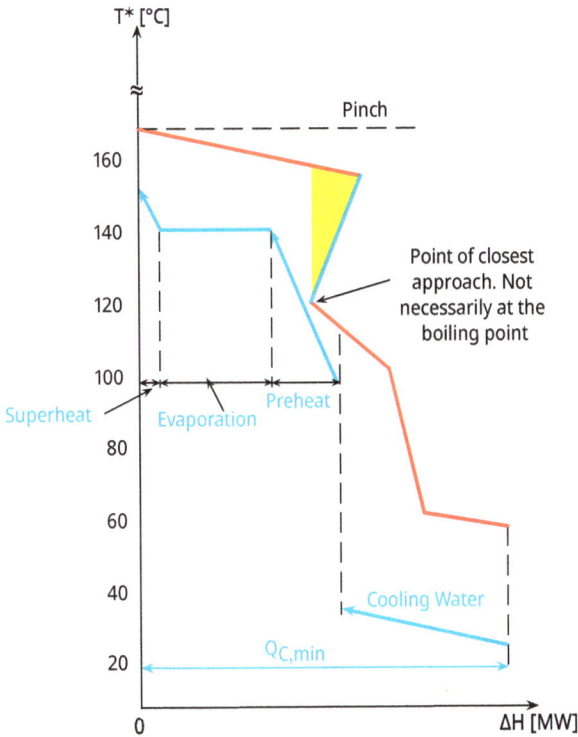

Fig. 2.41: Generating steam Below the Pinch.

When fuel is burned in a furnace or a boiler, the resulting flue gas becomes available to heat up the corresponding cold-stream medium (for steam generation or direct process duty). Transferring heat to the process causes the flue gas temperature to drop as it moves from the furnace to the stack. The stack temperature has to be above a specified value: the *minimum allowed stack temperature*, which is determined by limitations due to corrosion. If flue gas is used directly for heating, then the Pinch point, if it is higher, may become more limiting than the minimum allowed stack temperature. If the analysed process features both high- and moderate-temperature utility heating demands, then flue gas heating may not be appropriate for satisfying all those demands. If steam is cheaper, then combining it with flue gas reduces the latter's CP and the corresponding stack heat losses.

Another option for utility placement is to use part of the cooling demand of a process for generating steam. This is illustrated in Fig. 2.41, in which steam generation is placed Below the Pinch.

The GCC can reveal where utility substitution may improve energy efficiency; see Fig. 2.42. The main idea is to exploit Heat Recovery Pockets that span two or more utility temperature levels. The technical feasibility of this approach is determined by both the temperature span and the heat duty within the pocket, which should be large enough (in terms of both heat duty and temperature span) to make utility substitution worthwhile when weighed against the required capital cost.

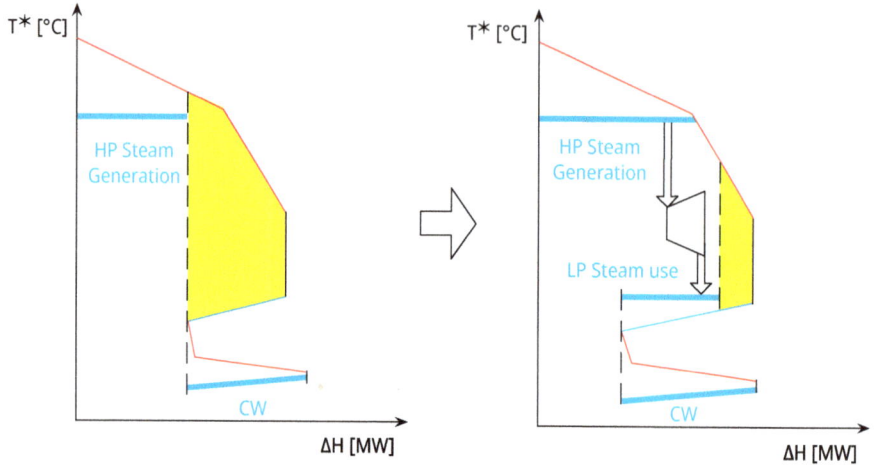

Fig. 2.42: Exploiting the pocket of the Grand Composite Curve for utility substitution.

Utility cooling below ambient temperatures may be required, a need that is usually met by refrigeration. Refrigerants absorb heat by evaporation, and pure refrigerants evaporate at a constant temperature. Therefore, refrigerants are represented, on the plot of T (or T*) vs ΔH, by horizontal bars, similarly to steam levels. On the GCC, refrigeration levels are placed similarly to steam levels; see Fig. 2.43.

When the level of a placed utility is between the temperatures of a Heat Recovery Pocket, the Utility Pinch cannot be located by using the GCC. In this case, Balanced Composite Curves (BCCs) are used. Figure 2.44 shows how the data about the placed utilities can be transferred from the GCC to the BCCs and enables the correct location of the Utility Pinch associated with LP steam.

The BCCs create a combined view in which all heat sources and sinks (including utilities) are in balance, and all Pinches are shown. Balanced Composite Curves are a useful additional tool for evaluating Heat Recovery, obtaining targets for specific utilities, and planning HEN design regions.

2.2.3.8 Investment and total cost targeting

In addition to estimating maximum Heat Recovery, it is also possible to estimate the required capital cost. The expressions for obtaining these estimates are derived from the relationship between the heat transfer area and the efficiency of a single heat exchanger. Methods for targeting capital cost and total cost were initially developed by Townsend and Linnhoff (1984) and further elaborated by others, e.g. using simplified

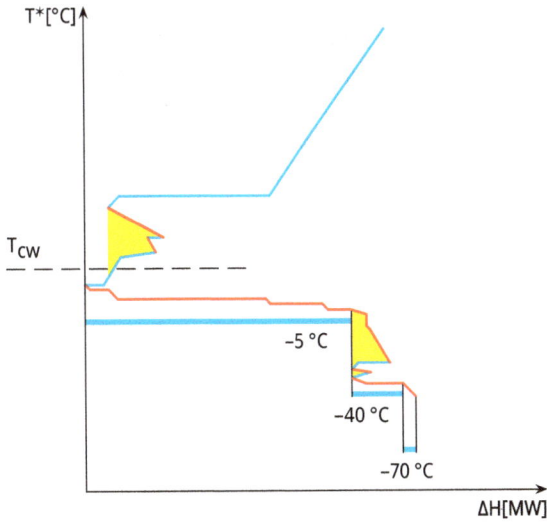

Fig. 2.43: Placing refrigeration levels for pure refrigerants.

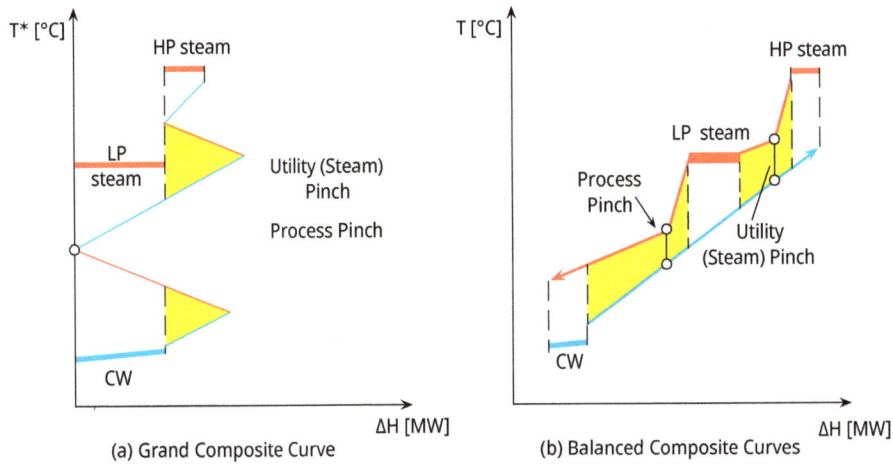

(a) Grand Composite Curve

(b) Balanced Composite Curves

Fig. 2.44: Locating the LP steam Utility Pinch.

capital cost models (Linnhoff and Ahmad, 1990); using detailed capital cost models (Ahmad et al., 1990); using NLP transhipment models and account for forbidden matches (Colberg and Morari, 1990); and integrated HEN targeting and synthesis within the context of the so-called "block decomposition" for HEN synthesis (Zhu et al., 1995).

The HEN capital cost depends on the heat transfer area, the number of heat exchangers, the number of shell-and-tube passes in each heat exchanger, construction materials, equipment type, and operating pressures. The heat transfer area is the most significant factor, and assuming one-shell-one-tube-pass exchangers, it is possible to estimate the overall minimum required heat transfer area; this value helps establish the lower bound on the network's capital cost. Estimating the minimum heat transfer area is based on the concept of an enthalpy interval. As shown in Fig. 2.45, an *enthalpy interval* is a slice constrained by two vertical lines with fixed values on the ΔH axis. This interval is characterised by its enthalpy difference, the corresponding temperatures of the CCs at the interval boundaries, its process stream population, and the heat transfer coefficients of those streams.

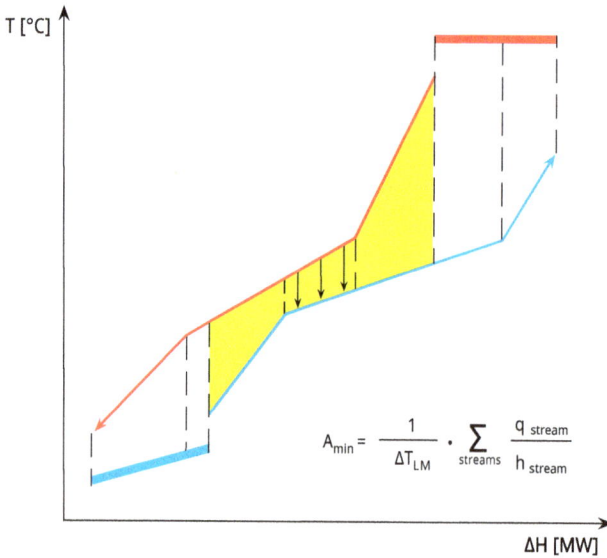

Fig. 2.45: Enthalpy intervals and area targeting.

The minimum heat transfer area target can be obtained by estimating it within each enthalpy interval of the BCC and then summing up the values over all intervals (Linnhoff and Ahmad, 1990):

$$A_{HEN, min} = \sum_{i=1}^{EI} \left[\frac{1}{\Delta T_{LM,i}} \cdot \sum_{s}^{NS} \frac{q_{s,i}}{h_{s,i}} \right]. \tag{2.12}$$

Here i denotes ith enthalpy interval; s, the sth stream; $\Delta T_{LM,i}$, the log-mean temperature difference in interval i (from the Composite Curve segments); q_s, the enthalpy change flow of the sth stream part in the current enthalpy interval; and h_s, the heat transfer coefficient of sth stream. The area targets can be supplemented by targets for the number of shells (Ahmad and Smith, 1989) and for the number of heat exchanger units, thus providing a basis for estimating the HEN capital cost and the total cost. This approach is known as *supertargeting* (Ahmad et al., 1989).

With supertargeting, it is also possible to optimise the value of ΔT$_{min}$ prior to designing the HEN. Proposed improvements to the capital cost targeting procedure of Townsend and Linnhoff (1984) mainly involve: 1) obtaining more accurate surface area targets for HENs with non-uniform heat transfer coefficients, covered by references previously assessed (Colberg and Morari, 1990; Jegede and Polley, 1992; Zhu et al., 1995; Serna-González et al., 2007); 2) accounting for construction materials, pressure rating, and different heat exchanger types (Hall et al., 1990); or 3) accounting for safety factors, including "prohibitive distance" (Santos and Zemp, 2000). Further information can be found in the paper by Taal et al. (2003), which summarises the common methods used to estimate the cost of heat exchange equipment and also provides sources of projections for energy prices. Further development includes the STEPS procedure for supertargeting by Sun et al. (2013), which applies simultaneous targeting and design, resulting in a significant improvement in the target accuracy and the robustness of the design procedure, as well as the HEN retrofit using different heat exchanger types (Wang et al., 2021).

2.2.3.9 Heat Integration of energy-intensive processes

Heat engines
Particularly important processes are the heat engines: steam and gas turbines as well as hybrids. They operate by drawing heat from a higher-temperature source, converting part of it to mechanical power; then (after some energy loss), they reject the remaining heat at a lower temperature; see Fig. 2.46. The energy losses are usually neglected for targeting purposes.

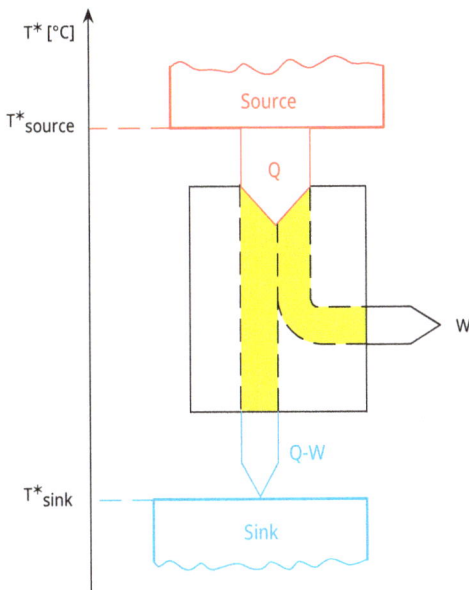

Fig. 2.46: Heat engine configuration.

Integrating a heat engine Across the Pinch is equivalent to a Cross-Pinch process to process heat transfer, and it results in a simultaneous increase of hot and cold utility. This usually also leads to excessive capital investment for the utility exchangers. If a heat engine is integrated Across the Pinch, then the hot utility requirement is increased by Q and the cold by Q–W (according to the notation in Fig. 2.47). Heat engines should be integrated in one of two ways:

Above the Pinch (Fig. 2.47(a)): This increases the hot utility for the main process by W, but all the extra heat is converted into shaft work;

Below the Pinch (Fig. 2.47(b)): This offers a double benefit. It saves on a cold utility, and the process heat Below the Pinch supplies Q to the heat engine (instead of rejecting it to a cooling utility).

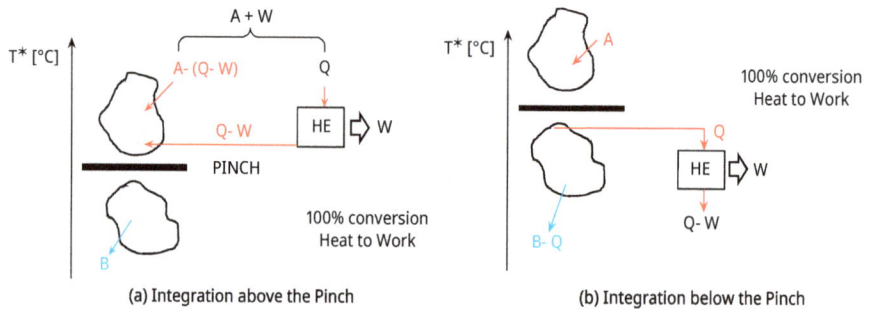

Fig. 2.47: Appropriate placement of heat engines.

From the perspective of the heat engines, their placement may differ. On the one hand, steam turbines may be placed either Below or Above the Pinch since they draw and exhaust steam. Figure 2.48 shows steam turbine integration Above the Pinch, which has the benefit of cogenerating extra power. Gas turbines, on the other hand, which use fuel as input, are typically used only as a utility heat source for the processes and can be placed only Above the Pinch.

Heat Pumps

Heat pumps present another opportunity for improving the energy performance of an industrial process. Their operation is the reverse of heat engines. That is, heat pumps take heat from a lower-temperature source, upgrade it by applying mechanical power, and then deliver the combined flow to a higher-temperature heat sink (Fig. 2.49).

An important characteristic of heat pumps is their coefficient of performance (COP). This metric for device efficiency is defined as the ratio between the heat delivered to the heat sink and the consumed shaft work (mechanical power):

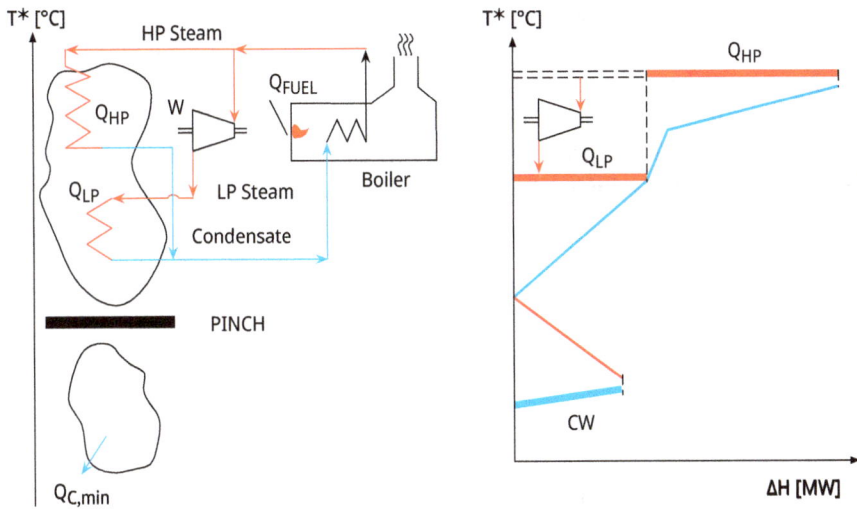

Fig. 2.48: Integrating a steam turbine Above the Pinch.

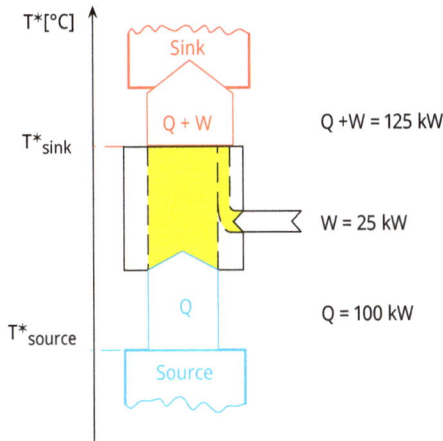

Fig. 2.49: Heat-pump configuration.

$$Q_{sink} = Q_{source} + W, \qquad (2.13)$$

$$COP = \frac{Q_{sink}}{W} = \frac{Q_{source} + W}{W}. \qquad (2.14)$$

The COP is a non-linear function of the temperature difference between the heat sink and the heat source (Laue, 2006); this difference is also referred to as *temperature lift*. Figure 2.50(a) shows the appropriate integration of a heat pump Across the Pinch, with the heat source located Below the Pinch and the heat sink above it. The GCC facilitates the sizing of the heat pump by evaluating the possible temperatures of the heat source and heat sink and their loads; see Fig. 2.50(b). Integrating entirely Above the Pinch results in the direct conversion of mechanical power to heat. This is a waste of primary resources because most of the power is generated at the expense of twice to three times the amount of primary fuel energy. The second alternative, placing the heat pump entirely Below the Pinch, results in the power flow consumed by the heat pump being added to the cooling utility demand Below the Pinch.

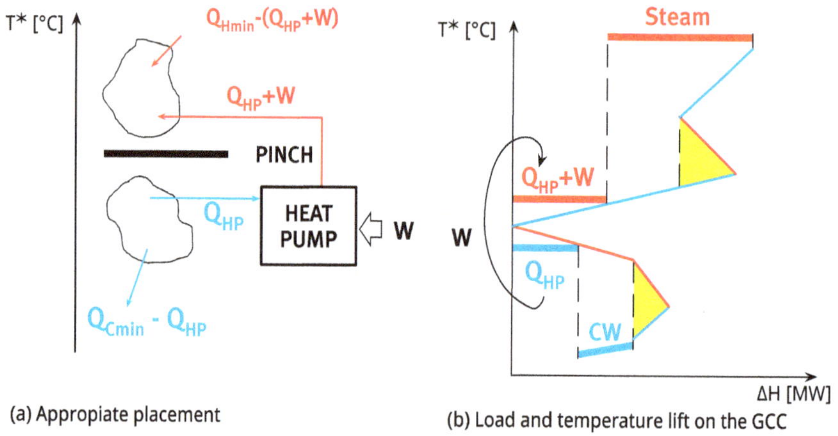

(a) Appropiate placement

(b) Load and temperature lift on the GCC

Fig. 2.50: Heat pump placement for a Heat Recovery problem.

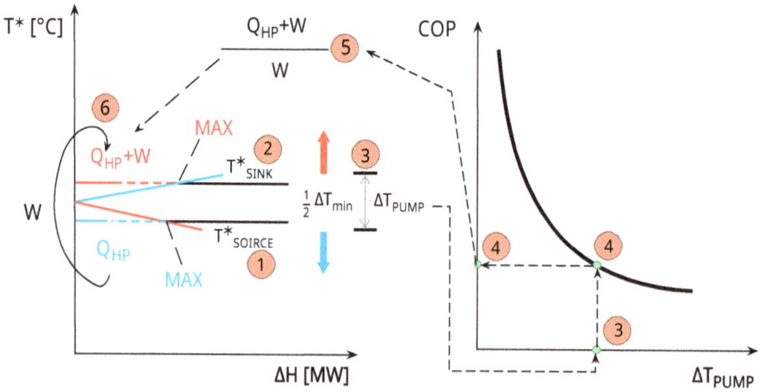

Fig. 2.51: Procedure for heat-pump sizing.

The procedure for sizing heat pumps to be placed Across a (process or utility) Pinch is illustrated in Fig. 2.51. First, temperatures are chosen for the heat source and the heat sink. Then, the horizontal projections spanning from the temperature axis to the GCC provide the maximum values for the heat source and sink loads. Recall that the GCC shows shifted temperatures. As true temperatures are used when calculating the actual heat-pump temperature lift, the GCC values must be modified by subtracting or adding $\Delta T_{min}/2$ (see Section 2.2.3.7). From the calculated ΔT_{pump} one can derive the COP, which is then used to calculate the necessary duties.

As a concrete example, assume that the GCC in Fig. 2.52 reflects an industrial process with $\Delta T_{min} = 20$ °C and a heat pump is available, described as follows:

$$COP = 100.18 \cdot DT_{pump}^{-0.874}. \tag{2.15}$$

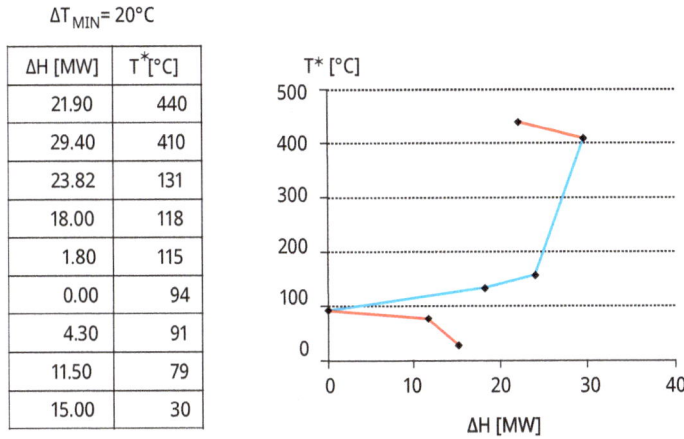

$\Delta T_{MIN} = 20°C$

ΔH [MW]	T^*[°C]
21.90	440
29.40	410
23.82	131
18.00	118
1.80	115
0.00	94
4.30	91
11.50	79
15.00	30

Fig. 2.52: Heat-pump sizing example: initial data.

Focusing on the Pinch "Nose" (a sharp nose provides a better integration option, but also watch the scale of both axes) allows choosing a shifted temperature for the heat source, in this example: $T^*_{source} = 85$ °C: see Fig. 2.53. Using this value allows one to extract a maximum of 6.9 MW from the GCC Below the Pinch. Setting $T^*_{sink} = 100$ °C results in an upper bound of 2.634 MW for the sink load. Transforming to real temperatures and taking the difference yields a temperature lift of $\Delta T_{pump} = 35$ °C. By eq. (2.15), the COP is thus equal to 4.4799. Given the upper bounds on the sink and source heat loads, the smaller one is chosen as a basis. Here, the sink bound is smaller, so the sink is sized to its upper bound: $Q_{sink} = Q_{sink,max} = 2.634$ MW. From this, the required pump power consumption is computed to be 0.588 MW. As a result, the actual source load for the heat pump is 2.046 MW. Comparing this value with the upper bound of 6.9 MW, it is evident that the source heat availability is considerably underutilised.

A different combination of source and sink temperatures is needed if the source availability is to be better utilised. As a second attempt, the sink temperature is increased from 100 to 110 °C. The maximum source heat remains 6.9 MW, but the maximum sink

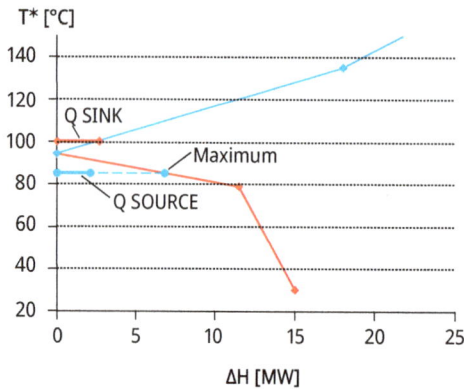

Fig. 2.53: Heat-pump sizing example: Attempt 1.

capacity increases from 2.634 to 7.024 MW. This results in the desired temperature lift of $\Delta T_{pump} = 45$ °C; the COP is now 3.5964, $W = 2.657$ MW, and the sink load $Q_{sink} = 9.557$ MW (Fig. 2.54).

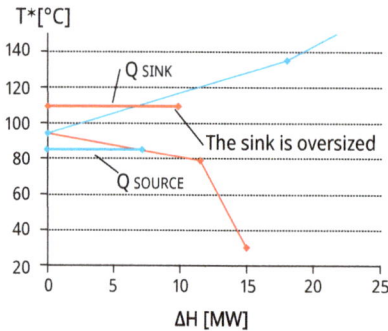

Fig. 2.54: Heat-pump sizing example: Attempt 2.

However, the heat sink is oversized in this new state (Fig. 2.54), so the heat source temperature must be shifted upward. An increase of 2.28 °C (see Fig. 2.55) yields the following values: $\Delta T_{pump} = 42.72$ °C; a maximum source load (also taken as the actual source load) of 5.152 MW; a maximum sink load equal to the actual sink load of $Q_{sink} = 7.016$ MW; COP = 3.7637; and $W = 1.864$ MW. Better results can be obtained by optimising the two temperatures simultaneously, while using overall utility cost as a criterion.

Once all utility levels are chosen, the heat pump can also be placed Across a Utility Pinch (Fig. 2.56). A special case of integration involves the placement of refrigeration levels (Fig. 2.57). Refrigeration facilities are actually heat pumps whose primary value is that their cold end absorbs heat. Utilising their hot ends for process heating can save considerable amounts of hot utility, especially when relatively low-temperature heating is needed.

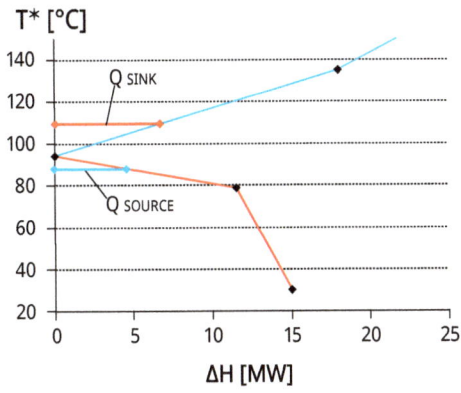

Fig. 2.55: Heat-pump sizing example: Attempt 3.

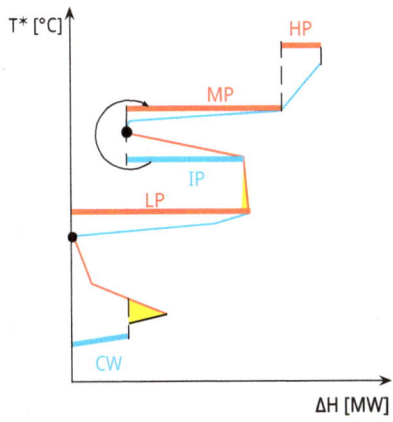

Fig. 2.56: Heat-pump placement Across the Utility Pinch.

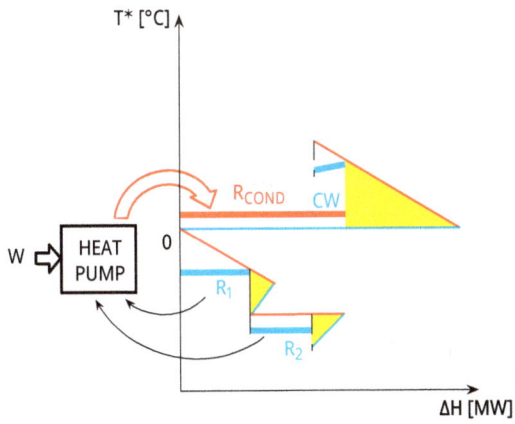

Fig. 2.57: Refrigeration systems.

Distillation processes and other separators

The simplest distillation column includes one reboiler and one condenser, and these components account for most of the column's energy demands. For purposes of Heat Integration, the column is represented by a rectangle: the top side denotes the reboiler as a cold stream, and the bottom side denotes the condenser as a hot stream; see Fig. 2.58.

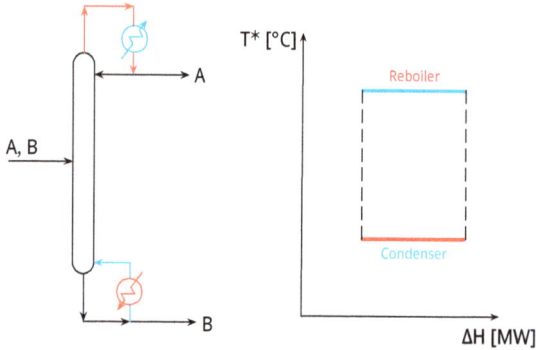

Fig. 2.58: Distillation column: T-H representation.

There are three options for integrating distillation columns: Across the Pinch, as shown in Fig. 2.59a; entirely Below; or entirely Above the Pinch, as shown in Fig. 2.59b. Integrating Across the Pinch only increases overall energy needs, so this option is not useful. The other two options result in net benefits by eliminating the need to use an external utility for supplying the distillation reboiler (Below the Pinch) or the condenser (Above the Pinch). The GCC is used to identify the appropriate column integration options. Figure 2.60 illustrates two examples of appropriately placed distillation columns.

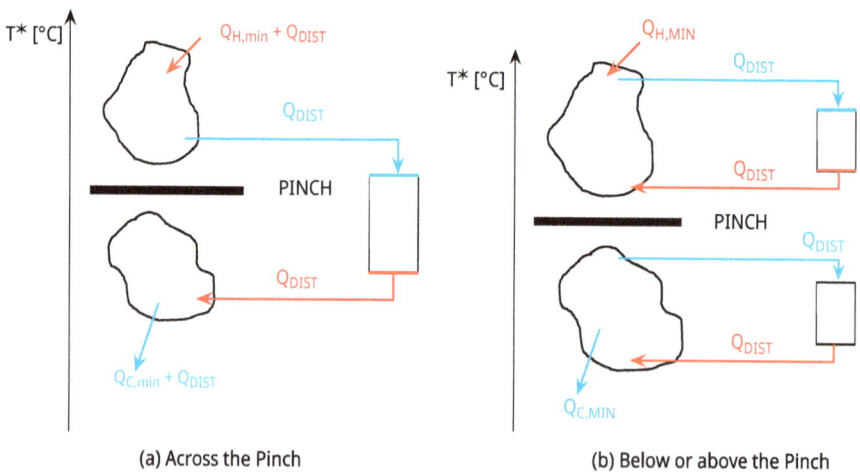

(a) Across the Pinch

(b) Below or above the Pinch

Fig. 2.59: Distillation column: integration options.

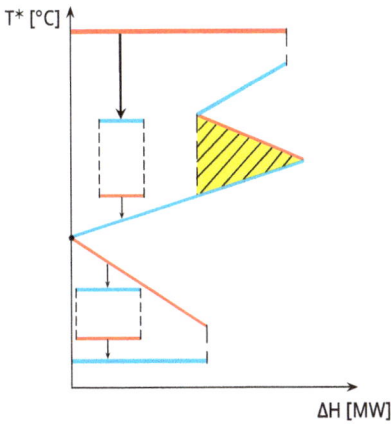

Fig. 2.60: Appropriate placement of distillation columns in terms of the GCC.

With regard to the operating conditions of a column, if placed Across the Pinch, several degrees of freedom can be utilised to change this and facilitate appropriate placement. One option is to change the operating pressure, which would shift the column with respect to the temperature scale until it fits Above or Below the Pinch. The second degree of freedom is to vary the reflux ratio, which results in simultaneous changes in the column temperature span and the duties of the reboiler and condenser. Increasing the reflux ratio yields a smaller temperature span and larger duties, whereas reducing the ratio has the opposite effect. It is also possible to split the column into two parts, thereby introducing a "double effect" distillation arrangement. In this approach, one of the effects is placed Below and the other Above the Pinch, which prevents internal thermal integration of the column effects.

Additional options are available, such as inter-reboiling and inter-condensing. When using available degrees of freedom, one has to keep in mind that the energy–capital trade-offs of the column and the main process are combined and thus become more complicated than the individual trade-offs. Another important issue is the controllability of the integrated designs: unnecessary complications should be avoided, and disturbance propagation paths should be discontinued. It is usually enough to integrate the reboiler or the condenser. If inappropriate column placement cannot be avoided, then it may be necessary to use condenser vapour recompression with a heat pump to heat up the reboiler.

Evaporators constitute another class of thermal separators. Because they also feature a reboiler and a condenser, their operation is similar to that of distillation columns. Therefore, the same integration principles can be applied to them. Absorber and stripper loops and dryers are similarly integrated.

Process modifications

The basic Pinch Analysis calculations assume that the core process layers in the onion diagram (Fig. 2.3, p. 20) remain fixed. However, it is possible, and in some cases beneficial, to alter certain properties of the process. Properties that can be exploited as degrees of freedom include: 1) the pressure, temperature, or conversion rate in reactors; 2) the pressure, reflux ratio, or pump-around flow rate in distillation columns; and 3) the pressure of feed streams to evaporators and pressure inside evaporators.

Such modifications will also alter the heat capacity flow rates and temperatures of the related process streams for Heat Integration; this will cause further changes in the shapes of the CCs and the GCC, thereby modifying the utility targets. The CCs are a very valuable tool for suggesting beneficial process modifications (Linnhoff et al., 1982). Figure 2.61 illustrates this application of the CCs in terms of the Plus-Minus Principle. The main idea is to alter the CC's slope in the proper direction in order to reduce the amount of utilities needed. This can be achieved by changing CP (e.g. by mass flow variation). Such decreases in utility requirements can be brought about by (Smith, 2016) 1) increases in the total hot stream duty Above the Pinch, 2) decreases in the total cold stream duty Above the Pinch, 3) decreases in the total hot stream duty Below the Pinch, and/or 4) increases in the total cold stream duty Below the Pinch.

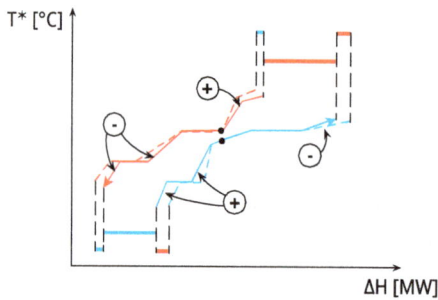

Fig. 2.61: The Plus-Minus Principle.

Another guide to modifying processes is the principle of *Keep Hot Streams Hot* (KHSH) and *Keep Cold Streams Cold* (KCSC). As illustrated in Fig. 2.62, increasing the temperature differences by process modification allows for more overlap of the curves and results in improved Heat Recovery. Energy targets improve if the heating and cooling demands can be shifted Across the Pinch. The principle suggests: 1) shifting hot streams from Below to Above the Pinch and/or 2) shifting cold streams from Above to Below the Pinch.

2.3 Summary

This chapter has provided an introduction to the most established methodology for evaluating Heat Recovery problems, combining as well as improving processes with the goal of minimising process utility demands. This is the methodology united under the term "Heat Integration". Applying Heat Integration allows the identification of the thermodynamic limitations of a Heat Recovery problem in the form of the targets for

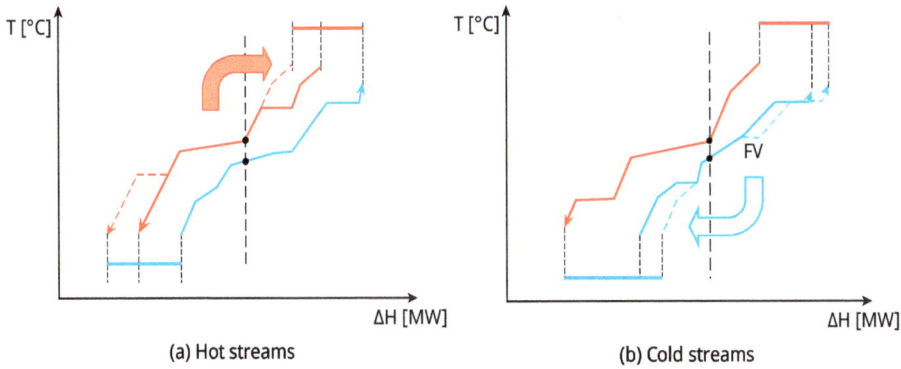

Fig. 2.62: Keep Hot Streams Hot (KHSH) / Keep Cold Streams Cold (KCSC).

Minimum Utility Heating, Minimum Utility Cooling, and the Heat Recovery Pinch. These targets can be further used for completing related tasks:

- Estimation of the maximum possible performance of a Heat Exchanger Network (HEN) before designing it. This can be done by using the Composite Curves, the Grand Composite Curve, or the Problem Table Algorithm.
- Synthesis and design of Heat Exchanger Networks (HENs). In this context, the utility targets alone can be used as lower bounds on utility demands in mathematical-based optimisation formulations. The utility targets, combined with the Pinch location, can be used for decomposing the Heat Recovery problem into several design areas, delimited by the process Pinch point as well the utilities at various levels defining Utility Pinches.
- Total Site Integration. This is an important extension of Heat Integration to complete industrial or combined sites, which include many industrial processes and/ or other energy users/generators, all of them linked by a common utility system. In this case, the obtained process-level Heat Recovery Targets can be used together to establish the site-wide utility targets and even evaluate the power cogeneration options.

Nomenclature

Acronyms and abbreviations

BFW	Boiler feed water
BCCs	Balanced Composite Curves
COND	Condenser
CW	Cooling water
CC	Composite Curve
CCC	Cold Composite Curve

GCC	Grand Composite Curve
HCC	Hot Composite Curve
HEN	Heat Exchanger Network
HI	Heat Integration
HP	(steam) High pressure (steam)
IP	(steam) Intermediate pressure (steam)
LP	(steam) Low pressure (steam)
MP	(steam) Medium pressure (steam)
PI	Process Integration
PTA	Problem Table Algorithm
REB	Reboiler
TB	Temperature boundary
XP	Cross-Pinch (heat flow)

Symbols

A	Heat transfer area, m^2
$A_{HEN,min}$	Minimum heat transfer area for a HEN, m^2
COP	Coefficient of performance for a thermodynamic machine: heat pump, refrigerator, etc.
CP	Heat capacity flowrate, kW/°C or MW/°C
$C_{p,stream}$	Heat capacity flowrate, kJ/(kg · °C)
H	Enthalpy flow, kW or MW
$h_{IN, COLD}$	Inlet enthalpy of a cold stream, kJ/kg
$h_{OUT,COLD}$	Outlet enthalpy of a cold stream, kJ/kg
$h_{IN,HOT}$	Inlet enthalpy of a hot stream, kJ/kg
h_s/h_{stream}	Heat transfer coefficient (film) of a process stream, kW/(m^2 · °C)
$h_{OUT,HOT}$	Outlet enthalpy of a hot stream, kJ/kg
m_{HOT}	Hot stream mass flow rate, kg/s
m_{COLD}	Cold stream mass flow rate, kg/s
m_{stream}	Stream mass flow rate, kg/s
Q	Heat load/duty, kW or MW
Q_C	Utility cooling demand/load, kW or MW
Q_{Cmin}	Minimum utility cooling, in the context of a process-level Pinch Analysis, kW or MW
Q_{DIST}	Heat load associated with a distillation column, kW or MW
Q_{HE}	Heat exchanger duty, kW or MW
Q_H	Utility heating demand/load, kW or MW
Q_{Hmin}	Minimum utility heating, in the context of a process-level Pinch Analysis, kW or MW
Q_{REC}	Heat recovery load, kW or MW
q_s/q_{stream}	Enthalpy change flow of a process stream or its part, kW or MW
Q_{sink}	Heat load supplied by a heat pump to a heat sink, kW or MW
Q_{source}	Heat load drawn by a heat pump from a heat source, kW or MW
$Q_{UC,above}$	Cold utility load Above the Pinch, kW or MW
$Q_{UH,below}$	Hot utility load Below the Pinch, kW or MW
T	Temperature, °C
T^*_{PINCH}	Pinch shifted temperature, °C
$T_{HOT\ PINCH}$	Pinch temperature for the hot streams, °C
$T_{COLD\ PINCH}$	Pinch temperature for the cold streams, °C

T^*	Shifted temperature – mainly used in the process-level Pinch Analysis, for the Grand Composite Curve, °C
T_S^*	Shifted starting temperature, °C
T_T^*	Shifted target (end) temperature, °C
T_{sink}	Temperature of a heat sink (of a heat pump), °C
T_{source}	Temperature of a heat source (of a heat pump), °C
T_S/T_{start}	Starting (supply) temperature, °C
T_T/T_{end}	Ending (target) temperature, °C
T_x	Temperature whose value has to be calculated, °C
U	Overall heat transfer coefficient, kW/(m² · °C)
W	Power, kW, MW
XP	Cross-Pinch heat transfer, kW or MW
ΔT_{LM}	Logarithmic-mean temperature difference (for a heat exchanger), °C
ΔT_{min}	Minimum allowed temperature difference, °C
ΔT_{PUMP}	Temperature lift of a heat pump, °C

References

Ahmad S, Linnhoff B, Smith R. (1989). Supertargeting: Different Process Structures for Different Economics, Journal of Energy Resources Technology, 111(3), 131–136.

Ahmad S, Linnhoff B, Smith R. (1990). Cost Optimum Heat Exchanger Networks: 2. Targets and Design for Detailed Capital Cost Models, Computers & Chemical Engineering, 14(7), 751–767.

Ahmad S, Smith R. (1989). Targets and Design for Minimum Number of Shells in Heat Exchanger Networks, Chemical Engineering Research and Design, 67(5), 481–494.

Atkins M J, Walmsley M R W, Neale J R. (2010). The Challenge of Integrating Non-continuous Processes – Milk Powder Plant Case Study, Journal of Cleaner Production, 18, 927–934.

Boldyryev S, Varbanov P S, Nemet A, Klemeš J J, Kapustenko P. (2013). Capital Cost Assessment for Total Site Power Cogeneration. In: Kraslawski A, Turunen I (eds.), Computer Aided Chemical Engineering, 23 European Symposium on Computer Aided Process Engineering. Elsevier, pp 361–366. doi: 10.1016/B978-0-444-63234-0.50061-0.

Boldyryev S, Varbanov P S, Nemet A, Klemeš J J, Kapustenko P. (2014). Minimum Heat Transfer Area for Total Site Heat Recovery, Energy Convers Management, 87, 1093–1097. doi: 10.1016/j.enconman.2014.04.029.

Cerdá J, Galli M R, Camussi N, Isla M A. (1990). Synthesis of Flexible Heat Exchanger Networks – I. Convex Networks, Computers & Chemical Engineering, 14(2), 197–211.

Colberg R D, Morari M. (1990). Area and Capital Cost Targets for Heat Exchanger Network Synthesis with Constrained Matches and Unequal Heat Transfer Coefficients, Computers & Chemical Engineering, 14(1), 1–22.

CPI (Centre for Process Integration). (2004). Heat Integration and Energy Systems (Msc Course, DPI UMIST), Manchester, UK.

Daichendt M M, Grossmann I E. (1997). Integration of Hierarchical Decomposition and Mathematical Programming for the Synthesis of Process Flowsheets, Computers & Chemical Engineering, 22(1–2), 147–175.

Foo D C Y, Chew Y H, Lee C T. (2008). Minimum Units Targeting and Network Evolution for Batch Heat Exchanger Network, Applied Thermal Engineering, 28(16), 2089–2099.

Friedler F. (2009). Process Integration, Modelling and Optimisation for Energy Saving and Pollution Reduction, Chemical Engineering Transaction, 18, 1–26.

Friedler F. (2010). Process Integration, Modelling and Optimisation for Energy Saving and Pollution Reduction, Applied Thermal Engineering, 30(16), 2270–2280.

Furman K C, Sahinidis N V. (2002). A Critical Review and Annotated Bibliography for Heat Exchanger Network Synthesis in the 20th Century, Industrial & Engineering Chemistry Research, 41, 2335–2370.

Gogenko A L, Anipko O B, Arsenyeva O P, Kapustenko P O. (2007). Accounting for Fouling in Plate Heat Exchanger Design, Chemical Engineering Transaction, 12, 207–212.

Hall S G, Ahmad S, Smith R. (1990). Capital Cost Targets for Heat Exchanger Networks Comprising Mixed Materials of Construction, Pressure Ratings and Exchanger Types, Computers & Chemical Engineering, 14(3), 319–335.

Jegede F O, Polley G T. (1992). Capital Cost Targets for Networks with Nonuniform Heat Exchanger Specifications, Computers & Chemical Engineering, 16(5), 477–495.

Klemeš J, Friedler F, Bulatov I, Varbanov P. (2010). Sustainability in the Process Industry – Integration and Optimization. McGraw-Hill, New York, USA.

Klemeš J J. (ed). (2022). Handbook of Process Integration (PI): Minimisation of Energy and Water Use, Waste and Emissions, 2nd updated edition. Woodhead/Elsevier, Cambridge, UK. ISBN: 978-0128238509.

Klemeš J J, Arsenyeva O, Kapustenko P, Tovazhnyanskyy L. (2015). Compact Heat Exchangers for Energy Transfer Intensification. CRC Press, Taylor & Francis Company, New York, USA, ISBN-13: 978-1482232592, 372 ps.

Klemeš J J, Varbanov P S, Kravanja Z. (2013). Recent Developments in Process Integration, Chemical Engineering Research and Design, 91, 2037–2053. doi: 10.1016/j.cherd.2013.08.019.

Klemeš J J, Varbanov P S, Walmsley T G, Jia X. (2018). New Directions in the Implementation of Pinch Methodology (PM), Renewable and Sustainable, Energy Reviews, 98, 439–468. doi: 10.1016/j.rser.2018.09.030.

Laue H J. (2006). Heat Pumps. In Clauser C, Strobl T, Zunic F (eds.). Renewable Energy, Vol. 3C. Springer, Berlin, Germany, DOI: 10.1007/10858992_21, Book DOI: 10.1007/b83039, ISBN 978-3-540-45662-9.

Linnhoff B, Ahmad S. (1990). Cost Optimum Heat Exchanger Networks. 1. Minimum Energy and Capital Using Simple Models for Capital Cost, Computers & Chemical Engineering, 14(7), 729–750.

Linnhoff B, Flower J R. (1978). Synthesis of Heat Exchanger Networks: I. Systematic Generation of Energy Optimal Networks, AIChE Journal, 24(4), 633–642.

Linnhoff B, Hindmarsh E. (1983). The Pinch Design Method for Heat Exchanger Networks, Chemical Engineering Science, 38(5), 745–763.

Linnhoff B, Townsend D W, Boland D, Hewitt G F, Thomas B E A, Guy A R, Marsland R H. (1982). A User Guide to Process Integration for the Efficient Use of Energy. IChemE, last edition, Rugby, UK, 1994.

Magege S R, Majozi T. (2021). A Comprehensive Framework for Synthesis and Design of Heat-Integrated Batch Plants: Consideration of Intermittently-available Streams, Renewable and Sustainable, Energy Reviews, 135, 110125. doi: 10.1016/j.rser.2020.110125.

Masso A H, Rudd D F. (1969). The Synthesis of System Designs II. Heuristic Structuring, AIChE Journal, 15(1), 10–17.

Pan M, Bulatov I, Smith R, Kim J K. (2013a). Optimisation for the Retrofit of Large Scale Heat Exchanger Networks with Different Intensified Heat Transfer Techniques, Applied Thermal Engineering, 53(2), 373–386.

Pan M, Jamaliniya S, Smith R, Bulatov I, Gough M, Higley T, Droegemueller P. (2013b). New Insights to Implement Heat Transfer Intensification for Shell and Tube Heat Exchangers, Energy, doi: 10.1016/j.energy.2013.01.017.

Santos L C, Zemp R J. (2000). Energy and Capital Targets for Constrained Heat Exchanger Networks, Brazil Journal of Chemical Engineering, 17(4–7), 659–670.

Seferlis P, Varbanov P S, Papadopoulos A I, Chin H H, Klemeš J J. (2021). Sustainable Design, Integration, and Operation for Energy High-performance Process Systems, Energy, 224, 120158. doi: 10.1016/j.energy.2021.120158.

Serna-González M, Jiménez-Gutiérrez A, Ponce-Ortega J M. (2007). Targets for Heat Exchanger Network Synthesis with Different Heat Transfer Coefficients and Nonuniform Exchanger Specifications, Chemical Engineering Research and Design, 85(10), 1447–1457.

Shah R K, Sekulić D P. (2003). Fundamentals of Heat Exchanger Design. Wiley, New York, USA.

Shilling R L, Berhhagen P M, Goldschmidt V M, Hrnjak P S, Johnson D, Timmerhaus K D. (2008). Heat-transfer equipment, In Green D W, Perry R H. (eds.), Perry's Chemical Engineers Handbook, 8th ed, chap. 11, New York, USA: McGraw-Hill, 11-1–11-121.

Smith R. (2016). Chemical Process Design and Integration, 2nd ed. John Wiley & Sons, Chichester, UK. ISBN: 978-1119990147.

Sun K N, Wan Alwi S R, Manan Z A. (2013). Heat Exchanger Network Cost Optimization Considering Multiple Utilities and Different Types of Heat Exchangers, Computers & Chemical Engineering, 49, 194–204. doi: 10.1016/j.compchemeng.2012.10.017.

Taal M, Bulatov I, Klemeš J, Stehlik P. (2003). Cost Estimation and Energy Price Forecast for Economic Evaluation of Retrofit Projects, Applied Thermal Engineering, 23, 1819–1835.

Ten Broeck H. (1944). Economic Selection of Exchanger Sizes, Industrial & Engineering Chemistry Research, 36(1), 64–67.

Townsend D W, Linnhoff B. (1983). Heat and Power Networks in Process Design. Part II: Design Procedure for Equipment Selection and Process Matching, AIChE Journal, 29(5), 748–771.

Townsend D W, Linnhoff B. (1984). Surface Area Targets for Heat Exchanger Networks, IChemE 11th Annual Research Meeting, Bath, U.K., April.

Varbanov P, Klemeš J. (2008). Analysis and Integration of Fuel Cell Combined Cycles for Development of Low-carbon Energy Technologies, Energy, 33(10), 1508–1517.

Varbanov P, Klemeš J, Shah R K, Shihn H. (2006). Power Cycle Integration and Efficiency Increase of Molten Carbonate Fuel Cell Systems, Journal of Fuel Cell Science and Technology, 3(4), 375–383.

Wan Alwi, S.R., Manan, Z.A., 2010. STEP—A new graphical tool for simultaneous targeting and design of a heat exchanger network. Chemical Engineering Journal 162, 106–121. https://doi.org/10.1016/j.cej.2010.05.009.

Wang B, Klemeš J J, Li N, Zeng M, Varbanov P S, Liang Y. (2021). Heat Exchanger Network Retrofit with Heat Exchanger and Material Type Selection: A Review and a Novel Method, Renewable and Sustainable, Energy Reviews, 138, 110479. doi: 10.1016/j.rser.2020.110479.

Zhu X X, O'Neill B K, Roach J R, Wood R M. (1995). Area-targeting Methods for the Direct Synthesis of Heat Exchanger Networks with Unequal Film Coefficients, Computers & Chemical Engineering, 19(2), 223–239.

Zhu X X, Zanfir M, Klemeš J. (2000). Heat Transfer Enhancement for Heat Exchanger Network Retrofit, Heat Transfer Engineering, 21(2), 7–18.

3 Synthesis of Heat Exchanger Networks

Having established how much Heat Recovery can be achieved at most and having realised the thermodynamic limitations posed by the process streams, that information can be used to synthesise Heat Recovery systems using heat exchangers – Heat Exchanger Networks (HEN). This procedure is frequently referred to as Heat Exchanger Network Synthesis (HENS). There are plenty of methods for performing this task – some heuristic, others based on Mathematical Programming as well as evolutionary ones using Genetic Algorithms and Simulated Annealing.

This chapter presents a systematic procedure for synthesising HENs using simple reasoning and calculations. This is a fundamental method – named the Pinch Design Method – that brings the necessary understanding of HENs, which allows experienced engineers as well as beginners to apply it and benefit from the offered insights. Other synthesis methods are also discussed as well as the fundamentals of HEN retrofit.

3.1 Introduction

In Chapter 2, "Setting energy targets and Heat Integration", Pinch Analysis was introduced. This is a powerful method based on sound thermodynamic concepts, applied to the overall Heat Recovery problem for a process. The targets obtained by Pinch Analysis can be used as lower bounds on utility demand in Mathematical Programming procedures. However, the real power of the method is revealed when it is realised that the Pinch divides the problem into a net heat sink Above the Pinch and a net heat source Below the Pinch. This provides a means for problem decomposition, i.e. replacing one large and complex problem with an equivalent combination of two or more problems that are an order of magnitude simpler. A second benefit of the Pinch division is that the thermodynamic bottleneck identified by the Pinch, combined with a requirement for maximum Heat Recovery, provides a set of rules for Heat Exchange match placement at the Pinch and further evolution of the design. This is the fundamental thinking behind the Pinch Design Method.

This chapter presents the Pinch Design Method in detail and then discusses other HEN synthesis methods, followed by a presentation of the basic features of HEN retrofit. Working sessions are also provided to help improve and consolidate the understanding of the material. At the end of the chapter, an additional section for the current edition is provided, which illustrates step-by-step one possible implementation of the Data Extraction procedure that is based on requirements for data integrity and continuity. This allows engineers from collaborating teams to back-trace the logic and the reasoning for certain Data Extraction decisions.

https://doi.org/10.1515/9783110782981-003

3.2 HEN synthesis

Most industrial-scale methods synthesise Heat Recovery networks under the assumption of a steady state. This assumption is also adopted in the current chapter.

3.2.1 The Pinch Design Method

The traditional Pinch Design Method (Linnhoff and Hindmarsh, 1983) has become popular owing to its simplicity and efficient management of complexity. The method has evolved into a complete suite of tools for Heat Recovery and design methods for energy efficiency, including guidelines for changing and integrating a number of energy-intensive processes. It has been developed over the years, and several overviews have been published. Let us mention at least the review of Klemeš et al. (2018) and the most recent handbook Klemeš (2022).

3.2.1.1 HEN representation

The representation of a Heat Exchanger Network by a general process flowsheet, as in Fig. 3.1, is not convenient. The reason is that this representation makes it difficult

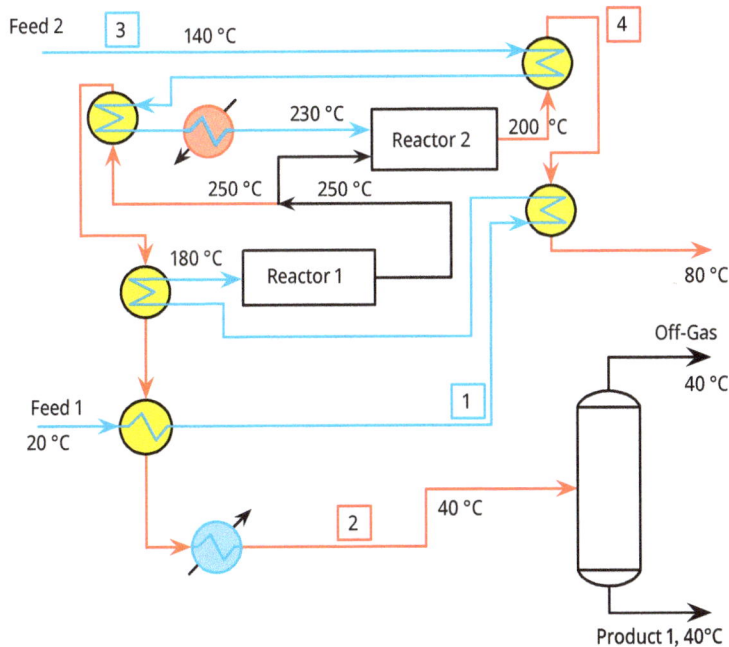

Fig. 3.1: Using a general process flowsheet to represent a HEN.

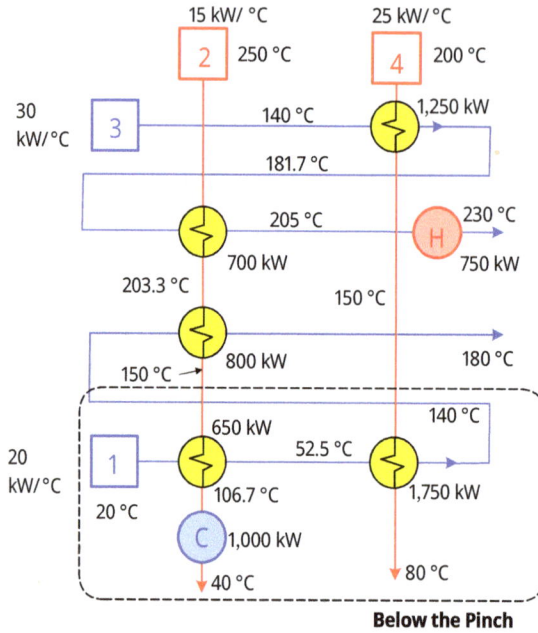

Fig. 3.2: Conventional HEN flowsheet.

to answer a number of important questions: "Where is the Pinch?", "What is the degree of Heat Recovery?", "How much cooling and heating from utilities is needed?"

The example given in Fig. 3.1 results in the streams data set from Tab. 3.1. The so-called conventional HEN flowsheet (Fig. 3.2) offers a small improvement. It shows only heat transfer operations and is based on a simple convention: cold streams are depicted horizontally and hot streams vertically. Although the Pinch Location can be marked for simple cases, it is still difficult to see. This representation makes it difficult to express the proper sequencing of heat exchangers and to represent clearly the network temperatures. Changing the positions of some matches often results in complicated path representations.

Tab. 3.1: PTA example: process streams data (ΔT_{min} = 10 °C).

No.	Type	T_S (°C)	T_T (°C)	CP (kW/°C)
1	Cold	20	180	20
2	Hot	250	40	15
3	Cold	140	230	30
4	Hot	200	80	25

The Grid Diagram, as shown in Fig. 3.3, provides a convenient and efficient representation of HENs by eliminating the problems just described. It has to be stressed that only heat transfer operations are shown. The process streams are drawn only horizontally. In the base version of the diagram – simplified Fig. 3.3(a) and full Fig. 3.3 (b), the hot streams are represented by arrows pointing from right to left, and the cold streams are represented by arrows pointing from left to right. The heat exchange matches are placed vertically. Each match is drawn as a pair of circles placed on one hot stream and one cold stream, and connected with a vertical line. A more recent version – named Shifted Retrofit Thermodynamic Grid Diagram – Shifted Temperature Range (Wang et al., 2020) is shown in Fig. 3.3(c). This advanced diagram shows explicitly the stream CP values and temperatures on a shifted scale. This type of diagram is especially useful for designing or retrofitting HENs while accounting for different heat exchanger types.

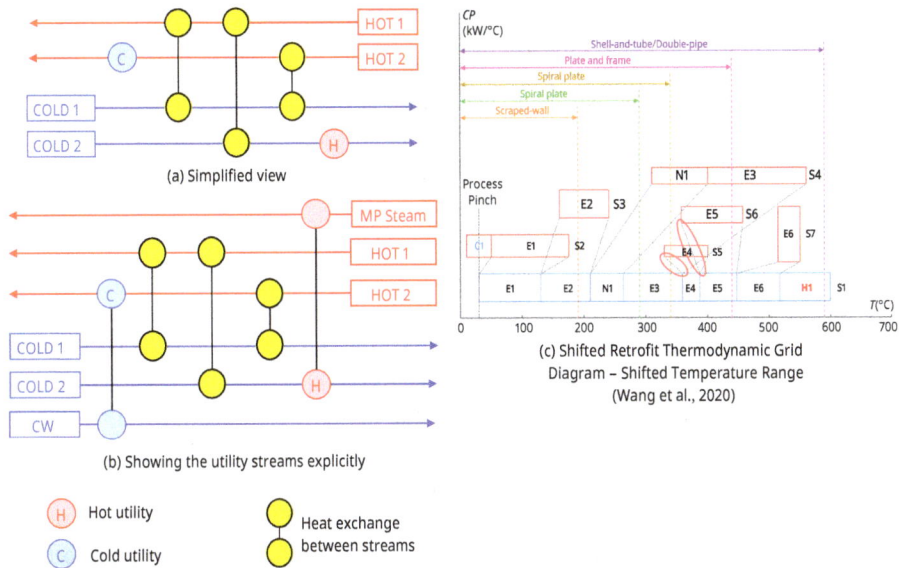

Fig. 3.3: The Grid Diagram for HENs.

The Grid Diagram has several advantages: the representation of streams and heat exchangers is clearer, it is more convenient for representing temperatures, and the Pinch Location (and its implications) are clearly visible; see Fig. 3.4. Temperature increases from left to right in the grid, which is intuitive, and (re)sequencing heat exchangers is straightforward.

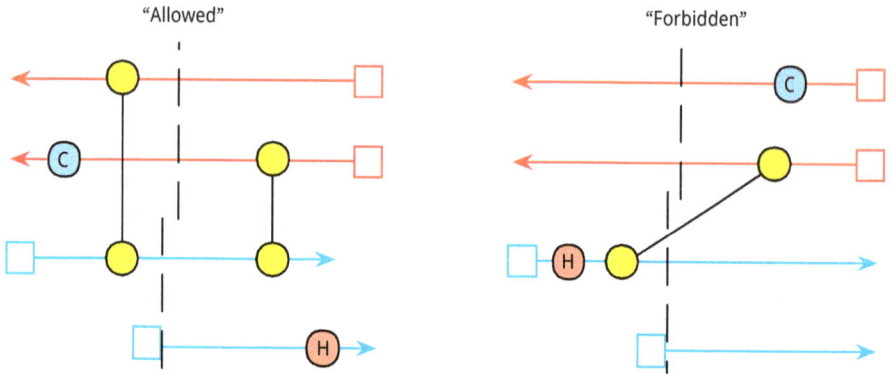

Fig. 3.4: The Grid Diagram and implications of the Pinch.

3.2.1.2 The design procedure

The procedure for designing a HEN follows several simple steps:

1. Specification of the Heat Recovery problem:
2. Identification of the Heat Recovery targets and the Heat Recovery Pinch (Chapter 2, "Setting Targets and Heat Integration");
3. Synthesis;
4. Evolution of the HEN topology.

The first two steps are discussed in Chapter 2, "Setting Targets and Heat Integration". The synthesis step begins by dividing the problem at the Pinch and then positioning the process streams, as shown in Fig. 3.5.

The Pinch Design Principle suggests starting the network design from the Pinch (the most restricted part of the design, owing to temperature differences approaching ΔT_{min}) and then placing heat exchanger matches, while moving away from the Pinch (Fig. 3.6). When placing matches, a few rules have to be followed in order to obtain a network that minimises utility use: i) no exchanger may have a temperature difference smaller than ΔT_{min}; ii) no process-to-process heat transfer may occur Across the Pinch, and iii) no inappropriate use of utilities is allowed.

At the Pinch, the driving force restrictions entail that certain matches have to be made if the design is to achieve the minimum utility usage without violating the ΔT_{min} constraint; these are referred to as essential matches. Above the Pinch, the hot streams may be cooled only by transferring the heat to cold process streams, not to utility cooling, and all hot streams Above the Pinch have to be matched up with cold streams. This means that all hot streams entering the Pinch have to be given priority when matches are made Above the Pinch. Conversely, cold streams entering the Pinch are given priority when matches are made Below the Pinch.

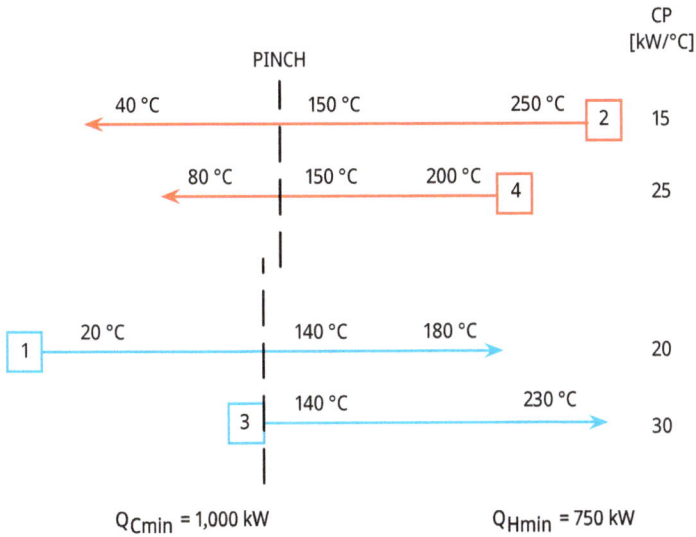

Fig. 3.5: Dividing at the Pinch for the streams from Tab. 3.1.

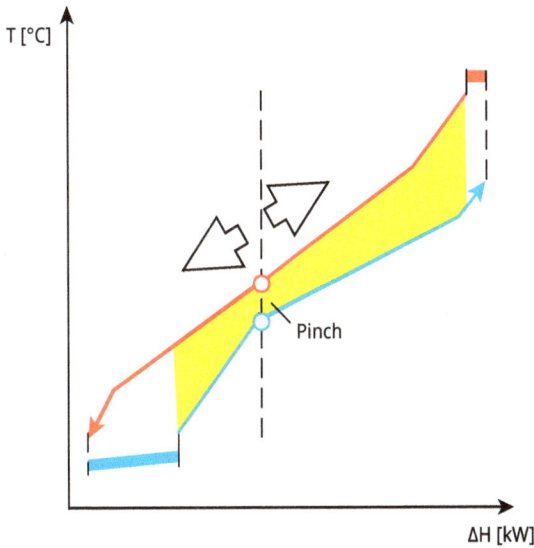

Fig. 3.6: The Pinch Design principle.

Now recall the example from Chapter 2, "Setting Targets and Heat Integration". It is repeated in Tab. 3.1. Fig. 3.5 shows the empty Grid Diagram, which is divided at the Pinch. The hot part (Above the Pinch) requires essential matches for streams 2 and 4 since they are entering the Pinch.

Consider Stream 4. One possibility is to match it against Stream 1, as shown in Fig. 3.7. Stream 4 is the hot stream, and its CP is greater than the CP of cold Stream 1. As shown in the figure, at the Pinch, the temperature distance between the two streams is exactly equal to ΔT_{min}. Moving away from the Pinch results in temperature convergence because the slope of the hot streamline is less steep, owing to its larger CP. Since ΔT_{min} is the lower bound on network temperature differences, the proposed heat exchanger match is infeasible and thus is rejected.

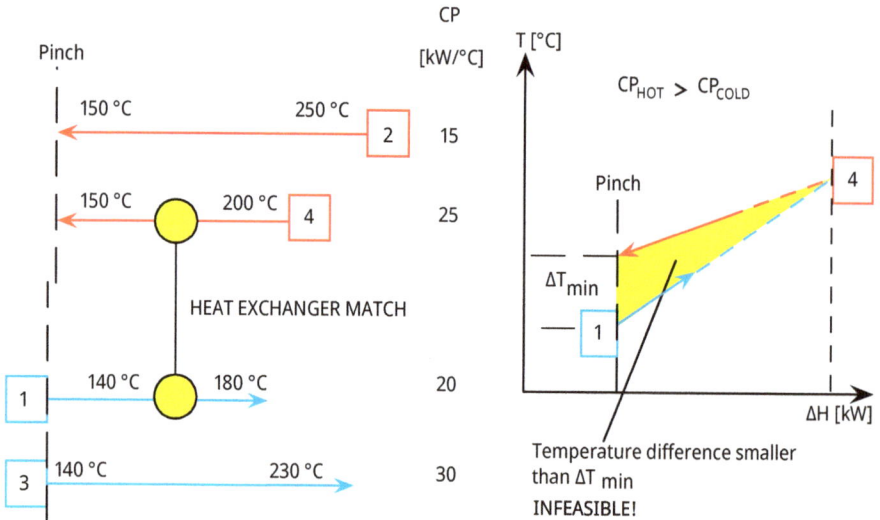

Fig. 3.7: An infeasible heat exchanger match Above the Pinch.

Another possibility for handling the cooling demand of Stream 4 is to implement a match with Stream 3, as shown in Fig. 3.8. The CP of Stream 3 is larger than the CP of Stream 4, resulting in divergent temperature profiles in the direction away from the Pinch. The rule for placing matches Above the Pinch may be expressed as follows:

$$CP_{\text{hot stream}} \leq CP_{\text{cold stream}} \tag{3.1}$$

The picture below the Pinch is symmetric. This part of the design is a net heat source, which means that the heating requirements of the cold streams must be satisfied by matching up with the hot streams. The same type of reasoning as before yields the following requirement: the CP value of a cold stream must not be greater than the CP value of a hot stream if a feasible essential match is to result. Generalising Equation (3.1) shows that the CP of the stream entering the Pinch has to be less than or equal to the CP of the stream leaving the Pinch:

$$CP_{\text{entering Pinch}} \leq CP_{\text{leaving Pinch}} \tag{3.2}$$

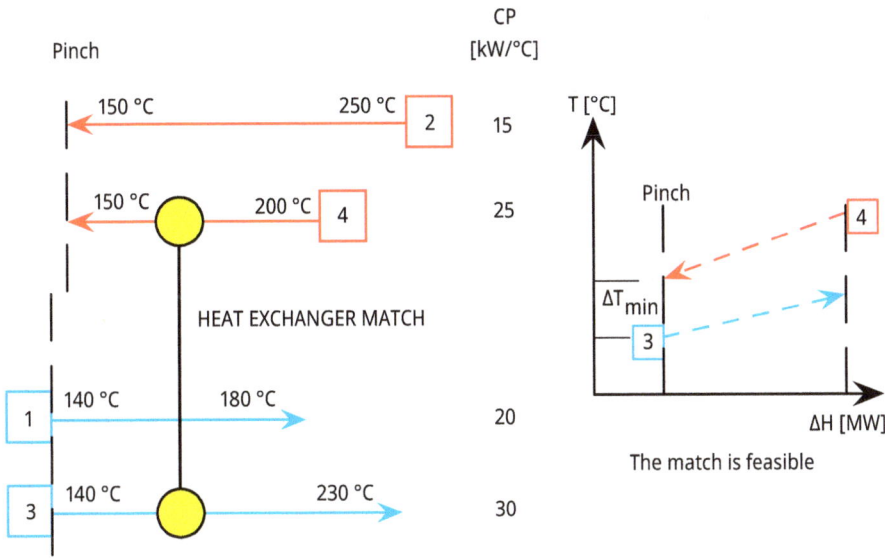

Fig. 3.8: A feasible heat exchanger match Above the Pinch.

The Pinch Design Method incorporates a special tool for handling this stage: CP tables – Fig. 3.9 (Linnhoff and Hindmarsh, 1983). Here the streams are represented by their CP values, which are sorted in descending order. This facilitates the identification of promising combinations of streams as essential match candidates.

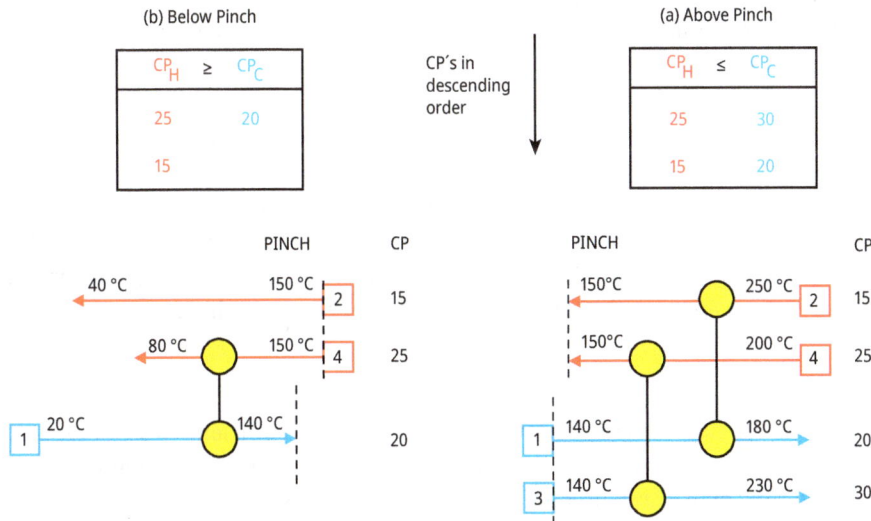

Fig. 3.9: CP tables.

Sizing the matches follows the tick-off heuristic, which stipulates that one of the streams in a Heat Exchange match should be completely satisfied and then "ticked off" from the design list; see Fig. 3.10.

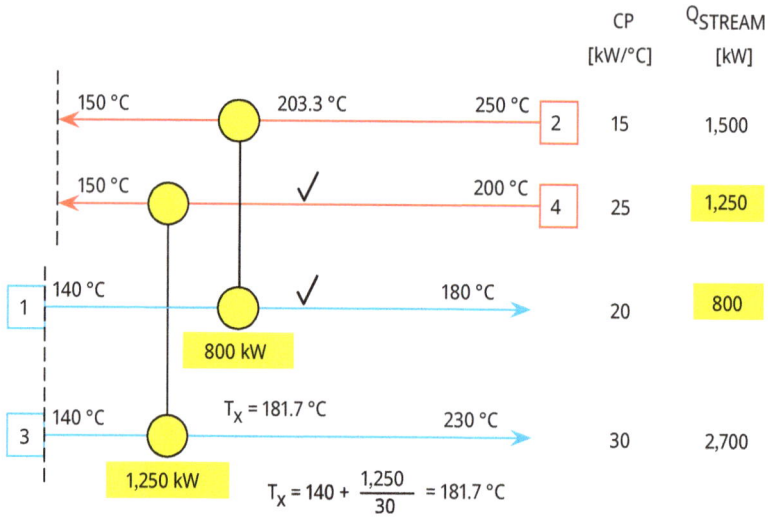

Fig. 3.10: The tick-off heuristic.

3.2.1.3 Completing the design

The HEN design Above the Pinch is illustrated in Fig. 3.11. The design Below the Pinch follows the same basic rules, with the difference that here it is the cold streams that define the essential matches.

Fig. 3.12 details the design Below the Pinch for the considered example. Firstly, the match between Streams 4 and 1 is placed and sized to the duty required by Stream 4. The other match, between Streams 2 and 1, formally violates the Pinch rule for placing essential matches. However, since Stream 1 already has another match at the Pinch, the current match (between 4 and 1) cannot be strictly termed essential. Up to the required duty of 650 kW, this match does not violate the ΔT_{min} constraint, which is the relevant one. The completed HEN topology is shown in Fig. 3.13.

3.2.1.4 Stream splitting

The basic design rules described previously are not always sufficient. In some cases, it is necessary to split the streams so that Heat Exchange matches can be appropriately placed. Splitting may be required in the following situations:

- Above the Pinch when the number of hot streams is greater than the number of cold streams ($N_H > N_C$); see Fig. 3.14;
- Below the Pinch when the number of cold streams is greater than the number of hot streams ($N_C > N_H$); see Fig. 3.15
- When the CP values do not suggest any feasible essential match, see Fig. 3.16.

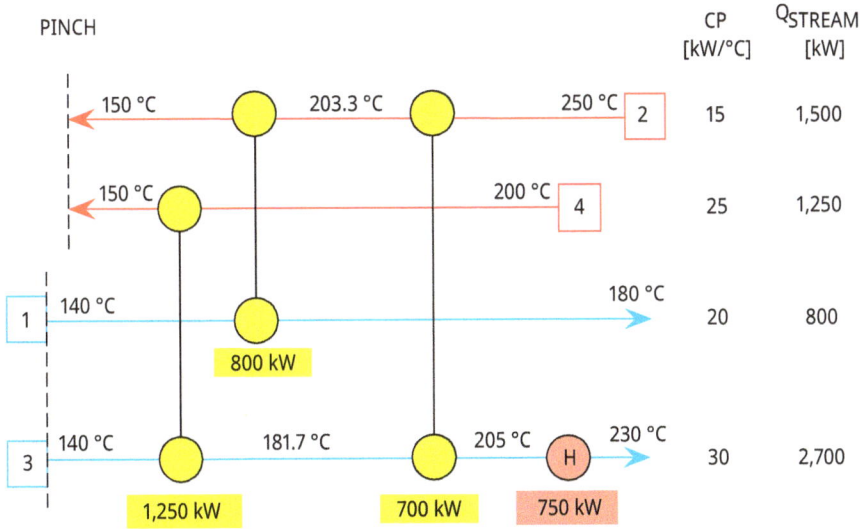

Fig. 3.11: Completing the HEN design Above the Pinch.

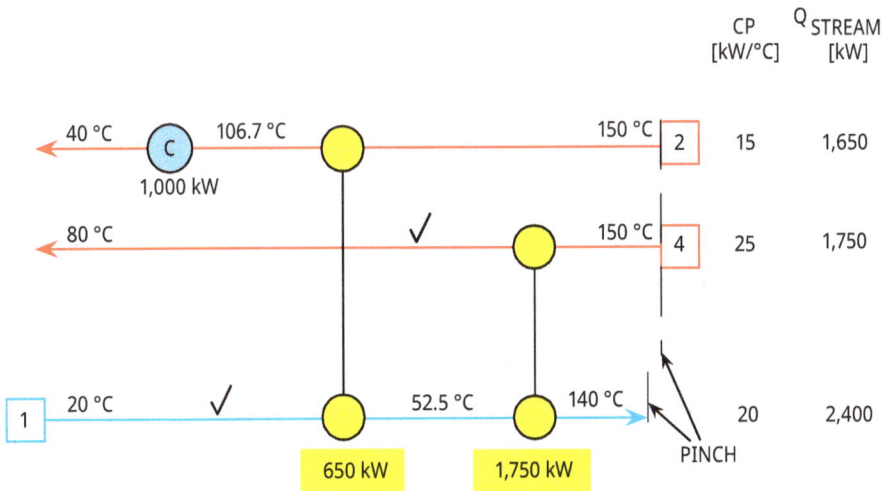

Fig. 3.12: The HEN design Below the Pinch.

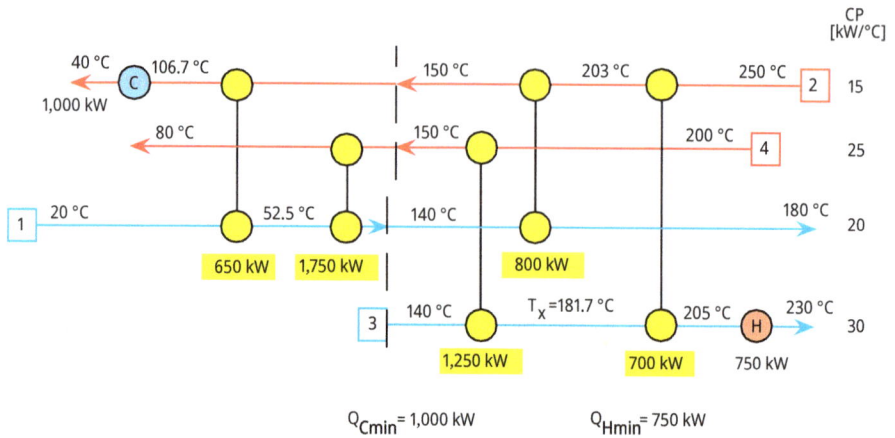

Fig. 3.13: The completed HEN design.

Loads of the matches involving stream branches are again determined using the tick-off heuristic. Because each stream splitter presents an additional degree of freedom, it is necessary to decide how to divide the overall stream CP between the branches. One possibility is illustrated in Fig. 3.17. The suggested split ratio there is 4: 3. This approach would completely satisfy the heating needs of the cold stream branches.

The arrangement in Fig. 3.17 is, in a way, trivial and can be improved on. Unless the stream combinations impose some severe constraints, there is an infinite number of possible split ratios, providing a useful degree of freedom. This fact can be exploited to optimise the network. In many cases, it is possible to save an extra heat exchanger unit by ticking off two streams with a single match, as illustrated in Fig. 3.18, which also satisfies hot Stream 2, in addition to both branches of cold Stream 3, thereby eliminating the second cooler.

The complete algorithm for splitting streams Below the Pinch is given in Fig. 3.19. The procedure for splitting Above the Pinch is symmetrical, with the cold and hot streams switching their roles.

3.2.1.5 Network evolution

The HEN obtained using the design guidelines described previously is optimal with respect to its energy requirements, but it is usually away from the optimum total cost. Observing the Pinch division typically introduces loops into the final topology and leads to a larger number of heat exchanger units. The final step in HEN design is an evolution of the topology: identifying heat load loops and open heat load paths; then using them to optimise the network in terms of heat loads, heat transfer area, and topology. During this phase, formerly rigorous requirements, for example, that all

(a) More Hot Streams than Cold Streams

(b) Split a cold stream

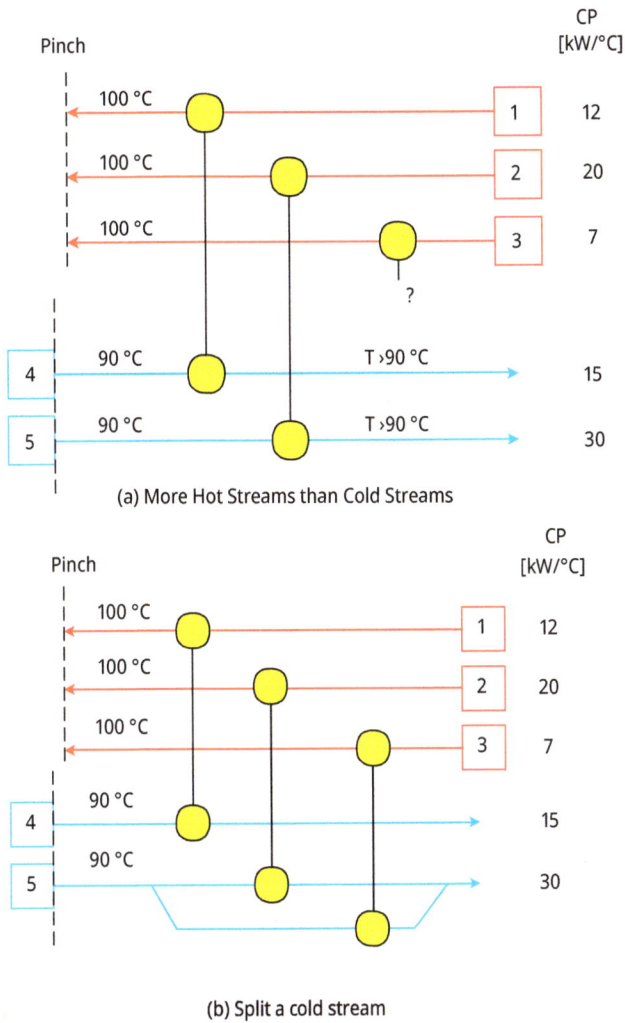

Fig. 3.14: Splitting Above the Pinch for $N_H > N_C$.

temperature differences exceed ΔT_{min} and that Cross-Pinch heat transfers be excluded, are usually relaxed.

The resulting optimisation formulations are typically non-linear and involve structural decisions; consequently, they are MINLP problems. Different approximations and simplifying assumptions can be introduced to obtain linear and/or continuous formulations. The design evolution step can even be performed manually by breaking the loops and reducing the number of heat exchangers.

Eliminating Heat Exchangers from the topology is done at the expense of shifting heat loads (from the eliminated Heat Recovery exchangers) to utility exchangers:

(a) More Cold Streams than Hot Streams

(b) Split a Hot Stream

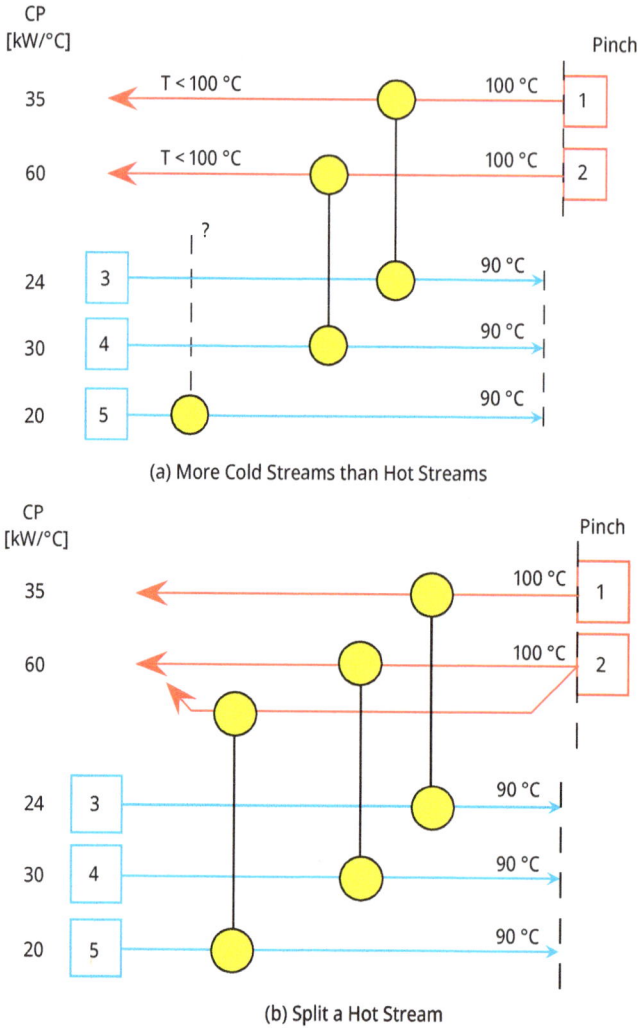

Fig. 3.15: Splitting Below the Pinch for $N_C > N_H$.

heaters and coolers. Topology evolution terminates when the resulting energy cost increase exceeds the projected savings in capital costs, which corresponds to a total cost minimum.

Network evolution is performed by shifting loads within the network toward the end of eliminating excessive heat exchangers and/or reducing the effective heat transfer area. It is necessary to exploit the degrees of freedom provided by loops and utility paths to shift loads. In this context, a loop is a circular closed path connecting two or more heat exchangers, and a utility path connects a hot with a cold utility or connects

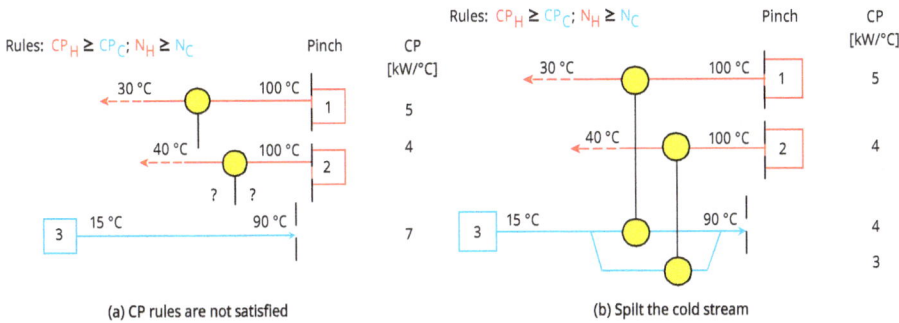

Fig. 3.16: Splitting to enable CP values for essential matches.

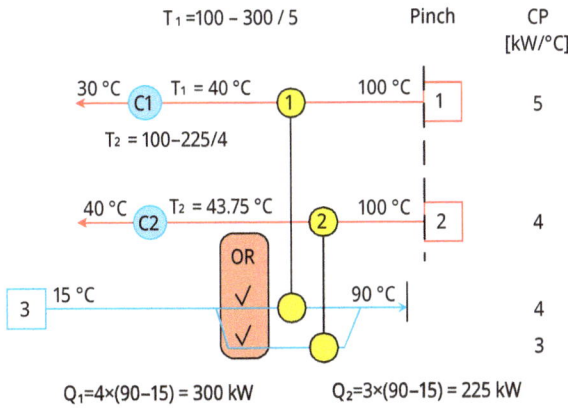

Fig. 3.17: Splitting and (trivial) tick-off.

two utilities of the same type. Figure 3.20 shows a HEN loop and a utility path. A network may contain many such loops and paths.

3.2.1.6 Working Session "HEN design for maximum Heat Recovery"

Assignment

The same flowsheet, as in the working sessions in Chapter 2, "Setting targets and Heat Integration", is considered (Fig. 3.21, Tab. 3.2). The process stream data is the same (Tab. 3.2), so are the remaining specifications of the Heat Recovery problem (ΔT_{min} = 10 °C, Steam at 200 °C, CW at 25 °C). The assignment is to design a network of heat exchangers for Maximum Energy Recovery (MER).

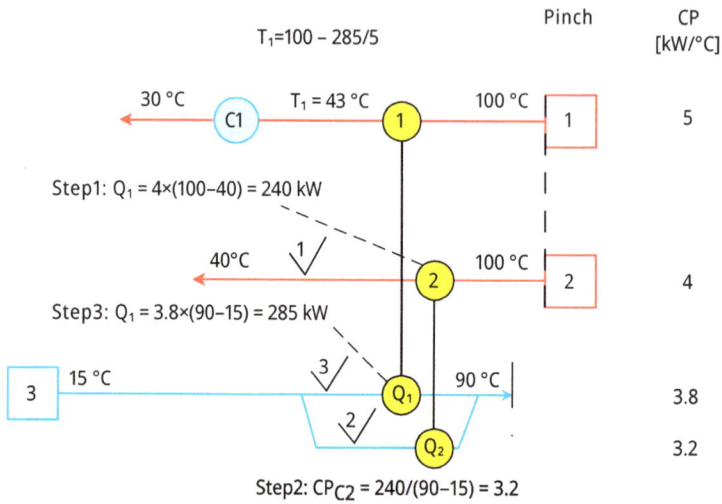

Fig. 3.18: Splitting and advanced tick-off.

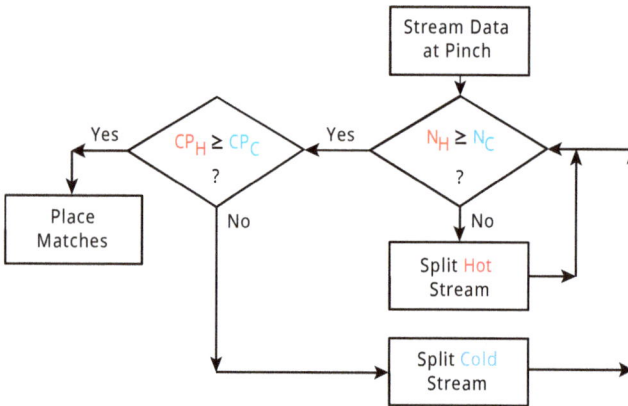

Fig. 3.19: Splitting procedure Below the Pinch.

Solution

The temperature boundaries are calculated first to prepare the construction of the Problem Table. This step is illustrated in Fig. 3.22. Further, the Problem Table is constructed, filled in and calculated, as shown in Fig. 3.23.

The heat cascade is obtained and made feasible, as shown in Fig. 3.24. The cascade also provides all the needed energy targets.

The targets are then used to sketch the empty design grid (Fig. 3.25)

The design Below the Pinch (Fig. 3.26) is simple, as there is only one hot and one cold stream. Since the cold stream (No. 4) has a smaller CP value, starting from the

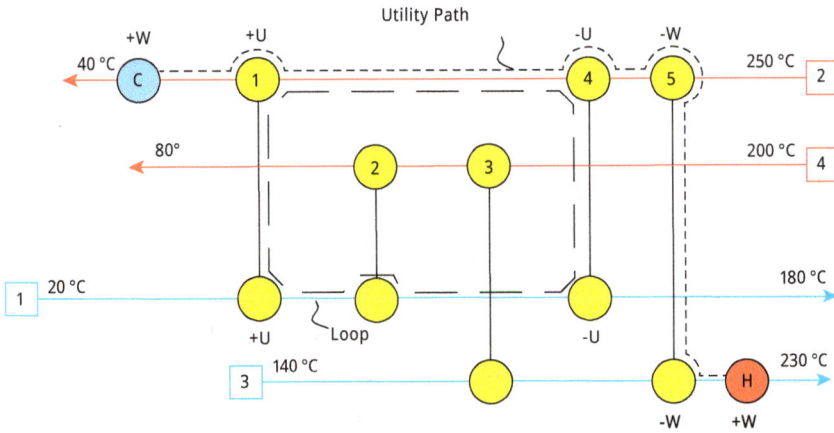

Fig. 3.20: Loop and path in a Heat Exchanger Network.

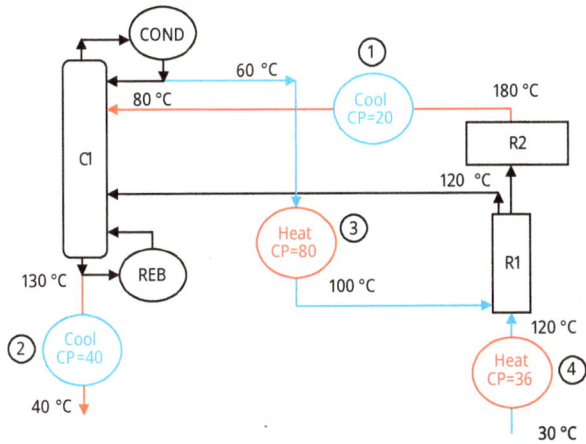

Fig. 3.21: Initial flowsheet for 3.2.1.5 Working Session "Introduction to Heat Integration".

Tab. 3.2: Heat Recovery problem for 3.2.1.5 Working Session "Introduction to Heat Integration".

Stream	T_S (°C)	T_T (°C)	CP (kW/°C)
1: Hot	180	80	20
2: Hot	130	40	40
3: Cold	60	100	80
4: Cold	30	120	36

	Shifted Temperatures				Candidate Boundaries	Eliminate Duplicates

Stream No. and type	T_S [°C]	T_T [°C]	T^*_S [°C]	T^*_T [°C]
1 Hot	180	80	175	75
1 Hot	130	40	125	35
3 Cold	60	100	65	105
4 Cold	30	120	35	125

$\Delta T_{min} = 10\ °C$

Candidate Boundaries: 175, 125, 125 ✕, 105, 75, 65, 35, 35 ✕

Eliminate Duplicates: 175, 125, 105, 75, 65, 35

Fig. 3.22: Obtaining the temperature boundaries for 3.2.1.6 Working Session "HEN design for maximum Heat Recovery".

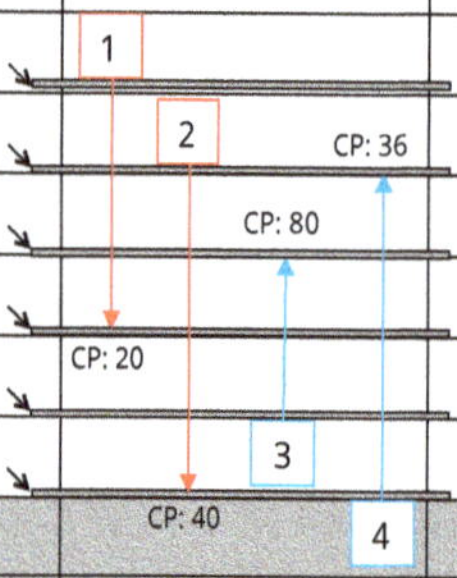

Interval Temperature	Stream Population				$\Delta T_{interval}$	$\Sigma CP_H -\Sigma CP_C$	Heat Balance	Surplus / Deficit
175	1							
125		2	CP: 36		50	20	1,000	Surplus
105			CP: 80		20	24	480	Surplus
75					30	−56	−1,680	Deficit
65	CP: 20				10	−76	−760	Deficit
35			3		30	4	120	Surplus
	CP: 40			4				

Fig. 3.23: The Problem Table for 3.2.1.6 Working Session "HEN design for maximum Heat Recovery".

Pinch and moving to lower temperatures opens the T-H profiles, which makes the match feasible. Stream 4 also has a smaller duty and defines the load of the exchanger match. The residual cooling requirement on the hot stream (No. 2) is satisfied by a cooler with cooling water, with a load of 120 kW.

The design Above the Pinch is more complicated (Fig. 3.27), as all four streams are present in that part. It takes several steps, as shown in that figure: first, the recovery heat exchangers 2 and 3 are placed, followed by heaters H1 and H2. The sum of the heater duties is 960 kW, which is exactly equal to the minimum heating target.

The complete HEN design, merging the Below- and Above-the-Pinch parts, is shown in Fig. 3.28. It is interesting to note that the design from Fig. 3.28 is not unique in terms of maximum Heat Recovery. It is also possible to obtain an alternative maximum Heat Recovery network, as shown in Fig. 3.29.

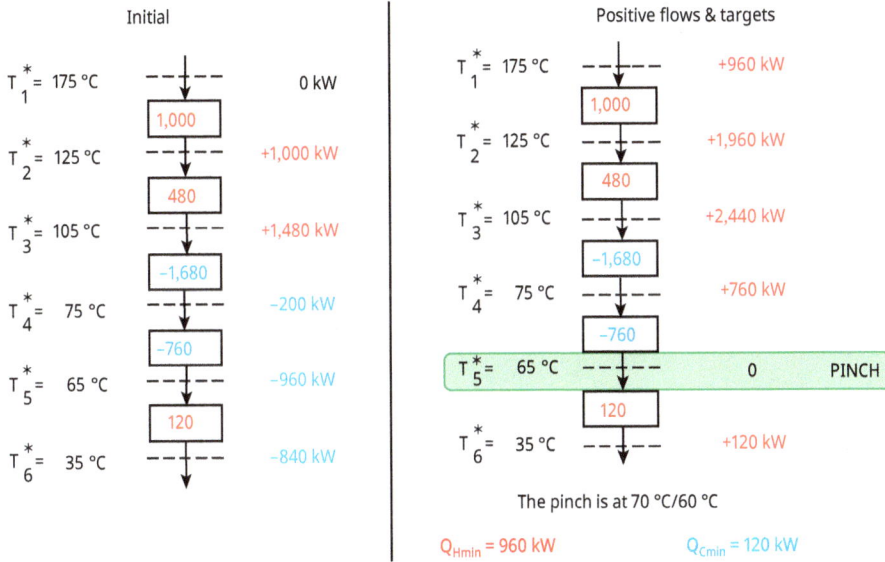

Fig. 3.24: The heat cascade for 3.2.1.6 Working Session "HEN design for maximum Heat Recovery".

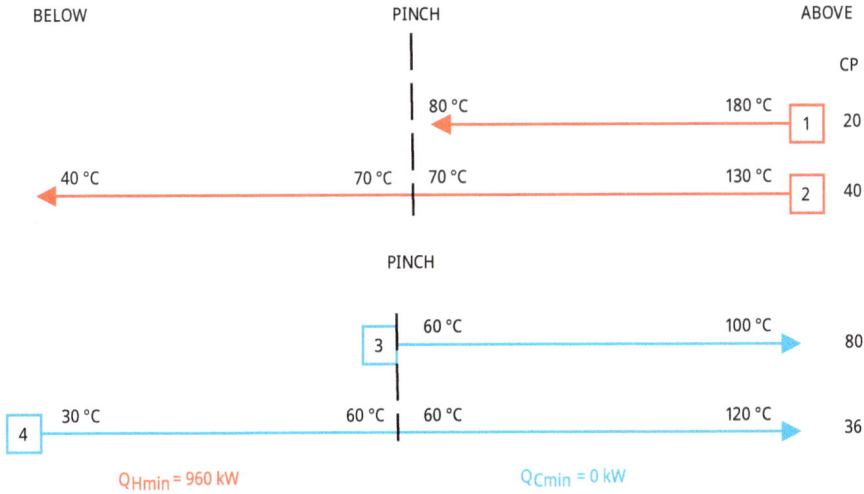

Fig. 3.25: The empty design grid for 3.2.1.6 Working Session "HEN design for maximum Heat Recovery".

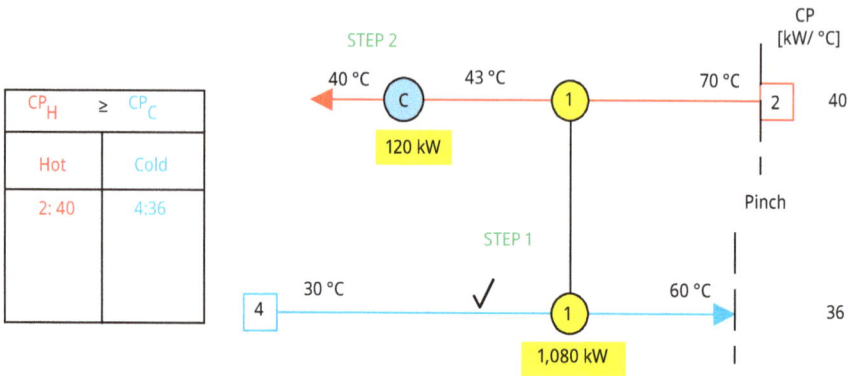

Fig. 3.26: Design Below the Pinch for 3.2.1.6 Working Session "HEN design for maximum Heat Recovery".

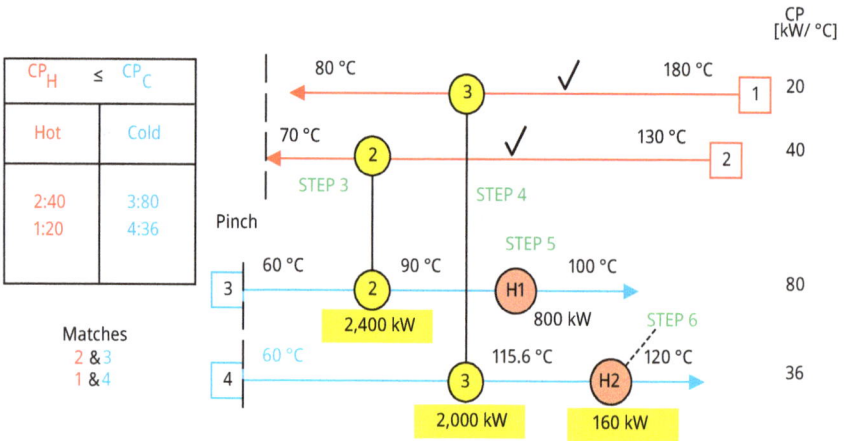

Fig. 3.27: Design Above the Pinch for 3.2.1.6 Working Session "HEN design for maximum Heat Recovery".

3.2.1.7 Working Session "An advanced HEN example"

Assignment

The stream data for a process are listed in Tab. 3.3. The hot utility is steam at 200 °C, the cold utility is water at 38 °C, and $\Delta T_{min} = 24$ °C.

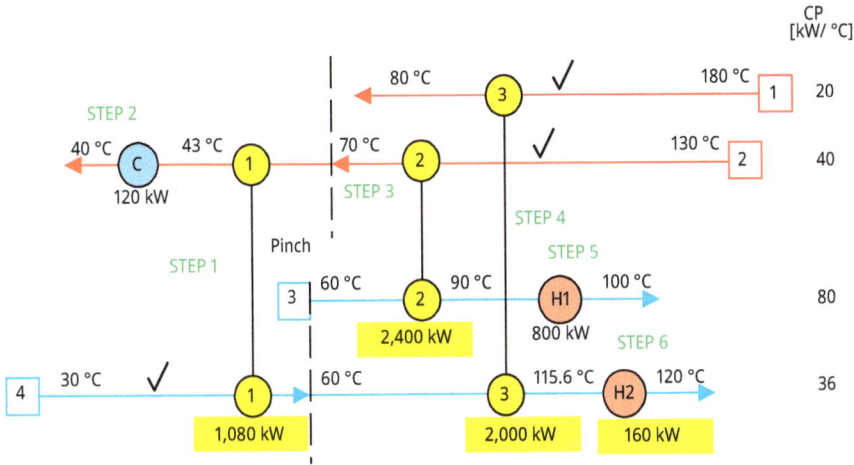

Fig. 3.28: The complete HEN design for 3.2.1.6 Working Session "HEN design for maximum Heat Recovery".

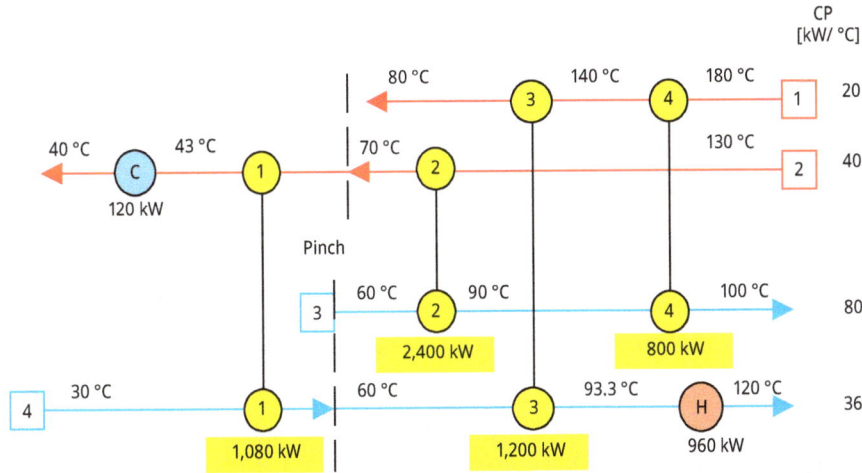

Fig. 3.29: An alternative HEN design for 3.2.1.6 Working Session "HEN design for maximum Heat Recovery".

Tab. 3.3: Process stream data for 3.2.1.7 Working Session "An advanced HEN example".

Stream		T_S (°C)	T_T (°C)	CP (kW/°C)
1	Cold	38	205	11
2	Cold	66	182	13
3	Cold	93	205	13
4	Hot	249	121	17
5	Hot	305	66	13

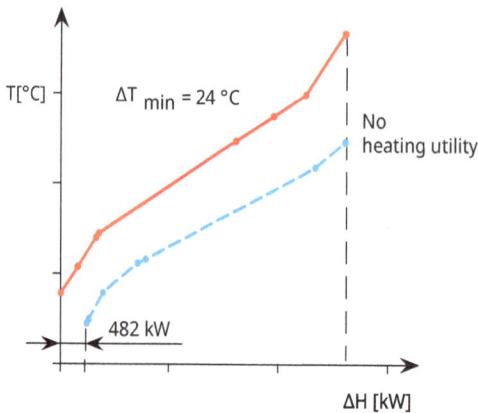

Fig. 3.30: Composite Curves for 3.2.1.7 Working Session "An advanced HEN example".

The following tasks are required to be performed:
(a) Plot the Composite Curves for this process.
(b) Determine the minimum heating requirement $Q_{H,min}$, the minimum cooling requirement $Q_{C,min}$, and the Pinch temperatures.
(c) Assuming that the cost of cooling water and steam is 18.13 and 37.78 $ kW^{-1} y^{-1}, plot the minimum annual cost for the utilities as a function of ΔT_{min} in the range 51–54 °C (in 1 ° increments).
(d) Design a network that features a minimum number of units and maximum energy recovery for ΔT_{min} = 24 °C.

Solution

Answer to (a). The Composite Curves for the process stream data are shown in Fig. 3.30.

Answer to (b). The position of the Composite Curves (CCs) in Fig. 3.30 indicates that for ΔT_{min} = 24 °C, this is a threshold problem with utility cooling only: $Q_{H,min}$ = 0 kW. On the cold end, there is an excess of hot streams, which means that some cold utility is required: $Q_{C,min}$ = 482 kW.

Answer to (c). The utility cost plot, given in Fig. 3.31, is based on the data in Tab. 3.4.

Answer to (d). A pair of networks that feature the minimum number of units and the maximum energy recovery is shown in Fig. 3.32. Please note that, besides being a threshold Problem, this is also one requiring stream splitting.

Tab. 3.4: Utility requirements and costs for various ΔT_{min} values for 3.2.1.7 Working Session "An advanced HEN example".

	ΔT_{min} (°C)			
	51	**52**	**53**	**54**
Q_H (kW)	0	0	27	57
Cost of heating ($/y)	0	0	1,020	2,153
Q_C (kW)	482	482	509	539
Cost of cooling ($/y)	8,739	8,739	9,228	9,772
Total utility cost ($/y)	8,739	8,739	10,248	11,926

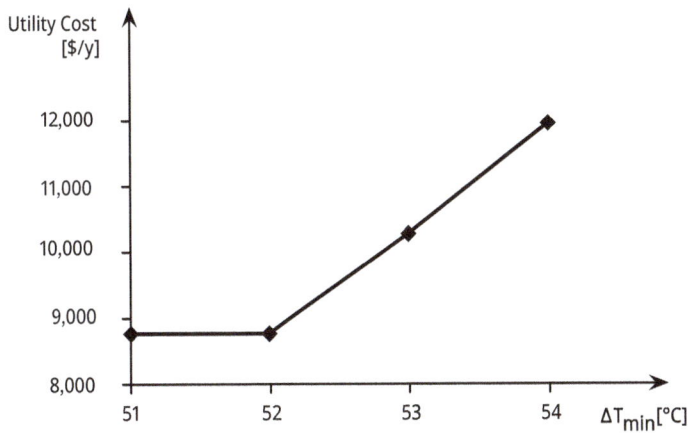

Fig. 3.31: Minimum allowed temperature difference ΔT_{min} versus annual utility cost for 3.2.1.7 Working Session "An advanced HEN example".

3.2.2 Methods using mathematical programming

3.2.2.1 MINLP approach

MINLP stands for "Mixed Integer Non-Linear Programming". This is one of the most used approaches for applying optimisation to chemical engineering and other industrial problems. As presented so far, the Pinch Design Method is based on a sequential strategy for the conceptual design of Heat Exchanger Networks. It first develops an understanding of the thermodynamic limitations imposed by the set of process streams, and then it exploits this knowledge to design a highly energy-efficient HEN. However, another approach has also been developed: the superstructure approach to HEN synthesis, which relies on developing a reducible structure (the superstructure) of the network under consideration and then reducing it using MINLP. An example of the spaghetti-type HEN superstructure fragments typically generated by such methods (Yee and Grossmann, 1990) is shown in Fig. 3.33.

(a) One possible design

(b) A second possible solution

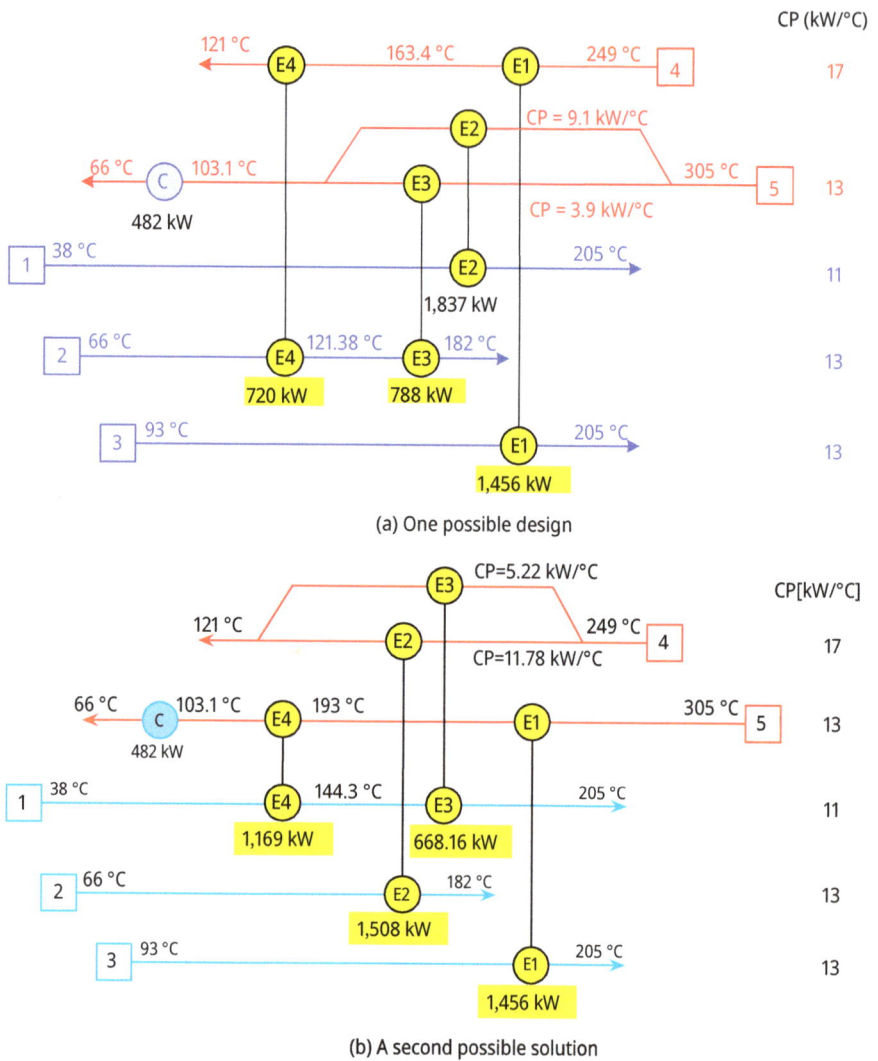

Fig. 3.32: Optimal Heat Exchanger Networks for 3.2.1.7 Working Session "An advanced HEN example".

This kind of superstructure is most often developed in stages – first formulated by Yee and Grossmann (1990) and later elaborated by Aaltola (2002) or in blocks (Zhu et al., 1995) and later (Zhu, 1997), each of which is a group of several consecutive enthalpy intervals. Zhu et al. (1995) investigated integrated HEN targeting and synthesis within the context of the so-called "block decomposition", which was further elaborated into an automated HEN synthesis method (Zhu, 1997). Within each block, each hot process stream is split into a number of branches, corresponding to the number of the cold streams presented in the block, and all cold streams are split similarly. This is followed

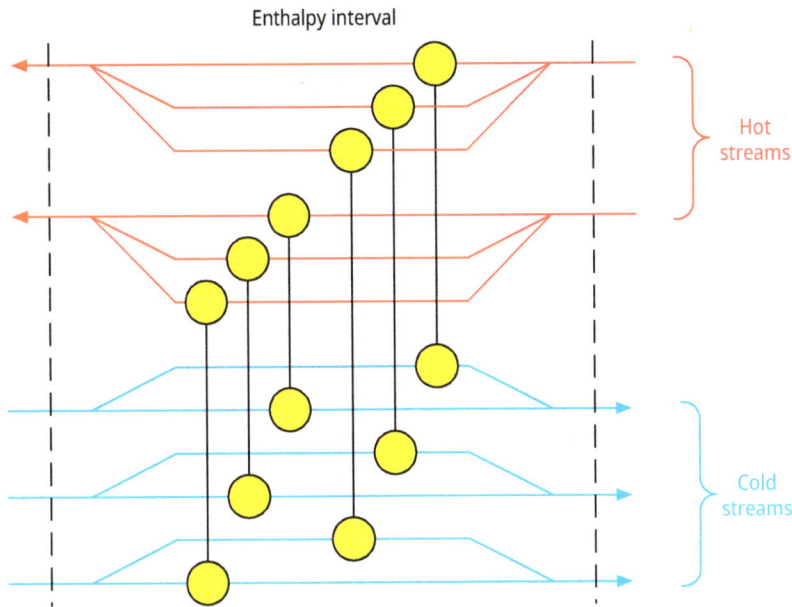

Fig. 3.33: Spaghetti superstructure fragment.

by matching each hot branch with each cold branch. The superstructure approach is still being developed. Examples of further improvements are the HEN synthesis superstructure, considering complex splits and cross-flows (Pavão et al., 2018) and extensions to the synthesis of large-scale HENs (Nemet et al., 2019), as well as for allowing intermediate placement of utilities (Liu et al., 2022).

Once developed, the superstructure is subjected to optimisation. The set of decision variables includes the existence of the different stream split branches and heat exchangers, the heat duties of the exchangers, and the split fractions or flow rates of the split streams. The objective function involves mainly the total annualised cost of the network. The objective function may be augmented by some penalty terms for dealing with difficult constraints. Because the optimisation procedure makes structural as well as operating decisions about the network being designed, it is called a structure-parameter optimisation.

"Structure-parameter optimisation" is a term widely used in Process Synthesis with a peculiar origin. It may be a little confusing to people familiar with mathematical optimisation only. In the context of optimisation, a parameter is an entity whose value is specified beforehand and is kept constant during the optimisation. However, in the early days of Process System Engineering, the term "parameter" was used more loosely, referring to any properties of the modelled system – including variables, which could be represented by real numbers.

Depending on which assumptions are adopted, it is possible to obtain both MILP and MINLP formulations. Linear formulations are usually derived by assuming isothermal mixing of the split branches and then using piecewise linearisation of the heat exchanger capital cost functions.

Within the framework of the superstructure approach, it is possible to include more than one type of heat exchanger, for example, direct heat transfer units (i.e., mixing) and further surface heat exchanger types (double-pipe, plate-fin, shell-and-tube). Soršak and Kravanja (2004) presented a method incorporating the different heat exchanger types into a superstructure block for each potential Heat Exchange match. Some other interesting works in this area are by Daichendt and Grossmann (1997) – using hierarchical decomposition (Zamora and Grossman, 1998) – emphasising global MINLP optimisation, Björk and Westerlund (2002) – exploring the non-isothermal mixing options, and Frausto-Hernández et al. (2003) – considering pressure drop effects.

3.2.2.2 A hybrid approach

The superstructure methodology offers some advantages in the synthesis of process systems and, in particular, of Heat Exchanger Networks. Among these advantages are: 1) the capacity to evaluate a large number of structural and operating alternatives simultaneously; 2) the possibility of automating (to a high degree) the synthesis procedure; and 3) the ability to deal efficiently with many additional issues, such as different heat exchanger types and additional constraints (e.g. forbidden matches) – see (Wang et al., 2021) for an elaboration.

However, these advantages also give rise to certain weaknesses. First, the superstructure approaches, in general, cannot eliminate the inherent non-linearity of the problem. They resort to linearisation and simplifying assumptions, such as allowing only the isothermal mixing of split streams. Second, the transparency and visualisation of the synthesis procedure are almost completely lost, excluding the engineer from the process. Third, the final network is merely given as an answer to the initial problem, and it is difficult to assess how good a solution it represents or whether a better solution is possible. Fourth, the difficulties of computation and interpreting the result grow dramatically with problem size. This is a consequence of the large number of discrete alternatives to be evaluated. Finally, the resulting networks often contain subnetworks that exhibit a spaghetti-type structure: a cluster of parallel branches on several hot and cold streams with multiple exchangers between them. Because of how superstructures are constructed, many such subnetworks usually cannot be eliminated by the solvers.

All these considerations highlight the fundamental trade-off between applying techniques that are based on thermodynamic insights, such as the Pinch Design Method, and relying on a superstructure approach. It would therefore be useful to follow a middle path, taking the best from these two basic approaches.

One such middle path is the class of hybrid synthesis methods that is described next. A method of this class first applies Pinch Analysis to obtain a picture of the thermodynamic limitations of the problem. Still, instead of continuing on to direct synthesis, it builds a reduced superstructure. At this point, the method follows the route of the classical superstructure approaches, including structure-parameter optimisation and topology simplifications. The cycle of optimisation and simplification is usually repeated several times before the final optimal network is obtained. All resulting networks feature a high degree of Heat Recovery, but not necessarily the maximum possible. A key component of this technique is avoiding the addition of unnecessary features to the superstructure, and this is an area where Pinch Analysis can prove helpful. A good example of a hybrid method for HEN synthesis is the block decomposition method (Zhu, 1997). A comparison and overview of both approaches have been presented by (Klemeš et al., 2013) and, more recently, by (Bogataj et al., 2022).

3.2.2.3 Comparison of HEN synthesis approaches

The networks obtained by the different synthesis methods have distinct features, which influence their total cost and their properties of operation and control. Because the Pinch Design Method incorporates the tick-off heuristic rule (Fig. 3.10), the networks synthesised by this method tend to have simple topologies with few streams splits. Both the tick-off rule and the Pinch principle dictate that utility exchangers be placed last; so they are usually located immediately before the target temperatures of the streams. However, the tick-off rule may also result in many process streams not having utility exchangers assigned to them, which reduces controllability. The Pinch Design Method may also reduce network flexibility because it relies on the Pinch decomposition of the problem (Fig. 3.6) and so, to a large degree, fixes the network behaviour.

Both the pure superstructure approach and the hybrid approach tend to produce more complex topologies. Their distinctive feature is the greater number of heat exchangers and stream splits, a result of how the initial superstructure is built. Spaghetti-type subnetworks also present a significant challenge to flexibility and control.

3.3 Grassroots and retrofits; the impact of economic criteria

The HEN design method discussed in Section 3.2 refers to the design of new plants. It is useful for developing a good understanding of the relevant heat- and energy-saving projects. However, most of the projects in the industry try to make the most of existing facilities, as opposed to completely replacing them with newly designed ones. Typically, they are geared toward improving the operation, removing bottlenecks, and improving efficiency with respect to energy and raw material utilisation.

The most commonly used term for such project types is *Retrofit*. Typical economic parameters or constraints are the maximum allowed values for Payback Time and

Investment Cost. The objective of a retrofit project is then to save as much energy as possible while satisfying these economic constraints. The strategy for retrofit problems needs to differ from that for new designs. Three different approaches seem to have been most popular historically (Kemp and Lim, 2020):

1. Synthesise a Maximum Energy Recovery (MER) – HEN for the existing plant, maximising the reuse of existing matches where possible. This implicitly puts the current HEN design as the procedure goal; therefore, bound to produce poor results.

2. Start with the existing network and attempt to modify it step by step to conform closer to the MER targets. The procedure usually starts with taking the specifications of the current process streams, utilities, and ΔT_{min}. The MER targets are obtained by Pinch Analysis (minimum heating and cooling by various utilities), as well as Pinch locations. The Grid Diagram is used to identify which heat transfer matches are the Pinch violators. Also, HEN topology modifications are proposed to correct the violations: addition of matches, shifting, and removal.

3. Another option, reasonably attractive to the industry, is to start from the existing HEN and identify the most critical changes to the network that would produce a substantial utility use reduction. This approach could be quite suitable if the current HEN is very far from the MER targets, and the required topology modifications to achieve the targets would involve too high complexity and cost. Asante and Zhu (1997) showed that it is also highly effective in other situations. A key insight is the *Network Pinch*, which shows the current bottleneck to improving the energy recovery imposed by the current network structure.

This chapter next provides a short overview of the Network Pinch Method introduced by Asante and Zhu (1997).

3.3.1 Network optimisation

The *Network Pinch Method* is based on identifying the Heat Recovery bottleneck, presented by the topology of the existing HEN. This is usually identified by performing optimisation of the HEN continuous properties: heat exchanger duties and stream temperatures, keeping the network topology fixed. In the course of this optimisation, the ΔT_{min} constraint may eventually be relaxed, even down to zero, leaving the total cost objective function to take care of identifying the best temperature differences for each heat exchanger.

3.3.2 The Network Pinch

Performing the mentioned optimisation on a HEN would eventually produce a combination of duties and temperatures, where at least one heat exchanger match will

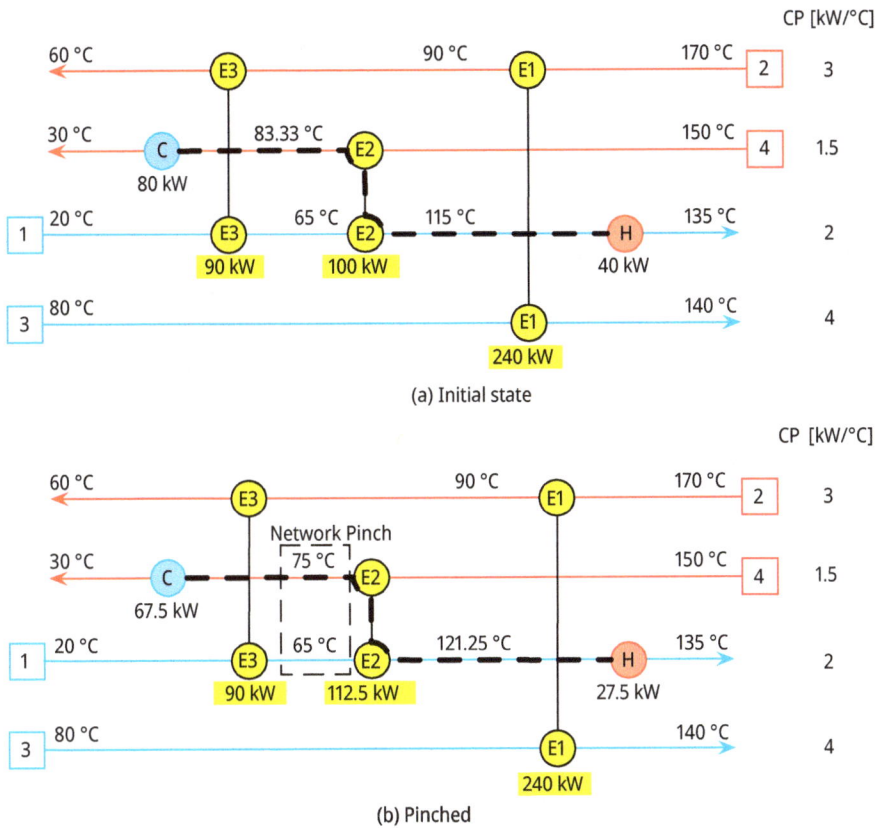

Fig. 3.34: An existing HEN.

feature ΔT_{min} at some of its ends, and the HEN utility demands are usually larger than the minima predicted by Pinch Analysis. This match is termed *the Pinching Match* and its end, where the stream temperatures approach each other at ΔT_{min}, is termed the *Network Pinch*.

Consider the network in Fig. 3.34(a), which has $\Delta T_{min} = 10$ °C. For the stream data, the targets are equal to 20 kW for hot utility and 60 kW for cold utility. There is a utility path from the heater **H**, through heat exchanger **E2**, to the cooler **C**. Exploiting the path, **E2** can be increased in size from 100 kW up to 112.5 kW. At this point, the temperature difference at the cold end of **E2** becomes equal to the 10 °C constraint, and the **E2** duty cannot be increased any further without violating the constraint (Fig. 3.34(b)). The load of the heater is 27.5 kW and that of the cooler is 67.5 kW, both 7.5 kW lower than the targets.

If ΔT_{min} is reduced to zero, a similar situation arises, where the targets are 0 kW for heating and 40 kW for cooling utility, presenting a threshold problem. For this case, the network from Fig. 3.34 can be theoretically pushed to a little larger Heat Recovery, with the heater load at 12.5 kW and cooler at 52.5 kW.

The cold end of exchanger **E2** represents the Network Pinch. After identifying the Network Pinch, there are several types of actions that can be applied to improve Heat Recovery. These are:

- **Resequence matches**. This means changing the order of matches and keeping the streams to which they connect.
- **Repiping**. This action would change one of the streams in the match.
- **Addition of a new match**. This adds a new heat exchanger at a strategically chosen location in the existing network.
- **Stream splitting**. One of the streams in the considered match can be split, allowing a better adjustment of the CP of the stream branch remaining in the match.

For the considered case, it happens that adding a new match produces a network capable of achieving the Pinch Targets (Fig. 3.35).

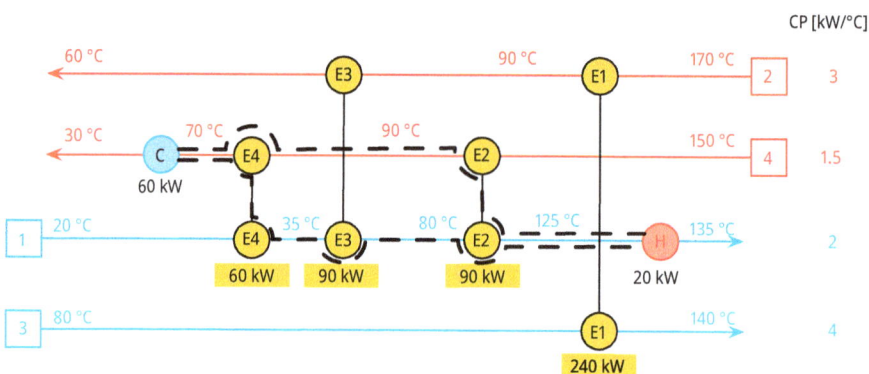

Fig. 3.35: A HEN with added heat exchanger (E4) achieving the targets.

Generally, after a retrofit step, it may happen that Heat Recovery would improve. However, more recovery would still be possible, and even the Network Pinch Location would most likely move to another match. In this case, the procedure of identifying the Network Pinch, followed by a new retrofit action, can be applied. More detailed instructions on the retrofit methodology of Network Pinch can be found in the work of Asante and Zhu (1997).

3.4 Advanced Data Extraction

A crucial activity for Heat Integration is Data Extraction. An introduction to the meaning of this activity, with a simple example, is given in Chapter 2. Overall, data acquisition is a complex process involving many stages and stakeholders. This section first discusses the treatment of Data Extraction in the published literature, showing that many publications just present the results from already performed Data Extraction, and others limit the provided reasoning only to a few key issues. Further, the importance of the preservation of data integrity and semantics is highlighted. One possible solution is presented based on the approach by Varbanov et al. (2019). This can be done by documenting the reasoning and the choices made during Data Extraction, as well as distributing the Data Extraction reasoning between the general discussion and stream-specific information placement into the Process Streams Table for Heat Integration, preserving the context. The approach is illustrated using a published example taken from (Varbanov et al., 2006).

HI is the foundational Process Integration (PI) part, for which Data Extraction has a crucial significance (Gundersen, 2022). The adequacy and quality of the obtained Heat Integration solutions depend on selecting the right data set, which captures all relevant heating and cooling demands of the considered process. If some of them are missed, the engineers would be solving a wrong HI problem (Rossiter, 2010). On the other hand, certain streams on a site should not be extracted for integration for safety reasons – for instance, parts with high explosion risk (Chew et al., 2013).

Another essential aspect of Data Extraction is the complex nature of the activities and the data flow (Matsuda et al., 2009). The activity involves several different stakeholder types exchanging data. Advanced software can be used to preserve data continuity – as can be seen from a case study on a crude oil refinery (Cui and Sun, 2017). However, a significant role in the data sets is played by semantics. All stakeholders have to reach a common understanding of the reasons for stream selection. Without that, mutual control and feedback are impossible, which reduces the level of confidence in the data quality. This brings the need to establish a common format and rules for composing datasets for HI. This format has to contain, in addition to the data, a set of descriptors of the reasoning for selecting and excluding the streams and their parameters. The procedure described here formulates a set of data requirements for DE systematically, enabling the development of an appropriate data format.

3.4.1 Review of Data Extraction treatment in the literature

Several key books have treated Data Extraction. The "User Guide" by Linnhoff et al. (1994) has been the cornerstone of teaching Heat Integration, providing complete instruction with conceptual discussions, Batch Heat Integration, Distillation Column Profiles, Low-Temperature Processes, Water/Wastewater Minimisation, Total Site, and Emissions Targeting. It discusses the Data Extraction principles, emphasising the importance of solving the right problem. The discussion starts with data accuracy, advising to strive for more accurate data only in the vicinity of the Pinch. It explains how to choose the Heat Integration Process Streams, including a selection of genuine heating and cooling requirements, "soft temperatures", and avoidance of non-isothermal mixing.

The Handbook of Process Integration (Klemeš, 2022) treats Data Extraction in Chapter 5. It focuses on explaining the Data Extraction rules applied in practice and the reasons for instituting them, elaborating on the stream mixing, practical constraints, and soft data. Chapter 37 of the PI Handbook (Klemeš, 2022) presents a focused discussion and a systematic consideration of the Data Extraction rules and their implications. Data Extraction is considered within the context of the overall workflow for performing a study for increasing the heat recovery of an industrial process. The core rules have been thoroughly examined and explained. These include when to extract heating/cooling requirements, data precision, heat capacity variation handling, temperature and load selections, and cost data. These traditional considerations have been supplemented by a short analysis of Data Extraction for the integration of renewable energy sources resulting from their intermittency.

Savulescu and Alva-Argaez (2008) investigated the additional opportunities for heat recovery improvement in Kraft paper mills that arise by including Direct Heat Transfer (DHT) in the analysis. The discussion starts with the classification and description of the DHT types in the process. It further analyses the process operations involving DHT – such as pulp dilution, questioning the need and extent of the heat transfer. The authors provide further analysis of the Data Extraction options, assuming a retrofit situation, aiming to minimise the DHT operations and eliminate Cross-Pinch heat transfer caused by the mixing.

Al-Riyami et al. (2001) have considered a real refinery case study, focusing on the Heat Integration of a Fluid Catalytic Cracking plant. They used a process simulator to establish the flowsheet, mass and energy balances. The paper provides a summary of the incurred Data Extraction issues, pointing out certain parts of the process excluded from the selection. However, as the focus of the article is the illustration of the Heat Exchanger Network (HEN) retrofit procedure, achieving 27% utility cost savings at 1.5 y payback time, no further details on the Data Extraction are provided.

A case study on the Heat Integration of a process for bio-ethanol production (Kravanja et al., 2013) provides a brief representation of Data Extraction reasoning, typical for published heat recovery studies. It lists the extracted process streams, with their temperatures and loads, and a summary of which heating or cooling demands have been included in the data set and which have been left out. The reasons for those choices are not discussed.

An evaluation of the Heat Integration opportunities for a palm oil processing plant (Lidu et al., 2016) starts with establishing the flowsheet and providing a description of the process and the main parameters in terms of temperatures and flowrates. The Data Extraction is then presented by listing the extracted process streams, briefly explaining the choice of temperature for one of the streams and documenting the exclusion of two other streams from the selection.

A Grid Diagram variation focusing on path tracing for HEN retrofit has been presented by (Nemet et al., 2015). The paper examines a HEN retrofit study, starting from the process identification, proceeding to balance, Data Extraction and Data Reconciliation

(DR), retrofit targeting, analysis of the current HEN for retrofit options, and recommending retrofit actions for improved heat recovery using the newly developed Retrofit Tracing Grid Diagram. The Data Extraction section of the article summarises the Data Extraction rules, as applied to the particular case study.

A biomass gasification case study is presented by (Pavlas et al., 2010). The process also has an integrated scrubber for the synthesis gas cleaning. The study evaluated the Heat Integration options, also considering a heat pump. The evaluation found an opportunity for economically justified electricity generation. Within the case study, the Data Extraction is represented only by the identified Heat Integration Process Streams. Pouransari et al. (2014) provide a discussion on how to perform Data Extraction for Heat Integration at various levels – from a single process up to the site level. They consider top-down and bottom-up options. While the considered options are valid and important, no clear interface between the process energy demands can be identified, which would allow performing a practical Heat Integration analysis.

The importance of safety considerations in Data Extraction is discussed in (Varbanov et al., 2016). Since Heat Integration increases the complexity of the process – in terms of management and control, certain process parts should be left out of the integration, allowing isolation and efficient management of the related process risks. Wan Alwi et al. (2014) have also added to the Heat Integration studies, consideration of heat losses and heat gains that take place due to differences in the stream temperatures and the ambient temperature – the reason the data to be extracted should also include the ambient conditions and heat transfer coefficients of the pipe and equipment casings.

Data integrity can be preserved within a single software suite. This can be seen in the example of the case study on a crude oil refinery (Cui and Sun, 2017), where the initial flowsheets were balanced and simulated in Aspen Plus (2019), and then the Data Extraction has been performed using Aspen Energy Analyzer (2019). While such an automated Data Extraction may capture most of the relevant process streams, it would still be missing the careful consideration by engineers and the relevant reasoning that should be kept together with the complete documentation in a Process Integration project.

3.4.2 Analysis of the Data Extraction workflow

A couple of patterns emerge from the review in the previous section. Rules and guidelines for Data Extraction have been well formulated for Heat Integration studies in the "User Guide" (Linnhoff et al., 1994) and, more recently, the PI Handbook (Klemeš, 2022). They have become a mandatory part of authoritative Process Integration books, including the book on the Process Integration approach to Chemical Process Design (Smith, 2016), as well as this current book. There have been other follow-up developments to the "User Guide" (Linnhoff et al., 1994). These include a systematic procedure for embedding Data Extraction into Heat Integration workflows (Klemeš and Varbanov, 2010) and an extension to handle cases of DHT (Savulescu and Alva-Argaez, 2008).

Some works have presented Data Extraction to a varying degree in their Heat Integration case studies. In some, only the results of the Data Extraction are presented in the typical tabular form, as in (Pavlas et al., 2010). Others provide Data Extraction reasoning, limited to specific issues that lead to excluding process streams from selection (Al-Riyami et al., 2001) or provide complete reasoning in the case analysis description (Kravanja et al., 2013).

The discussed sources do provide useful information and examples. However, for industrial-scale case studies, the data acquisition process should be considered systematically – preserving the continuity and the semantics of the performed evaluation. Such a consideration was not found in those works and the literature associated with the term "Data Extraction". This section fills in the gap by proposing one way of preserving the continuity and integrity of the extracted data, together with its associated semantics and related reasoning.

3.4.3 Site- and process-level procedures

The data acquisition for a HI evaluation most often starts at the site level and follows the steps:
(i)   Site description
(ii) Identification of potential interfaces
(iii) Scoping site processes
(iv) Data acquisition for each selected process

The first step involves obtaining a description of the site, including an overall diagram of process connections and the utility system, main parameters – as processing capacities, process energy demand, primary energy demand, heat and power generation, and waste energy flows. The potential interactions of the site with the surrounding regional and municipal entities can be evaluated for increasing the efficiency of the energy resource utilisation via any waste heat reuse. A typical scheme of this type is the potential to supply district heating to nearby residential demands using industrial waste heat. The further step is to perform process scoping – determining which site processes to include for further consideration. This selection can be made based on site steam system evaluation for potential cost savings – as done in Top-Level Analysis (Varbanov et al., 2004) or safety considerations (Varbanov et al., 2016). The last stage is performed for each of the selected processes. It identifies the process flowsheets, followed by the stream properties, and the data are processed with Data Reconciliation to ensure consistency, followed by the Data Extraction, proceeding in two parts. First, the current (existing) duties of the heat exchangers, heaters, and coolers are identified. This provides a baseline for the follow-up heat recovery evaluation. The second part of this procedure is the Process Streams selection for Heat Integration.

In acquiring the data for each process, it is vital to ensure several properties of the treated data set: continuity, clear links between objects that are linked in the real process, data consistency and adequacy for calculation and decision-making, and conformance to the interfaces of the applied methods. The data acquisition workflow at the process level is shown in Fig. 3.36. The first step is to establish the process flowsheet – the set of process operations involved, their connections, and the streams implementing the connections. This should be accompanied by the appropriate mass and energy flow measurements and approximate balancing. The next step is Data Reconciliation. This activity ensures the consistency of the measured data (Veverka and Madron, 1997). It uses the measured data for flowrates, temperatures, and pressures, together with the appropriate process constraints, to reconcile the data set, and ensure its consistency and fidelity. The next step is the Data Extraction for Heat Integration, which uses the Reconciled Data Set.

Considering Fig. 3.36, it can be noticed that the workflow refers to several data sets: (a) the flowsheet showing the operating units, flowsheet streams and their properties; (b) the Reconciled Data Set; and (c) the Heat Integration problem data set. Data continuity and consistency have to be kept among all these data sets. Using appropriate software tools – e.g. Aspen Energy Analyzer (2019), these properties can be ensured.

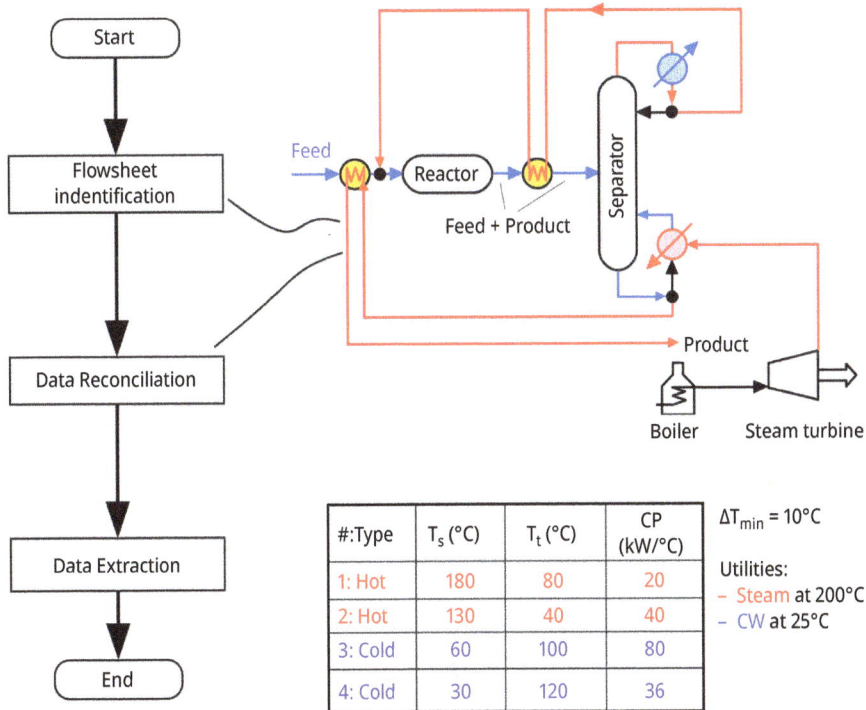

#:Type	T_s (°C)	T_t (°C)	CP (kW/°C)
1: Hot	180	80	20
2: Hot	130	40	40
3: Cold	60	100	80
4: Cold	30	120	36

ΔT_{min} = 10°C

Utilities:
– Steam at 200°C
– CW at 25°C

Fig. 3.36: Data acquisition for a process.

However, there is one more property that is important to preserve across all data sets in the workflow; it is the continuity of the data semantics, which includes the chain of reasoning applied to obtain each following data set from the previous. While Data Reconciliation is automated to a large degree, applying an agreed and unified mathematical model at the Data Extraction stage is subject to engineering judgement. The chain of reasoning should be recorded, linking the Reconciled Data Set and the data set for the HI Process Streams. An example of subtler reasoning, which requires judgement, is the case of the high-temperature waste heat, usually available from hydrogen production units in petroleum refineries. Some engineers may suggest using this waste heat, as it is also available in substantial amounts. However, refinery site engineers are reluctant to consider such an option due to the high risk of explosion, which leaves this type of waste heat out of the process/stream selection for HI.

One way to allow for such continuity and integrity is to provide the appropriate reasoning for the selection of the Process Streams in two places:

(a) The general discussion should be placed in the main text of the document – report or article. This should also include a table of streams excluded from the HI, with explanatory notes.

(b) The stream-specific information should be provided in the Process Streams Table (Tab. 3.5).

This arrangement provides the right balance and a hierarchy for recording the course of the DE, and allows one to easily trace any decisions made and to efficiently make changes to those choices during project iterations. In Tab. 3.5, the unit for temperature is K because the illustrative example uses this. Using °C is usually preferred.

Tab. 3.5: Template for an enhanced Data Extraction format.

#	Name	Flowsheet IDs and links	T_S (K)	T_T (K)	ΔH (kW)	CP (kW/K)
Number	An informative process stream name for the HI context	An identifier of a flowsheet stream or identifier of a linked group of streams	Supply Temperature Value	Target Temperature Value	Enthalpy Change (flow) Value	Heat-Capacity Flowrate Value
	Description of the selected context and reasoning					

3.4.4 Illustrative example

To illustrate the potential implementation of the proposed integrity preservation approach, an example from the Heat Integration of a Fuel Cell system is used. It has been sourced from (Varbanov et al., 2006). That study considered a Molten Carbonate Fuel Cell (MCFC) and investigated the opportunities for maximising the degree of utilisation of the fuel. It combines the MCFC with a HEN for recovering the maximum amount of heat and generating steam, which is then passed to a steam turbine, generating additional power. This allows for increasing the power generation efficiency from that of the MCFC (46.38%) up to the Fuel Cell Combined Cycle – to nearly 70% (between 60.23% and 69.38%). The Heat Integration study starts with the Data Extraction from the MCFC flowsheet (Fig. 3.37), assuming that the data are reconciled.

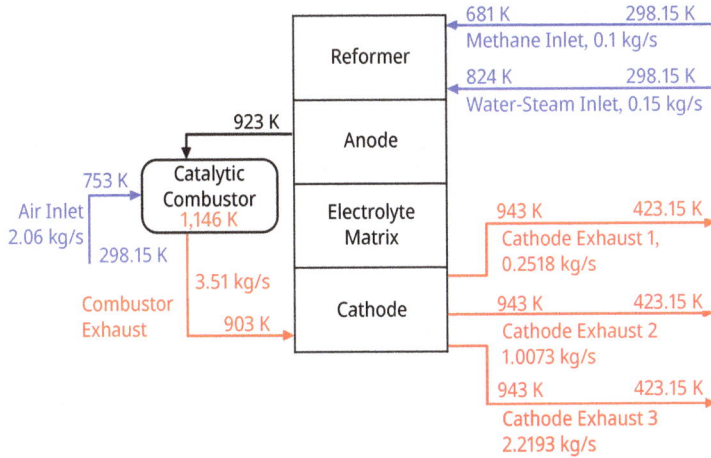

Fig. 3.37: MCFC Flowsheet, adapted from (Varbanov et al., 2006).

The Data Extraction reasoning starts with an overview of the flowsheet streams and reasoning for selecting each of them for Heat Integration or not. The anode exhaust, with a temperature of 923 K, is sent to the catalytic combustor to complete the combustion of the excess fuel. It does not undergo heat exchange; so it is left out of the selection. The rest of the streams have been found to need heat exchange, and have been selected for Heat Integration. The specific reasoning for selecting and descriptions of the streams are given in Tab. 3.6. In the table, the flowsheet identifier (column "Flowsheet ID") and the reasoning for the stream inclusion and parameter selection are added to the traditional HI table format.

In summary, information integrity and traceability are essential for industrial-scale projects, as they usually involve a large number of stakeholders from at least two categories – an industrial operator (client) and a consultant. The stakeholders work in a distributed way and play different but linked parts. Integrity preservation is important to ensure that all parties reach a common understanding. The proposed data set format illustrates one possible way of ensuring data integrity and traceability for selecting the Heat Integration Process Streams in Data Extraction. Further improvements are possible if workflows for Process Integration and resource efficiency optimisation are analysed to develop a better understanding of the need for information integrity at all stages. This follows from the iterative nature of industrial projects,

Tab. 3.6: Heat Integration Process Streams selected and the reasoning for the case study.

#	Name	Flowsheet ID	T_S (K)	T_T (K)	ΔH (kW)	CP (kW/K)
1	C1	Methane Inlet	298.15	681	109.5	0.2858

The methane gas/natural gas is fed to a reformer to undergo indirect reforming. It is taken from the ambient temperature of 298.15 K. It needs to be preheated to 681 K before it is injected to react with the steam feed.

#	Name	Flowsheet ID	T_S (K)	T_T (K)	ΔH (kW)	CP (kW/K)
2	C2	Water-Steam Inlet	298.15	824	282.7	0.5374

The water from the supply system enters the fuel cell reformer as steam and reacts with methane to undergo a steam reforming reaction. It is supplied at ambient temperature (298.15 K) and preheated to 824 K.

#	Name	Flowsheet ID	T_S (K)	T_T (K)	ΔH (kW)	CP (kW/K)
3	C3	Air Inlet	298.15	753	976.7	2.1465

The exhaust from the anode (mainly CO, CO_2 and H_2O) is burned with the presence of air in the catalytic combustor. The air is heated from the ambient temperature (298.15 K) to 753 K before entering the combustor.

#	Name	Flowsheet ID	T_S (K)	T_T (K)	ΔH (kW)	CP (kW/K)
4	H1	Combustor Exhaust	1,146	903	−1,279.1	5.2638

The combustor exhaust (mainly N_2, CO_2, unreacted CO and O_2) has to be cooled before it is introduced into the cathode. At high temperatures, evaporation of electrolytes and corrosion of material are more likely to occur. In this case, it is required to be cooled from 1,146 K to 903 K.

#	Name	Flowsheet ID	T_S (K)	T_T (K)	ΔH (kW)	CP (kW/K)
5	H2	Cathode Exhaust 1	943	423.15	−130.9	0.2518

The cathode exhaust consists mainly of N_2, CO_2, and O_2, with traces of SO_2. It is vented to the ambient through a stack. It should be cooled from 943 K to 423.15 K in the particular device design, and it is split into three branches. This is the first branch

#	Name	Flowsheet ID	T_S (K)	T_T (K)	ΔH (kW)	CP (kW/K)
6	H3	Cathode Exhaust 2	943	423.15	−523.6	1.0073

This is the second branch of the cathode exhaust.

#	Name	Flowsheet ID	T_S (K)	T_T (K)	ΔH (kW)	CP (kW/K)
7	H4	Cathode Exhaust 3	943	423.15	−1,153.7	2.2193

This is the third branch of the cathode exhaust.

Convention: The heating demands are assigned positive values for ΔH, and the cooling demands are negative.

where the results of one iteration often reveal the need to alter certain assumptions or measurement values or run re-calculation or re-optimisation loops.

3.5 Summary

In this chapter, the main concepts, representations, procedures, and diagrams of HEN synthesis using the Pinch Design Method have been described and illustrated with working sessions. The most important topics covered in this chapter are:
- Grid Diagram for HEN synthesis and optimisation;
- Pinch Decomposition;
- Pinch Design Method with Matching Rules;

- Design Evolution with Heat Load Loops and Paths;
- Overview of other HEN synthesis methods and basics of retrofit;
- An extended Data Extraction workflow and data format is also discussed to aid engineers in preserving the integrity of data, reasoning and semantics for multi-stakeholder projects.

References

Aaltola J. (2002). Simultaneous Synthesis of Flexible Heat Exchanger Network, Applied Thermal Engineering, 22, 907–918.

Al-Riyami B A, Klemeš J, Perry S. (2001). Heat Integration Retrofit Analysis of a Heat Exchanger Network of a Fluid Catalytic Cracking Plant, Applied Thermal Engineering, 21, 1449–1487. doi: 10.1016/S1359-4311(01)00028-X.

Asante N D K, Zhu X X. (1997). An Automated and Interactive Approach for Heat Exchanger Network Retrofit, Chemical Engineering Research and Design, 75(part A), 349–360.

Aspen Energy Analyzer. (2019). <https://www.aspentech.com/en/products/pages/aspen-energy-analyzer>, accessed 21.03.2019.

Aspen Plus. (2019). <https://www.aspentech.com/en/products/engineering/aspen-plus>, accessed 21.03.2019.

Chew K H, Klemeš J J, Wan Alwi S R, Z A Manan. (2013). Industrial Implementation Issues of Total Site Heat Integration, Applied Thermal Engineering, 61(1), 17–25. doi: 10.1016/j.applthermaleng.2013.03.014.

Björk K M, Westerlund T. (2002). Global Optimization of Heat Exchanger Network Synthesis Problems with and without the Isothermal Mixing Assumption, Computers & Chemical Engineering, 26(11), 1581–1593.

Bogataj M, Klemeš J J, Kravanja Z. (2022). Fifty Years of Heat Integration: Pinch Analysis and Mathematical Programming. In Klemeš J J (ed.)., Handbook of Process Integration (PI): Minimisation of Energy and Water Use, Waste and Emissions. 2nd updated edition. Woodhead/Elsevier, Cambridge, UK, doi: 10.1016/B978-0-12-823850-9.00020-7.

Chew K H, Klemeš J J, Wan Alwi S R, Manan Z A. (2013). Industrial Implementation Issues of Total Site Heat Integration, Applied Thermal Engineering, 61(1), 17–25. doi: 10.1016/j.applthermaleng.2013.03.014.

Cui C, Sun J. (2017). Coupling Design of Interunit Heat Integration in an Industrial Crude Distillation Plant Using Pinch Analysis, Applied Thermal Engineering, 117, 145–154. doi: 10.1016/j.applthermaleng.2017.02.032.

Daichendt M M, Grossmann I E. (1997). Integration of Hierarchical Decomposition and Mathematical Programming for the Synthesis of Process Flowsheets, Computers & Chemical Engineering, 22(1–2), 147–175.

Frausto-Hernández S, Rico-Ramírez V, Jiménez-Gutiérrez A, Hernández-Castro S. (2003). MINLP Synthesis of Heat Exchanger Networks considering Pressure Drop Effects, Computers & Chemical Engineering, 27(8–9), 1143–1152.

Gundersen T. (2022). Heat Integration E Performance Targets and Heat Exchanger Network Design. In Klemeš J J (ed.)., 2022 Handbook of Process Integration (PI): Minimisation of Energy and Water Use, Waste and Emissions. 2nd updated edition. Woodhead/Elsevier, Cambridge, UK, doi: 10.1016/B978-0-12-823850-9.00008-6.

Kemp I C, Lim J S. (2020). Pinch Analysis for Energy and Carbon Footprint Reduction: User Guide to Process Integration for the Efficient Use of Energy, 3rd ed. Elsevier / Butterworth-Heinemann, Oxford United Kingdom.

Klemeš J J (ed)., (2022). Handbook of Process Integration (PI): Minimisation of Energy and Water Use, Waste and Emissions, 2nd updated edition. Woodhead/Elsevier, Cambridge, UK, ISBN: 9780128238509 doi: 10.1016/B978-0-12-823850-9.00019-0.

Klemeš J J, Varbanov P S, Kravanja Z. (2013). Recent Developments in Process Integration, Chemical Engineering Research and Design, 91(10), 2037–2053.

Klemeš J, Varbanov P. (2010). Process Integration – Successful Implementation and Possible Pitfalls, Chemical Engineering Transaction, 21, 1369–1374.

Klemeš J J, Varbanov P V, Walmsley T G, Jia X X. (2018). New Directions in the Implementation of Pinch Methodology (PM), Renewable and Sustainable, Energy Reviews, 98, 439–468.

Kravanja P, Modarresi A, Friedl A. (2013). Heat Integration of Biochemical Ethanol Production from Straw – A Case Study, Applied Energy, 102, 32–43. doi: 10.1016/j.apenergy.2012.08.014.

Lidu S R, Mohamed N A, Klemeš J J, Varbanov P S, Yusup S. (2016). Evaluation of the Energy Saving Opportunities for Palm Oil Refining Process: Sahabat Oil Products (Sop) in Lahad Datu, Malaysia, Clean Technologies and Environmental Policy, 18, 2453–2465. doi: 10.1007/s10098-016-1252-6.

Linnhoff B, Hindmarsh E. (1983). The Pinch Design Method for Heat Exchanger Networks, Chemical Engineering Science, 38(5), 745–763.

Linnhoff B, Townsend D W, Boland D, Thomas B E A, Guy A R, Marsland R H. (1994). A User Guide on Process Integration for the Efficient Use of Energy, revised first edition. IChemE, Rugby, UK.

Liu Z, Yang L, Yang S, Qian Y. (2022). An Extended Stage-wise Superstructure for Heat Exchanger Network Synthesis with Intermediate Placement of Multiple Utilities, Energy, 248, 123372. doi: 10.1016/j. energy.2022.123372.

Matsuda K, Hirochi Y, Tatsumi H, Shire T. (2009). Applying Heat Integration Total Site Based Pinch Technology to a Large Industrial Area in Japan to Further Improve Performance of Highly Efficient Process Plants, Energy, 34, 1687–1692. doi: 10.1016/j.energy.2009.05.017.

Nemet A, Isafiade A J, Klemeš J J, Kravanja Z. (2019). Two-step MILP/MINLP Approach for the Synthesis of Large-scale HENs, Chemical Engineering Science, 197, 432–448. doi: 10.1016/j.ces.2018.06.036.

Nemet A, Klemeš J J, Varbanov P S, Mantelli V. (2015). Heat Integration Retrofit Analysis – An Oil Refinery Case Study by Retrofit Tracing Grid Diagram, Frontiers of Chemical Science and Engineering, 9, 163–182. doi: 10.1007/s11705-015-1520-8.

Pavão L V, Costa C B B, Ravagnani M A S S. (2018). A New Stage-wise Superstructure for Heat Exchanger Network Synthesis considering Substages, Sub-splits and Cross Flows, Applied Thermal Engineering, 143, 719–735. doi: 10.1016/j.applthermaleng.2018.07.075.

Pavlas M, Stehlík P, Oral J, Klemeš J, Kim J-K, Firth B. (2010). Heat Integrated Heat Pumping for Biomass Gasification Processing, Applied Thermal Engineering, 30, 30–35. doi: 10.1016/j. applthermaleng.2009.03.013.

Pouransari N, Bocquenet G, Maréchal F. (2014). Site-scale Process Integration and Utility Optimization with Multi-level Energy Requirement Definition, Energy Convers Management, 85, 774–783.

Rossiter A P. (2010). Improve Energy Efficiency via Heat Integration, Chemical Engineering Progress, 106, 33–42.

Savulescu L E, Alva-Argaez A. (2008). Direct Heat Transfer Considerations for Improving Energy Efficiency in Pulp and Paper Kraft Mills, Energy, 33, 1562–1571. doi: 10.1016/j.energy.2008.07.015.

Smith R. (2016). Chemical Process Design and Integration, 2nd ed. Wiley, Chichester, West Sussex, UK.

Soršak A, Kravanja Z. (2004). MINLP Retrofit of Heat Exchanger Networks Comprising Different Exchanger Types, Computers & Chemical Engineering, 28(1–2), 235–251.

Varbanov P S, Klemeš J, Shah R K, Shihn H. (2006). Power Cycle Integration and Efficiency Increase of Molten Carbonate Fuel Cell Systems, Journal of Fuel Cell Science and Technology, 3(3), 375–38. doi: 10.1115/1.2349515.

Varbanov P S, Klemeš J J, Liu X. (2016). Process Integration Contribution to Safety and Related Financial Management Issues, Chemical Engineering Transaction, 53, 241–246. doi: 10.3303/CET1653041.

Varbanov P, Perry S, Makwana Y, Zhu X X, Smith R. (2004). Top-level Analysis of Site Utility Systems, Chemical Engineering Research and Design, 82, 784–795. doi: 10.1205/026387604774196064.

Varbanov P S, Yong J Y, Klemeš J J, Chin H H. (2019). Data Extraction for Heat Integration and Total Site Analysis: A Review, Chemical Engineering Transaction, 76, 67–72. doi: 10.3303/CET1976012.

Veverka V, Madron F. (1997). Material and Energy Balancing in the Process Industries: From Microscopic Balances to Large Plants, Elsevier Science B.V, Amsterdam, The Netherlands, ISBN: 9780080535869.

Wan Alwi S R, Lee C K M, Lee K Y, Manan Z A, Fraser D M. (2014). Targeting the Maximum Heat Recovery for Systems with Heat Losses and Heat Gains, Energy Convers Management, 87, 1098–1106. doi: 10.1016/j.enconman.2014.06.067.

Wang B, Klemeš J J, Varbanov P S, Zeng M. (2020). An Extended Grid Diagram for Heat Exchanger Network Retrofit considering Heat Exchanger Types, Energies, 13, 2656. doi: 10.3390/en13102656.

Wang B, Klemeš J J, Varbanov P S, Zeng M, Liang Y. (2021). Heat Exchanger Network Synthesis considering Prohibited and Restricted Matches, Energy, 225, 120214. doi: 10.1016/j.energy.2021.120214.

Yee T F, Grossmann I E. (1990). Simultaneous Optimization Models for Heat Integration – II. Heat Exchanger Networks Synthesis, Computers & Chemical Engineering, 14, 1165–1184.

Zamora J M, Grossmann I E. (1998). A Global MINLP Optimization Algorithm for the Synthesis of Heat Exchanger Networks with No Stream Splits, Computers & Chemical Engineering, 22(3), 367–384.

Zhu X X. (1997). Automated Design Method for Heat Exchanger Networks Using Block Decomposition and Heuristic Rules, Computers & Chemical Engineering, 21(10), 1095–1104.

Zhu X X, O'Neill B K, Roach J R, Wood R M. (1995). Area-targeting Methods for the Direct Synthesis of Heat Exchanger Networks with Unequal Film Coefficients, Computers & Chemical Engineering, 19(2), 223–239.

4 Total Site Integration

Maximising Heat Recovery at the process level is a good step toward the better performance of industrial facilities. However, industrial processes rarely operate in isolation. They are usually organised in larger areas termed sites and are served by a common utility system. The processes interact with each other via the utility system. There are significant benefits to be gained from considering the complete sites as integrated energy systems, evaluating and optimising the energy generation, distribution, use, and recovery. This chapter introduces a systematic framework and the associated algorithmic and visual tools for performing such an analysis.

This framework, with its tools, is widely known as Total Site Integration. Targeting for Total Sites is an extension of the established Pinch Technology Targeting methodologies and has been used extensively in the industry. The Total Site targeting methodology includes Data Extraction methods, construction of Total Site Profiles, Total Site Composite Curves, and the Site Utility Grand Composite Curve.

4.1 Introduction

Total Site Integration of industrial systems, based on the concept of the Site Heat Source and Heat Sink Profiles, was introduced by Dhole and Linnhoff (1993). Klemeš et al. (1997) made further advances in the field by adding targets for power cogeneration. The concept has been applied to industrial sites – in the beginning to refinery and petrochemical processes. These processes generally operate as parts of large sites or factories. The Heat Integration on such Total Sites is performed through a set of energy carriers, usually steam at several pressure levels, plus hot water and cooling water. The site processes are serviced by a central utility system providing the thermal energy carriers and power.

Total Site Integration has been applied to a number of chemical industrial sites (Matsuda et al., 2009) and even to a heterogeneous Total Site involving a brewery and several commercial energy users. It has also been used as the targeting stage for the synthesis of utility systems, first by Shang and Kokossis (2004) and then elaborated for choosing steam header pressure levels by Varbanov et al. (2005). In recent years the methodology has been extended into geothermal energy heating and cooling (Wang et al., 2021), material recycling networks (Chin et al., 2021b), Total Site Hydrogen Integration (Gai et al., 2022), Spatial Total Site Heat Integration Targeting Using Cascade Pinch (Wahab et al., 2022), and Integrating Trigeneration Systems at Total Sites (Jamaluddin et al., 2022). Total Site Centralised Water Integration (Fadzil et al., 2017) and Total Site Water Mains – prototyped in (Jia et al., 2020) and fully developed in (Chin et al., 2021a) are both covered in focusing on Water Integration – see Chapter 10.

The current chapter provides a guide through the basics of Total Site Integration. It starts with a definition, aided by a description of the reasons for and the benefits of

https://doi.org/10.1515/9783110782981-004

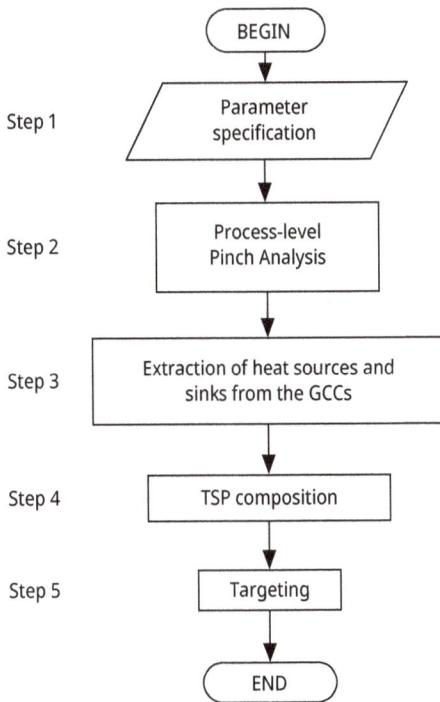

Fig. 4.1: Procedure for Total Site targeting.

applying the concept. The Data Extraction from process-level Heat Integration results is considered, which is one of the necessary steps for analysing new system designs. Next, follow the steps of the basic Total Site Analysis – construction and use of the Total Site Profiles, the Total Site Composite Curves, and the Site Utility Grand Composite Curve. The next section touches on the subject of cogeneration targeting and its extended use for analysing the best power-to-heat ratio for a site. Advanced developments of the Total Site Methodology are briefly reviewed after that, including more flexible specifications of the minimum allowed temperature differences, numerical tools for Total Site Heat Integration and Hybrid Power Systems. The chapter is concluded with a summary of the potential sources of further information.

The analysis of a Total Site almost invariably involves collecting information about the heating and cooling requirements on the site, thermodynamic analysis of these demands based on composite profiles, and identification of the targets for minimum fuel use, power cogeneration, and the amount of power import. The most frequently used procedure is shown in Fig. 4.1.

4.2 What is a Total Site, and what are the benefits?

It is rare to have single isolated industrial processes. The usual situation is to have clusters of processes interlinked with one another by various lines for exchanging raw materials, side products, intermediates and, most importantly, energy carriers. The latter usually represent the most significant links.

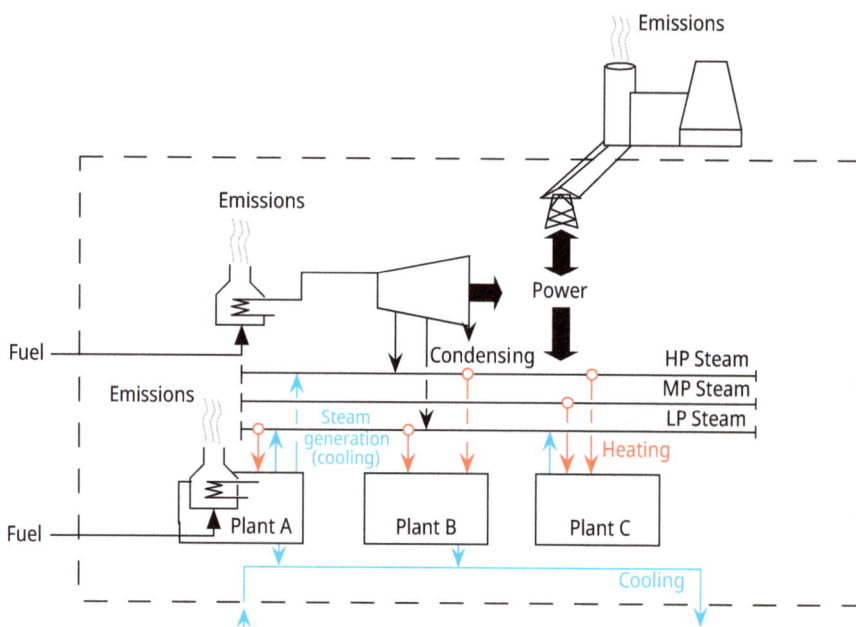

Fig. 4.2: Typical Total Site in the industry (the arrow colours reflect the viewpoint of the plants).

4.2.1 Total Site definition

A typical chemical or another industrial site (Fig. 4.2) usually consists of a number of production and auxiliary processes. These processes require the supply of different utilities to carry out their functions. Such utilities are the following.

- **Process heating**. Steam is usually the preferred heating medium because of its high specific heat content in the form of latent heat and superheat. High-temperature processes, however, may require heating with hot oil or directly with flue gas in furnaces.
- **Process cooling**. This is performed by using cooling water, ambient air, or refrigeration.

- **Power demands**. These arise from the need for driving process equipment, such as pumps, compressors, mills, etc., and also lighting and electric heating (where high precision and responsiveness are necessary).
- **Water supply and disposal**. This is also an important utility. It includes mainly the supply of freshwater, eventually treated to satisfy the water quality requirements, as well as the wastewater treatment, recycling and disposal.

The utility system is considered to supply the site processes with heating and cooling utilities and satisfy their power demands. This follows from the two primary components of Total Site Integration, which are closely related: heat recovery and power cogeneration using the utility system.

In summary, an industrial site is a cluster of production processes interlinked by a common utility system. There are many benefits of having utilities provided by a central system. To a large extent, a great benefit is an opportunity to share the large investment for the utility system between more processes and lower the specific cost of utility supply.

Another advantage – a crucial one – is that a common utility system allows the processes to exchange utilities: the high-temperature cooling demand of a process may be used for utility (e.g. steam) generation instead of serving the demand with utility cooling. Lower temperature cooling demands in other processes can sometimes be served by the process-generated utility instead of applying boiler steam, utility hot oil or furnace flue gas. In this way, a central utility provides a marketplace for exchanging utilities and enabling Heat Integration between separate processes via intermediate energy carriers.

Having the common marketplace as an important degree of freedom in inter-process energy integration allows for considering various options. When all processes on a site are considered together, the industrial site is termed a Total Site, implying that this is not just a collection of elements but an integrated system.

4.2.2 Total Site Analysis interfaces

The Total Site Methodology defines very clear interfaces to perform an analysis for Interprocess Energy Integration. For analysing the Heat Exchange, Heat Sinks, and Heat Sources are used.

A Heat Sink is a representation of heating demand. It is similar to a cold Process Stream. By the same analogy, it is defined by a starting temperature (T_S), a final/target temperature (T_T), and an enthalpy change (ΔH). Specifying these properties is significant to calculate all other information. Similarly, a Heat Source is defined with a starting temperature (T_S), a final/target temperature (T_T), and an enthalpy change (ΔH), which is analogous to a hot Process Stream.

Heat Sources and Sinks can represent individual Process Streams, including residual heating or cooling demands, which process managers would like to expose to the exchange via the utility system. However, the reason for defining the separate concepts of Heat Sources and Sinks is to also provide the opportunity for extending process-level Heat Integration Analysis (i.e. Pinch Analysis) to the site level. In this way, Process Integration applies the known Information Technology principle of information encapsulation to provide more analytical flexibility to the site engineers.

4.3 HI extension for Total Sites: Data Extraction for Total Sites

When obtaining the Heat Sources and Sinks from process-level Pinch Analysis, the Grand Composite Curve (GCC) of each analysed process is used. GCC, as discussed previously, is one of the representations of the process-utility interface for a single process. The GCC represents the remaining heating and cooling demands of the process after Heat Recovery has taken place within the process. The temperature scale of the GCC is in shifted temperature T^*. The shifted temperatures are produced by shifting downwards (to lower values) the supply and target temperatures of hot streams by $0.5 \cdot \Delta T_{min}$, and shifting upwards (to higher values) the supply and target temperatures of the cold streams by $0.5 \cdot \Delta T_{min}$, building feasible Heat Exchange into the curve. The GCC shows the Pinch, the heating (Q_{Hmin}), and cooling (Q_{Cmin}) requirements to be supplied by external sources.

The individual GCCs can be used to identify the potential for inter-process Heat Recovery via steam mains. When a site houses several production processes, the GCC of each process may indicate steam levels suitable for the given process without a guarantee that they will be suitable for the other processes. This suggests that trade-offs would be needed among the energy demands of the various processes on a site to minimise the overall utility demand.

4.3.1 The algorithm

The overall procedure for obtaining site-wide Heat Recovery targets is based on thermal profiles for the entire site: the Total Site Profiles (TSPs). Constructing the profiles requires data supplied in the form of Heat Sources and Heat Sinks (i.e. the input interfaces), plus specifications of the utilities: types and temperatures. Obtaining the data for the Heat Source and Sink specifications is referred to as "Total Site Data Extraction".

The Data Extraction procedure is shown in Fig. 4.3.

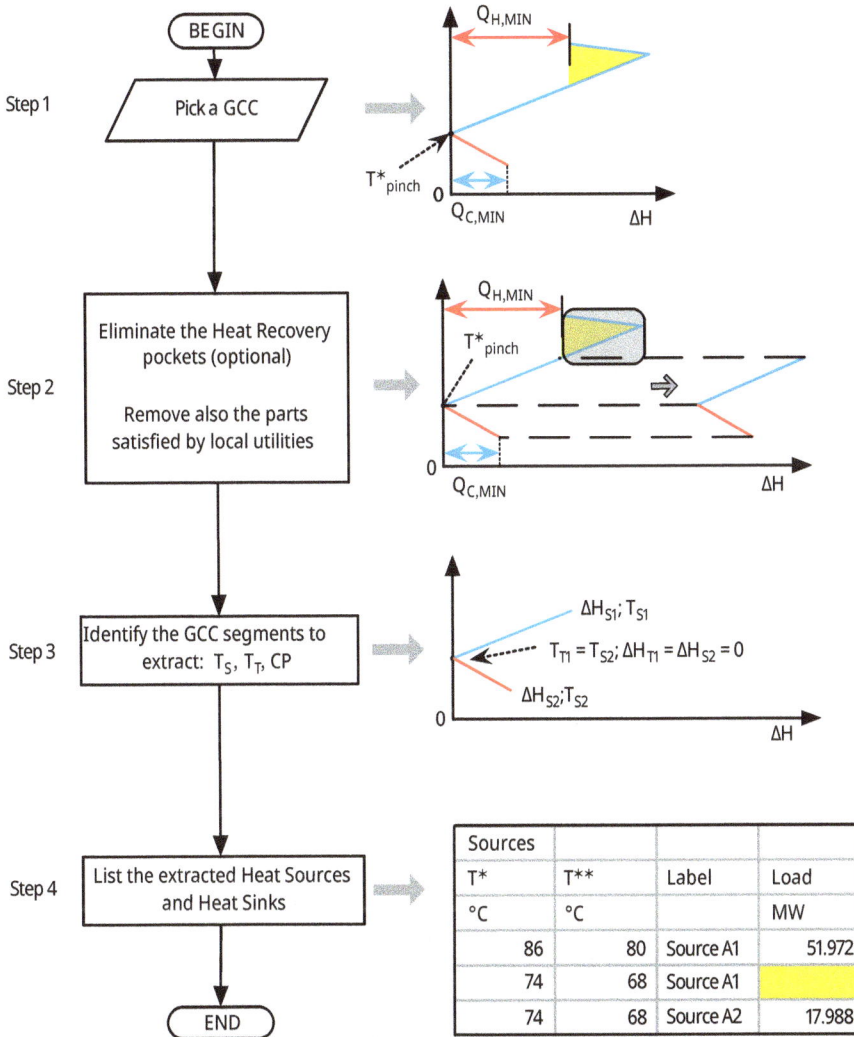

Fig. 4.3: Total Site Data Extraction procedure.

4.3.2 Step-by-step guide

Consider a Total Site comprising four processes (Fig. 4.4, Tabs. 4.1–4.5). This will be labelled "Total Site Example 1". The minimum allowed temperature difference for all processes on the site is specified as $\Delta T_{min} = 12\ °C$.

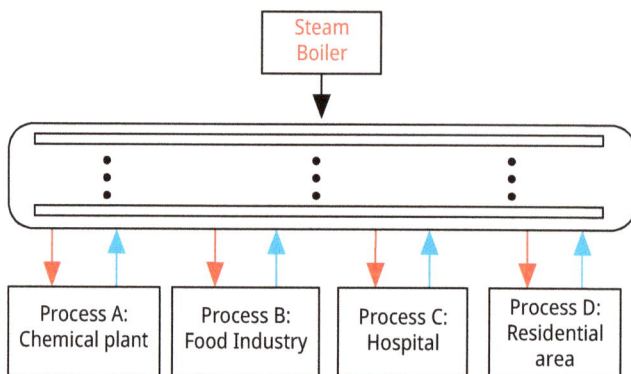

Fig. 4.4: A simple site.

Tab. 4.1: Stream specifications for Process A (Chemical Plant).

No.	Stream	Type	T_S (°C)	T_T (°C)	ΔH (MW)	CP (MW/°C)
1	A1	Hot	110	80	40.0	1.333
2	A2	Hot	150	149	180.0	180.000
3	A3	Cold	50	135	104.4	1.228
4	A4	Cold	85	100	82.3	5.487
5	A5	Cold	62	100	130.0	3.421
6	A6	Hot	92	55	282.939	7.647

Tab. 4.2: Stream specifications for Process B (Food Processing Plant).

No.	Stream	Type	T_S (°C)	T_T (°C)	ΔH (MW)	CP (MW/°C)
1	B1	Hot	200	195	160.0	32.000
2	B2	Cold	20	54	10.0	0.294
3	B3	Cold	50	85	107.3	3.066
4	B4	Cold	100	120	130.0	6.500
5	B5	Hot	150	40	83.5	0.759
6	B6	Cold	80	95	48.0	3.200
7	B7	Hot	95	25	80.0	1.143

Tab. 4.3: Stream specifications for Process C (Hospital).

No.	Stream	Type	T_S (°C)	T_T (°C)	ΔH (MW)	CP (MW/°C)
1	Soapy water	Hot	85	40	23.85	0.53
2	Condensate	Hot	80	40	96.4	2.41
3	Sanitary water	Cold	25	55	17.3	0.576
4	Laundry	Cold	55	85	18.0	0.600

Tab. 4.3 (continued)

No.	Stream	Type	T_S (°C)	T_T (°C)	ΔH (MW)	CP (MW/°C)
5	BFW	Cold	33	60	12.0	0.444
6	Sanitary water	Cold	25	60	15.0	0.429
7	Sterilisation	Cold	82	121	34.1	0.874
8	Swimming pool water	Cold	25	28	23.1	7.700
9	Cooking	Cold	80	100	32.0	1.600
10	Heating	Cold	18	25	41.1	5.871
11	Bedpan washers	Cold	21	121	5.0	0.050

The extraction of Heat Sources and Heat Sinks from process GCCs are illustrated in the example of Process A from Fig. 4.4 and Tab. 4.1. Its Grand Composite Curve is shown in Fig. 4.5, with the heat cascade data as point coordinates. For information on how to perform Pinch Analysis and obtain the GCC, please refer to Chapter 2.

Tab. 4.4: Stream specifications for Process D (Residential Area).

No.	Stream	Type	T_S (°C)	T_T (°C)	ΔH (MW)	CP (MW/°C)
1	Space heating	Cold	15	25	88.0	8.800
2	Hot water base	Cold	15	45	25.0	0.833
3	Hot water 2	Cold	15	45	65.0	2.167

Tab. 4.5: Available site utilities.

Type	Temperature range	Unit cost ($/(kW·y))
Cooling water	From 13 °C to 25 °C	150
LP steam	150 °C	260
Hot water	From 50 °C to 70 °C	–

The data in Fig. 4.5 are interpreted in the following way. The minimum heating demand of Process A (Q_{Hmin}) is 14.3 MW, and the minimum cooling demand (Q_{Cmin}) is 200.5 MW. The Pinch is located at interval temperature $T^* = 86$ °C, translating to 92 °C for the hot streams and 80 °C for the cold streams.

The next step (Step 2) of the extraction algorithm from Fig. 4.3, the elimination of Heat Recovery pockets, is optional. The idea is to eliminate the parts of GCC where additional intraprocess Heat Recovery takes place and expose to interprocess exchange only the net heating or cooling demands. When Heat Recovery is the only goal, then pocket elimination results in minimising capital costs for the utility piping and utility heat exchangers and therefore is usual. The characteristic of "optional" for pocket elimination comes from the eventual additional goal of Total Site Integration: cogeneration.

T* [°C]

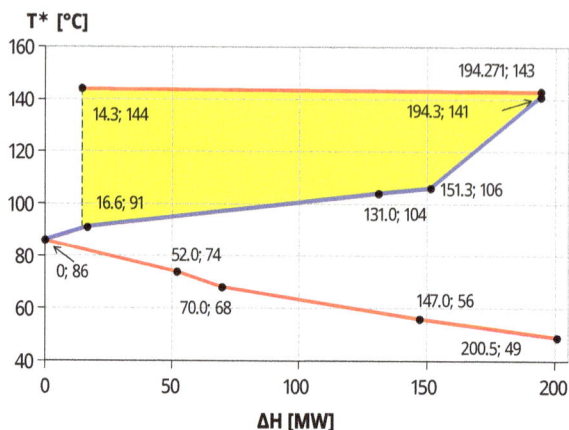

Fig. 4.5: GCC for Process A.

If there exist GCC pockets with large temperature spans and heat load (altogether – large area) and at sufficiently high temperatures, then it may be beneficial to export the GCC parts of the pockets. In such cases, steam at higher pressure may be generated and supplied by streams at the pocket top to the common utility system; in return, the streams of the pocket bottom take lower-pressure steam. In-between, the utility system can expand the higher-pressure steam to the lower-pressure cogenerating power.

If extraction of the pocket segments is to be worthwhile, the following two conditions should be met:

– The pocket should span a considerable temperature difference to enable fitting a steam turbine between the upper and lower saturation temperatures forming the pocket. In this way, the pocket would support generating a hot utility (steam) by the heat source in the pocket and using a lower temperature hot utility (steam) to supply the cold sink.
– The amount of the heat source should be sufficiently large – larger than a few hundred kW – to justify the eventual additional investment cost required to transform the heat source to a hot utility that can be used elsewhere on the Total Site and/or install a steam turbine.

For the current discussion, it is assumed that the temperature difference for the pocket in Fig. 4.5 (between 143 °C and 106 °C) is insufficiently large to accommodate a steam turbine at a reasonable cost, and the pocket can be eliminated. The elimination is performed by first finding the vertical projection of the pocket top-left point of the GCC (in this section with coordinates ΔH = 14.271 MW; T = 144 °C) onto the pocket bottom part of the curve. It is necessary to calculate the temperature at which the pocket bottom part of the GCC is at ΔH = 14.271 MW. This operation can be performed by linear interpolation inside the corresponding GCC segment. This sequence of operations is illustrated in

Fig. 4.6, where the unknown temperature is identified as $T^*_{pocket\ bottom}$ = 90.3 °C. After this calculation, the GCC segments are removed from further consideration, which in Fig. 4.6 is denoted by greying that part of the GCC.

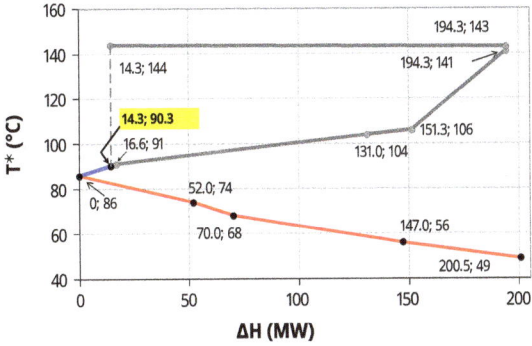

Fig. 4.6: Eliminating the pocket from the GCC of Process A.

Tab. 4.6: Heat Sources for Process A.

	T*	T**	Label	Load
	°C	°C		MW
Start	86	80	Source A1	51.972
End	74	68		
Start	74	68	Source A2	17.988
End	68	62		
Start	68	62	Source A3	77.028
End	56	50		
Start	56	50	Source A4	53.529
End	49	43		

Tab. 4.7: Heat Sinks for Process A.

	T*	T**	Label	Load
	°C	°C		MW
Start	86	92	Sink A1	14.271
End	90.3	96.3		

Next, follows Step 3 of the extraction algorithm from Fig. 4.3: identification of the GCC segments to extract. This operation involves reading the temperatures and ΔH values off the GCC or the related process heat cascade. This can be easily done by looking at Fig. 4.6.

Step 4 of the extraction algorithm is to list the extracted Heat Sources and Sinks. This also involves the additional calculation of the double-shifted temperature for each of the list entries (T^{**}). For a Heat Source, T^{**} is calculated by shifting its T^* values by $0.5 \cdot \Delta T_{min}$ downward (colder), and for a Heat Sink, the shift is upward (hotter), which is completely analogous to the shifting for hot and cold Process Streams in process-level Pinch Analysis. After this calculation, the Heat Sinks are grouped into one list and the Heat Sources into another. Both lists are sorted in descending order of the starting T^{**} values.

Additional information can be indicated in these lists for better documentation and comprehension. For instance, since entries from many processes have to be listed in the same lists, the type and origin of the respective curve segment can be indicated. One useful option is to state that the curve segment is a Heat Source by adding the label "Source" and that it is the first Heat Source from Process A (A1), which yields a combined label "Source A1".

After applying the described actions to the GCC of Process A after pocket elimination, the resulting Heat Sources and Sinks are listed in Tabs. 4.6 and 4.7.

4.3.3 Working session

4.3.3.1 Assignment

The assignment for this exercise is for each of Processes B, C, and D to:
1. Perform Process-Level Pinch Analysis and draw the relevant Grand Composite Curves;
2. Identify site-level Heat Sources and heat sinks by extracting them from the process of Grand Composite Curves. Note: remove the GCC pockets;
3. Calculate the double-shifted temperatures T^{**} of each Heat Source and Heat Sink and list the Heat Sources and Heat Sinks from Processes B, C, and D, combined with those from Process A, in two tables: one for Sources and another for Sinks.

4.3.3.2 Solution

The Process-Level Pinch Analysis has been performed for Processes B, C, and D. The resulting GCC plots are shown in Figs. 4.7–4.9.

For Process B, the minimum heating demand is $Q_{Hmin} = 21.97$ MW, and the minimum cooling demand (Q_{Cmin}) is 50.17 MW. The Pinch is located at interval temperature $T^* = 56$ °C, translating to 62 °C for the hot streams and 50 °C for the cold streams. Processes C and D have no Pinch Points, and their targets can be read in the same way.

After removing the GCC pockets, the Heat Sources and Heat Sinks for Processes B, C, and D have been identified, and the T^{**} temperatures have been calculated. After combining these with the results from Process A, the full lists are given in Tabs. 4.8 and 4.9.

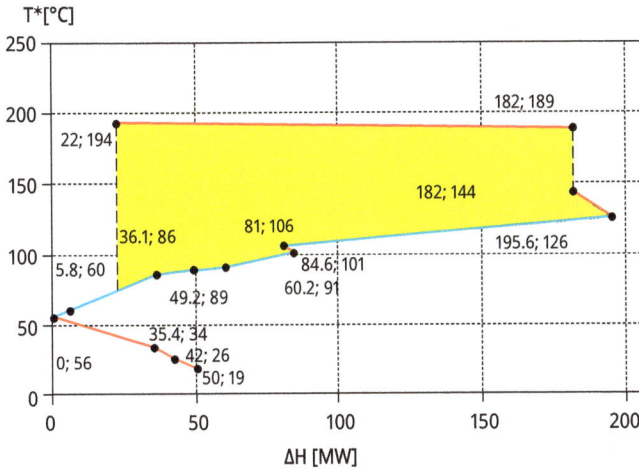

Fig. 4.7: GCC for process B.

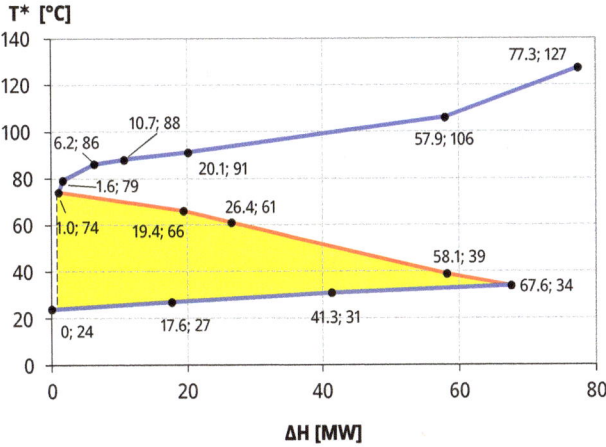

Fig. 4.8: GCC for Process C.

4.4 Total Site Profiles and Total Site Composite Curves

The next stage in the Total Site targeting procedure involves the combination of the Heat Sources from all site processes (A, B, C, D) into a composite Site Source Profile. The combination is performed among all Heat Sinks to produce the Site Sink Profile and among all Heat Sources to produce the Site Source Profile. The procedure is

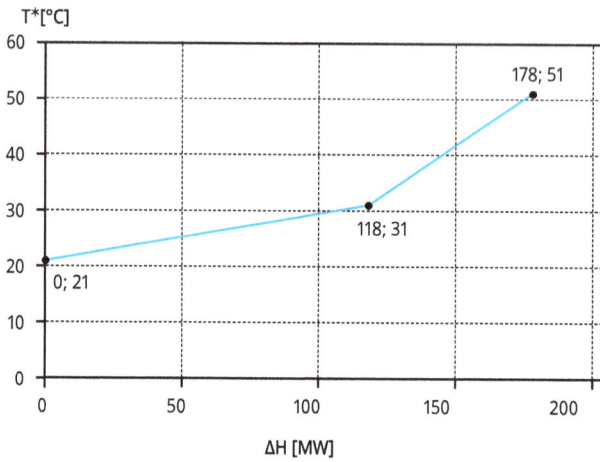

Fig. 4.9: GCC for Process D.

Tab. 4.8: Heat Sources for the example site.

	T*	T**	Label	Load
	°C	°C		MW
Start	86	80	Source A1	51.972
End	74	68		
Start	74	68	Source A2	17.988
End	68	62		
Start	68	62	Source A3	77.028
End	56	50		
Start	56	50	Source A4	53.529
End	49	43		
Start	56	50	Source B1	35.38
End	34	28		
Start	34	28	Source B2	6.8
End	26	20		
Start	26	20	Source B3	8
End	19	13		

Tab. 4.9: Heat Sinks for the example site.

	T*	T**	Label	Load
	°C	°C		MW
Start	86	92	Sink A1	14.271
End	90.3	96.3		

Tab. 4.9 (continued)

	T*	T**	Label	Load
	°C	°C		MW
Start	56	62	Sink B1	5.83
End	60.0	66.0		
Start	60	66	Sink B2	16.14
End	73.9	79.9		
Start	24	30	Sink C1	1.084
End	24.2	30.2		
Start	74	80	Sink C2	0.6
End	79.0	85.0		
Start	79	85	Sink C3	4.55
End	86.0	92.0		
Start	86	92	Sink C4	4.5
End	88.0	94.0		
Start	88	94	Sink C5	9.372
End	91.0	97.0		
Start	91	97	Sink C6	37.86
End	106.0	112.0		
Start	106	112	Sink C7	19.404
End	127.0	133.0		
Start	21	27	Sink D1	118
End	31.0	37.0		
Start	31	37	Sink D2	60
End	51.0	57.0		

completely analogous to the one for constructing the Composite Curves in Process-Level Pinch Analysis (Chapter 2).

The procedure is illustrated using two simple data sets termed "Total Site Example 2". Consider a simple site with two processes, A and B. The specifications are:
- $\Delta T_{min} = 10$ °C;
- Boiler (high-pressure, HP) steam at saturation temperature 270 °C (pressure 55.0 bar);
- Low-pressure (LP) steam at saturation temperature 140 °C (pressure 3.6 bar);
- Cooling water was supplied at 15 °C and returned to the utility system at 20 °C.

The stream data for Process A are given in Tab. 4.10 and for Process B in Tab. 4.11, and the resulting GCC plots are in Figs. 4.10 and 4.11.

The Heat Sources and the Heat Sinks from both processes are listed in Tabs. 4.12 and 4.13.

The composition of the Site Source Profile starts with the identification of temperature intervals based on the T** scale and then mapping the Heat Sources from Tab. 4.12 to the intervals. Also, from the T** temperatures and the heat load of each

Tab. 4.10: Total Site Example 2: Process A streams.

Stream	Type	T_S (°C)	T_T (°C)	CP (MW/°C)
A1	Cold	50	140	5
A2	Hot	100	30	6
A3	Cold	100	140	2

Tab. 4.11: Total Site Example 2: Process B streams.

Stream	Type	T_S (°C)	T_T (°C)	CP (MW/°C)
B1	Hot	190	120	6
B2	Cold	105	245	4
B3	Hot	80	60	2

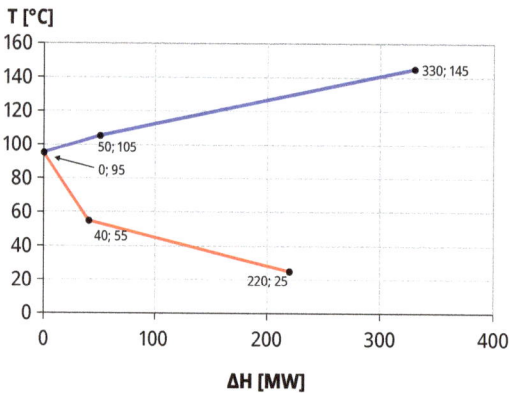

Fig. 4.10: Total Site Example 2: GCC for process A.

segment, the corresponding CP values are calculated and added to the picture. The resulting diagram is shown in Fig. 4.12.

After that, for each interval in Fig. 4.12, the sum of the CP values of the Heat Sources present in it is taken, and the interval enthalpy change is calculated as the product of the latter sum and the temperature difference in the interval.

The data from Tab. 4.13 are also processed in a similar way, and the diagram in Fig. 4.13 is obtained.

The next step is building the Site Source Profile and Site Sink Profile. For the Site Source Profile, the T^{**} values temperature intervals in Fig. 4.12 are taken and listed in descending order forming the second column of Tab. 4.14 (T^{**}, °C). In the first column of that table are calculated cascaded heat flows, starting with zero and subtracting the balance of each consecutive T^{**} interval. This algorithm produces the data points

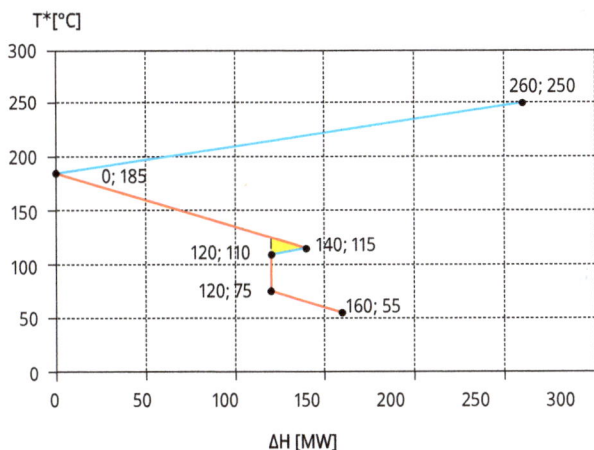

Fig. 4.11: Total Site Example 2: GCC for Process B.

Tab. 4.12: Total Site Example 2: heat sources.

texSegment	T^*_{start} (°C)	T^*_{end} (°C)	ΔH (MW)	T^{**}_{start} (°C)	T^{**}_{end} (°C)
Source A1	95	55	40	90	50
Source A2	55	25	180	50	20
Source B1	185	125	120	180	120
Source B2	75	55	40	70	50

Tab. 4.13: Total Site Example 2: heat sinks.

Segment	T^*_{start} (°C)	T^*_{end} (°C)	ΔH (MW)	T^{**}_{start} (°C)	T^{**}_{end} (°C)
Sink A1	95	105	50.00	100.00	110.00
Sink A2	105	145	280.00	110.00	150.00
Sink B1	185	250	260.00	190.00	255.00

for the Site Source Profile in such a way that it provides an opportunity for steam generation at the highest possible pressure level (represented here by saturation temperature).

In a similar but mirror way, the T^{**} values temperature intervals in Fig. 4.13 are listed in ascending order forming the second column of Tab. 4.15 (T^{**}, °C), and the ΔH column is filled with cascaded heat flow values from top to bottom (from lower to higher temperature).

The next step is to plot the data points from Tabs. 4.14 and 4.15, which yields the pair of Total Site Profiles: Site Source Profile and Site Sink Profile (Fig. 4.14).

TB** °C	Source A1 CP MW / °C	Source A2 CP MW / °C	Source B1 CP MW / °C	Source B2 CP MW / °C	ΔT interval °C	ΣCP interval MW / °C	ΔH interval MW / °C
180							
			2		60	2	120
120							
					30	0	0
90							
	1				20	1	20
70							
	1			2	20	3	60
50							
		6			30	6	180
20							

Fig. 4.12: Combining the site Heat Sources (Total Site Example 2).

TB** °C	Sink A1 CP MW / °C	Sink A2 CP MW / °C	Sink B1 CP MW / °C	ΔT interval °C	ΣCP interval MW / °C	ΔH interval MW / °C
255						
			4.00	65	4	260
190						
				40	0	0
150						
		7.00		40	7	280
110						
	5.00			10	5	50
100						

Fig. 4.13: Combining the site Heat Sinks (Total Site Example 2).

Tab. 4.14: Data for the Site Source Profile (Total Site Example 2).

ΔH (MW)	T** (°C)
0	180
−120	120
−120	90
−140	70
−200	50
−380	20

Tab. 4.15: Data for the Site Sink Profile
(Total Site Example 2).

ΔH (MW)	T** (°C)
0	100
50	110
330	150
330	190
590	255

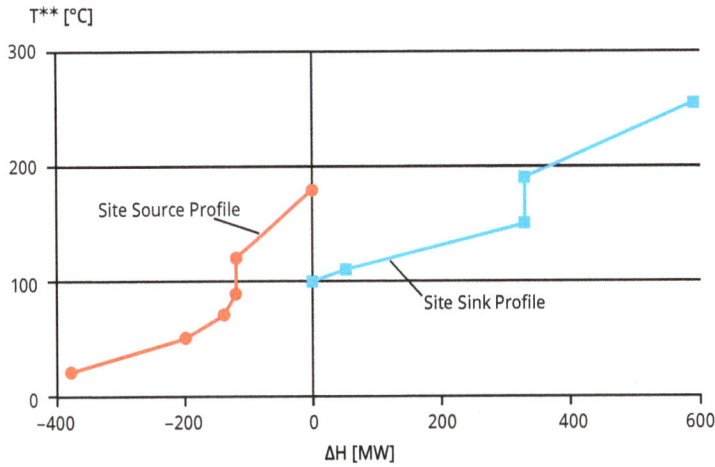

Fig. 4.14: Total Site Profiles (Total Site Example 2).

The evaluation so far has revealed that the total heating requirement of the Site Heat Sinks is 590 MW, and the total cooling requirement of the Site Heat Sources is 380 MW. Without any interprocess Heat Recovery, these would be the final amounts requested from the utility boiler and from the cooling water. Interprocess Heat Recovery can be performed using a utility as a carrier. It is important to underline that such an evaluation requires that the temperature levels of the utilities are specified before executing the procedure.

From all site utilities, the ones intersecting both Total Site Profiles can be used for such Heat Recovery. For Total Site Example 2, it can be seen that only LP steam intersects both the Heat Sink and Source Profiles, as illustrated in Fig. 4.15. So, for the example under consideration, the only vehicle for inter-process Heat Recovery is the LP steam.

The maximum possible Heat Recovery through a utility system can be targeted by using the Site Sink Profile and the Site Source Profile in combination with the steam header saturation temperatures. Site-level Composite Curves for utility generation and use are constructed, accounting for feasible heat transfer.

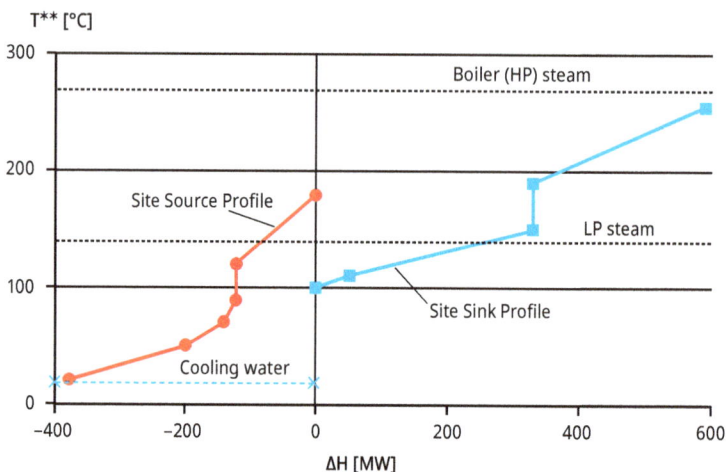

Fig. 4.15: Total Site Profiles with utilities (Total Site Example 2).

It is necessary to describe how the process-utility interface functions to understand the set of Site-Level Curves. Process Heat Sources can generate utilities, in most cases, steam at various levels. This is a heat transfer stage. The generated utility is represented by the Hot Utility Generation Composite Curve (HUGCC), also known as Site Source Composite Curve from previous publications, e.g. the pivotal Total Site work by Klemeš et al. (1997) also summarised in (Klemeš et al., 2010). At the next stage, the generated utility is supplied to a system of vessels, usually termed a "main" or a "header". The header system then routes the utility flows to the potential users. There the utility can be directly consumed, e.g. live steam used in reactors and distillation columns. More often, however, the utility is used for process heating via another heat transfer stage. The supply of heating utility from headers is represented by the Hot Utility Use Composite Curve (HUUCC), also known as Site Sink Composite Curve (Klemeš et al., 1997, 2010).

The two heat transfer stages described previously imply the need to ensure minimum temperature differences between the Site Source Profile and the HUGCC, as well as between the HUUCC and the Site Sink Profile, via the double shifting of the process Heat Source and Sink temperatures to the T* scale, as discussed previously. The two site CCs are analogous to the individual process CCs. The Source CC is built starting from the highest possible steam level. The steam generation at each level is maximised before the next lower temperature levels are considered. This ensures maximum utilisation of the temperature potential of the Heat Sources. The remainder of the Site Heat Source Profile, which does not overlap the Source CC, is served by cooling water. Building the Sink CC follows a symmetrical procedure, starting from the lowest possible steam level. The utility use of this level is maximised before moving up to the next higher temperature level and so on until steam with the highest possible level is used, including boiler-generated steam.

The construction of the Site CC for Example 2 is shown in Fig. 4.16. Let us start with the Utility Generation CC (Fig. 4.16 a). At Step 1, it can be seen that no HP steam can be generated by the Site Heat Sources, and therefore the curve stays at ΔH of zero MW. Moving toward the LP steam level (Step 2), it is obvious that some LP steam can be generated. The exact amount is identified by intersecting the LP steam temperature level with the Site Heat Source Profile (Step 3), and this is equal to 80 MW. The remaining Heat Sources have too low a temperature at the current utility specifications and are therefore served by 300 MW cooling water. The construction of the Utility Use

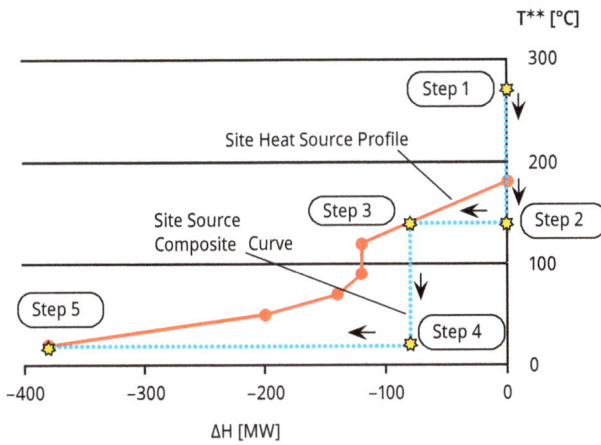

(a) Site Source Composite Curve

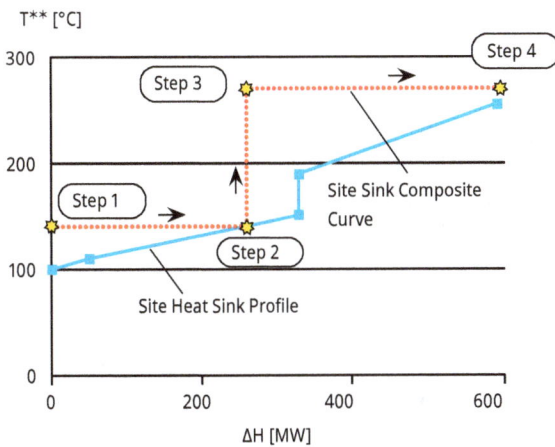

(b) Site Sink Composite Curve

Fig. 4.16: Construction of Site Source CC and Site Sink CC (Total Site Example 2).

Composite Curve is made in a similar way, identifying that the LP steam demand is 260 MW and the genuine HP steam demand is 330 MW.

After the construction of the Site CC, to target the site Heat Recovery via the intermediate utilities, the Source CC (Utility Generation CC) is shifted towards the Sink CC (Utility Use CC) to overlap, much in the same way as for the Composite Curves in process-level Pinch Analysis, thereby illustrating the Total Site Heat Recovery possible through the steam system. The amount of Heat Recovery for the Total Site is indicated by the amount of overlap between the CCs. Hence, Heat Recovery is maximised when the two curves touch and cannot be shifted further. The area where the curves touch, which is usually confined between two steam levels, is the Total Site Pinch. It is characterised by opposite utility load balances above and below. Above the Total Site Pinch, there is a net deficit of heat, and below it is a net excess of heat. For Example 2, the shift of the Total Site Composite Curves is illustrated in Fig. 4.17. There at the LP steam level, there is 80 MW LP steam generation resulting from the Site Source CC and 260 MW LP steam use resulting from the Site Sink CC, making a net deficit of 180 MW of LP steam to be cascaded from the HP steam (boiler).

At the lower utility level, that of the cooling water, there is a net excess of heat coming entirely from the Heat Source Profile, totalling 300 MW. As a result, the Total Site Pinch for Example 2 is located at the utility levels LP steam: cooling water.

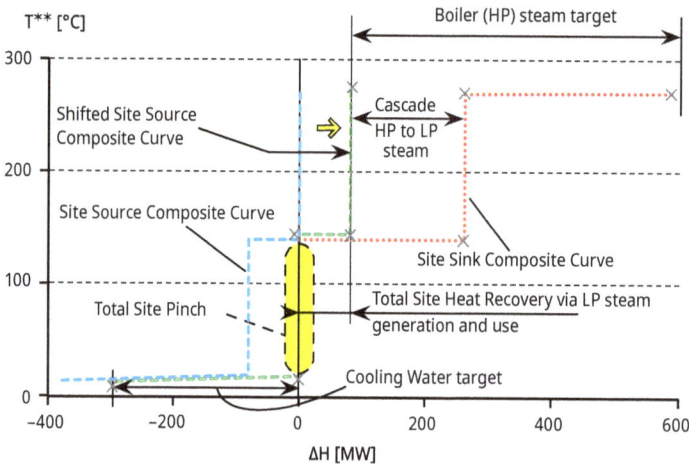

Fig. 4.17: Shift of the HUGCC toward the HUUCC and identification of the Total Site Pinch (Total Site Example 2).

4.5 Site Utility Grand Composite Curve (SUGCC)

Another targeting tool available for Total Sites is the Site Utility Grand Composite Curve (SUGCC), which evaluates the potential cogeneration that is available in the Site Utility system due to the distribution of steam to the site process Heat Sinks and the generation of steam from site boilers and process heat sources. The SUGCC can be derived from the targeting information already obtained in the form of the Total Site Profiles and Total Site Composite Curves.

Visually, the SUGCC can be obtained geometrically by subtracting the segments of the Site Sink CC from those of the Site Source CC. The algorithm for constructing the SUGCC includes the following.

(i)   Plot the highest temperature utility from zero ΔH to the right (data source: Site Source CC).

(ii) At the temperature reached in Step 1, plot to the left the utility use at the same utility level (data source: Site Sink CC). If the current utility is steam, its generation and use will have the same temperatures.

(iii) If there is a lower temperature utility, continue the curve vertically down to the next lower temperature utility level and select it as the current one. The SUGCC ends when there are no other lower-temperature utilities. The vertical line is drawn for the continuity of the curve and does not represent any heat transfer directly.

(iv) Repeat the above steps for the new "current" utility level.

Figure 4.18 illustrates the described procedure and its result when applied to the case of Total Site Example 2.

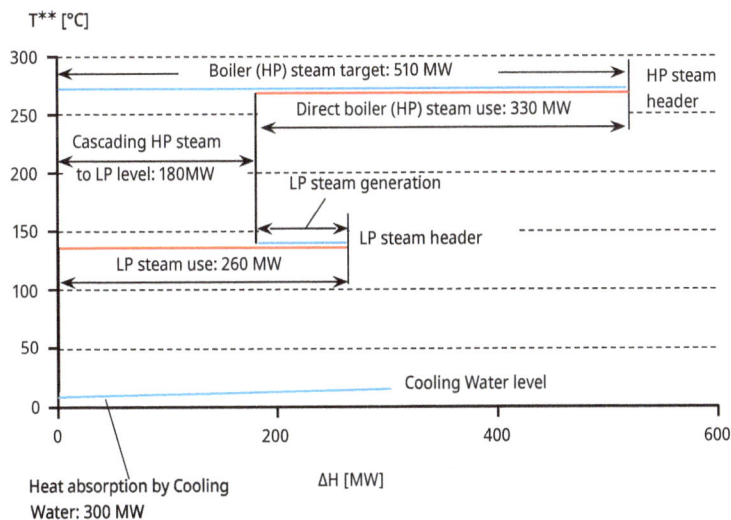

Fig. 4.18: The Site Utility Grand Composite Curve (SUGCC): Total Site Example 2.

4.6 Modelling of utility systems

Utility systems are an important part of most processing sites. Cogeneration is usually an important feature of such systems, particularly from steam turbines, gas turbines, or a combination of both. On large processing sites, the cost of fuel and power can be very high and better management of the utility system can lead to significant cost savings. A comprehensive textbook on Utility Systems is available from De Gruyter (Varbanov et al., 2020). That book provides detailed discussions on Utility System phenomena, thermodynamics, components, machines, networks and overall modelling using process simulators. The current chapter provides an introduction to the basics related to Total Sites. For existing sites, such cost savings can often be accomplished without capital expenditure by more effective day-to-day operational management. A flexible but sufficiently simple model of utility systems is an important tool in determining the strategy for energy efficiency improvement, e.g. to pursue improved Heat Integration or modifications to the utility system.

Tab. 4.16: Summary of steam turbine models suitable for network simulation and optimisation.

Sources	Features	Intended uses
(Raissi, 1994) and (Klemeš et al., 1997)	Based on heat flows intended specifically for cogeneration targeting	Cogeneration targeting only
(Bruno et al., 1998)	Non-linear model, relating efficiency to steam flow	Utility system synthesis, MINLP model
(Mavromatis and Kokossis, 1998)	Backpressure steam turbine model based on the Willans Line. Performance related to size and part-load. Willans Line intercept assumed identical with energy losses.	Steam network modelling, cogeneration targeting
(Shang and Kokossis, 2004)	Extension of (Mavromatis and Kokossis, 1998) to condensing steam turbines.	Steam network modelling, cogeneration targeting
(Varbanov et al., 2004a)	Complete overhaul of (Mavromatis and Kokossis, 1998) and later (Shang and Kokossis, 2004) models; accounting for the pressure at both inlets and outlets	Steam network modelling, cogeneration targeting
(Aguilar et al., 2007)	Customisation of the model in (Varbanov et al., 2004a) accounting for flexibility	Steam network modelling, cogeneration targeting
(Sun and Smith, 2015)	Refinement of the (Varbanov et al., 2004a) model accounting for more turbine types	Steam network modelling, cogeneration targeting
(Jiménez-Romero et al., 2022)	Further development of the model in (Sun et al., 2015), producing variations for the different stages of the steam network optimisation, interfacing with the steam mains model	Steam network synthesis

4.6.1 A flexible steam turbine model for cogeneration evaluation

A number of models have been used for evaluating the performance of steam turbines within the context of utility systems. Mavromatis and Kokossis (1998) proposed a simple model of back-pressure steam turbine performance. In this model, the performance of a steam turbine is related to its size (in terms of maximum shaft power) and to part-load performance. The shaft power is modelled as a function of the steam mass flow, a function that is known as the Willans Line (Willans, 1888). This model was later extended by Shang (2000) to condensing steam turbines. All these works follow the same model structure and employ the same equations; however, they use different values for the turbine regression coefficients.

The intercept of Willans line was mapped by Mavromatis and Kokossis (1998), as well as by Shang (2000), as identical to the turbine energy losses, also assuming a fixed loss rate. Varbanov et al. (2004a) introduced improvements to those models by: i) recognising that Willans line intercept has no direct physical meaning and is simply the intercept of a linearisation; and ii) accounting for both inlet and outlet pressures of the steam turbines. The improved steam turbine models have been incorporated into methodologies for simulating and optimising steam networks and have also been used to target the cogeneration potential by assuming a single large steam turbine for each expansion zone between two consecutive steam headers (Varbanov et al., 2004b). A concise summary can be seen in Tab. 4.16.

A flexible cogeneration model, suitable both for steam network modelling and targeting based on the SUGCC information, is the one (Varbanov et al., 2004a) described here. The elaborations, listed in Tab 4.16, are recommended as well, but for the purposes of learning this model is the starting point.

4.6.1.1 Factors determining steam turbine performance

Steam turbine performance is affected by a number of factors. The most significant among them are:
- Thermodynamic limitations and efficiency;
- Turbine size in terms of maximum power load;
- Pressure drop across the turbine;
- Current load.

Thermodynamic limitations, pressure drop and efficiency

The turbine thermodynamics and design as factors are best understood from the representation of the expansion of steam in a turbine on a Mollier Diagram (Fig. 4.19). The isentropic efficiency for a given turbine load is defined as

$$\eta_{is} = \frac{\Delta h_{\text{turbine}}}{\Delta h_{is}}.$$ (4.1)

The expansion process transforms part of the energy of the inlet steam into power. The magnitude if this part depends on the turbine design. Different ways of expanding steam would result in different isentropic efficiency. The total power from the expansion is further split (Fig. 4.19) into useful power, delivered to the shaft and energy losses. Losses occurring in steam turbines are mechanical friction losses, causing heat losses and kinetic energy losses with the turbine exhaust.

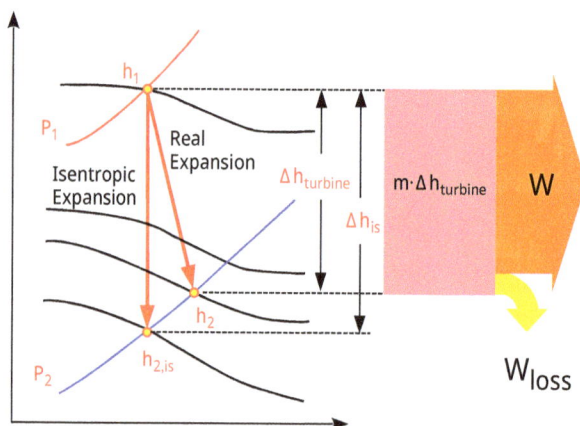

Fig. 4.19: Expansion of steam in a turbine.

Current load

For a particular steam turbine and a specified maximum steam flow, it is possible to calculate its maximum shaft power. During operation, however, many turbines work at part load. It is necessary to estimate the actual enthalpy change, isentropic efficiency, and the energy losses with changing turbine load. From operating practice, it is known that steam turbine efficiency varies with part load (Mavromatis and Kokossis, 1998). At lower loads, it is lower and gradually increases with the turbine reaching its maximum load. This trend is illustrated in Fig. 4.20.

The overall turbine efficiency can be represented as consisting of two components: the isentropic efficiency and the machine efficiency. The machine efficiency is generally higher than the isentropic efficiency (Siddhartha et al., 1999) and changes in a relatively narrow range. In contrast, the isentropic efficiency changes substantially with load:

$$\eta_{st} = \eta_{is} \cdot \eta_m, \eta_m \gg \eta_{is}.$$ (4.2)

The variation of the overall efficiency is non-linear (Fig. 4.20 (a)). It can also be interpreted as the equivalent relationship of the shaft power vs the mass flow of steam

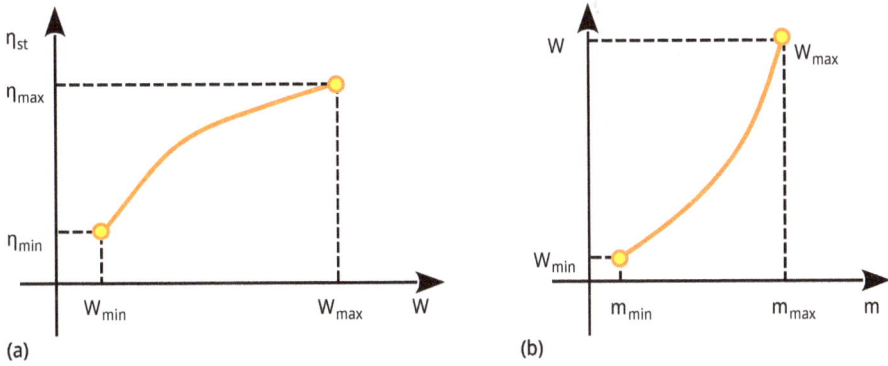

Fig. 4.20: Typical steam turbine performance trend.

(Fig. 4.20 (b)). The latter tends to be less non-linear and, when approximated with piecewise-linear segments, usually leads to better accuracy.

To model the turbine in a way suitable for efficient optimisation modelling requires deriving linear relationships and characterising the line coefficients. The slopes are related to the actual enthalpy change across the turbine. For any load of a steam turbine with fixed inlet and back pressures and fixed inlet temperature, the isentropic enthalpy change remains constant. Because of changing isentropic efficiency, the actual enthalpy drop across the turbine varies with the load, remaining smaller than the isentropic enthalpy drop. The actual non-linear power–steam flow relationship in Fig. 4.20 (b) can be approximated by linear segments (Fig. 4.21), usually termed Willans Lines (Willans, 1888). The actual steam turbine performance curve is replaced with one or more straight lines with different but fixed coefficients.

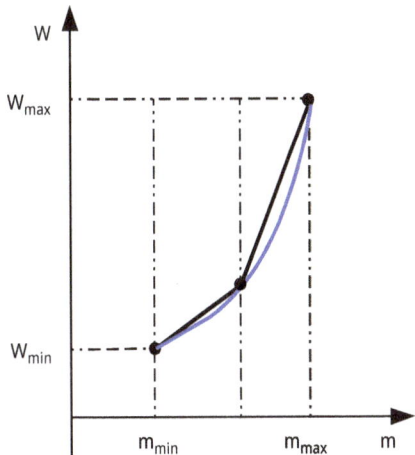

Fig. 4.21: Linear approximation segments (Willans Lines) superimposed on the performance curve.

The actual steam turbine performance is given by

$$W = \Delta h_{\text{turbine}} \cdot m - W_{\text{Loss}} \tag{4.3}$$

Where $\Delta h_{\text{turbine}}$ (see also Fig. 4.19) is the actual change in steam enthalpy across the turbine, and W_{Loss} designates the energy losses. The expression in eq. (4.3) has a typical linear form. However, it should be noted that the loss term also depends on the current load, which makes the resulting plot a curve.

The performance approximation in each modelling interval in Fig. 4.21 has the form of Willans Line:

$$W = n_{\text{ST}} \cdot m_{\text{STEAM}} - W_{\text{INT}}. \tag{4.4}$$

For existing utility systems, often the values of the steam parameters – pressures, temperatures – are fixed. It is sufficient to apply Willans Line to model steam turbines in such cases.

Turbine size

For designing new systems, including Total Site Targeting, evaluation of potential configuration changes, and buying new turbines, it is necessary to also model the turbine performance more broadly, including both design-point and off-design performance, across many steam turbines.

In the case of Total Site Targeting, the current steam turbine load is equal to the maximum turbine load and eq. (4.4) takes the following form:

$$W_{\text{MAX}} n_{\text{ST}} \cdot m_{\text{MAX}} - W_{\text{INT}}. \tag{4.5}$$

The Willans Line slope n_{ST} is estimated based on the isentropic enthalpy drop across the turbine and the maximum steam flowrate:

$$n_{\text{ST}} = \frac{L+1}{B_{\text{ST}}} \cdot \left(\Delta h_{\text{is}} - \frac{A_{\text{ST}}}{m_{\text{max}}} \right). \tag{4.6}$$

In eq. (4.6), L is referred to as the intercept ratio, A_{ST} and B_{ST} are regression parameters. The Willans Line intercept is

$$W_{\text{INT}} = \frac{L}{B_{\text{ST}}} \cdot (\Delta h_{\text{is}} \cdot m_{\text{MAX}} - A_{\text{ST}}). \tag{4.7}$$

The steam turbine parameters A_{ST} and B_{ST}, used in eqs. (4.6) and (4.7), are calculated from the following regression relationships:

$$A_{\text{ST}} = b_0 + b_1 \cdot \Delta T_{\text{sat}} \tag{4.8}$$

$$B_{\text{ST}} = b_2 + b_3 \cdot \Delta T_{\text{sat}} \tag{4.9}$$

The intercept ratio L from eqs. (4.6) and (4.7) has also been found to depend on the saturation temperature drop across steam turbines:

$$L = a_L + b_L \cdot \Delta T_{\text{sat}}. \tag{4.10}$$

The total energy extracted from the expanding steam can be estimated using a fixed specification for the mechanical efficiency of the steam turbine and generator:

$$W_{\text{total}} = \frac{W}{\eta_{\text{mech}}}. \tag{4.11}$$

Alternatively, the evaluation can be done from a linear regression model if regression data are available:

$$W_{\text{total}} = a_{\text{total}} + b_{\text{total}} \cdot W. \tag{4.12}$$

The latter allows estimation of the enthalpy of the turbine exhaust as follows:

$$h_{\text{ST, EX}} = h_{\text{ST, IN}} - \frac{W_{\text{total}}}{m_{\text{STEAM}}}. \tag{4.13}$$

The values of the performance regression parameters derived in (Varbanov et al., 2004a) for a given set of steam turbines are listed in Tab. 4.17. Additionally, further simplifications of this model may be used, where some of the calculated coefficients are fixed based on additional sensitivity analysis.

Tab. 4.17: Regression coefficients for turbine performance estimation.

Coefficient	Unit	Value
b_0	MW	$-2.080 \cdot 10^{-8}$
b_1	MW/°C	0.000297
b_2	–	1.60200
b_3	1/°C	−0.001600
a_L	–	−0.01
b_L	1/°C	0.000326
a_{tot}	MW	0.1422
b_{tot}	MW/°C	1.0174

The full derivation and reasoning of the steam turbine model, as well as models for the other utility system components, can be found in (Varbanov et al., 2004a). Other steam turbine models are also possible to apply for simplified evaluations. These include:

– Assuming certain isentropic efficiency of the turbine and neglecting the mechanical losses, the power generation can be estimated as the product of the isentropic enthalpy drop Δh_{is}, the steam mass flow rate m_{STEAM} and the isentropic efficiency. This also includes the case of assuming isentropic expansion ($\eta_{\text{is}} = 1$);

– Accounting for the mechanical losses in the isentropic model can be performed by adding mechanical efficiency to the product.

It is necessary to emphasise that the coefficients given in Tab. 4.17 have been obtained from published data (Peterson and Mann, 1985) and may not necessarily be appropriate to predict the performance of all steam turbines. Therefore, if possible, it is recommended that an independent regression for particular steam turbine designs be performed if data from equipment manufacturers are available or, better still, operating data. The data required to determine the coefficients are.

W (MW) – the shaft power;

$T_{sat,in}$ (°C) – the inlet steam saturation temperature;

T_{in} (°C) – the inlet steam actual temperature;

$T_{sat,out}$, (°C) – the outlet steam saturation temperature;

m_{ST} (t/h) – the mass flow of the expanding steam.

4.6.2 Utility network modelling: simulation and optimisation

With a basic understanding of steam turbines, it is possible to create simple utility network models for evaluating and optimising their performance. It is necessary to identify the degrees of freedom and trade-offs in utility plants to understand the system-level features.

4.6.2.1 Firing machines

Starting from the higher-temperature parts of the system, the first degrees of freedom relate to firing machines: gas turbines, boilers (fuel-based steam generators), and Heat Recovery steam generators (Fig. 4.22). Firing in utility systems can be done in gas turbines or boilers. A gas turbine can be connected to a heat recovery steam generator, where some supplementary firing can be applied. Every firing machine can use a

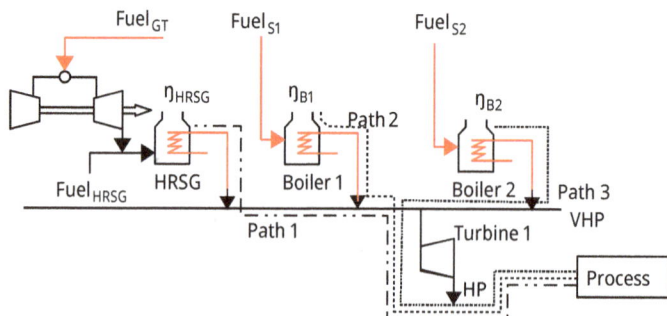

Fig. 4.22: Typical configuration of firing machines in a utility system.

different fuel or, perhaps, a different combination of fuels and has its own efficiency, varying with the load.

4.6.2.2 Steam distribution system

Steam can be transferred between headers via letdown stations or steam turbines (Fig. 4.23), moving down the temperature scale. Generally, steam turbines have different efficiencies, varying with a part load. If there are two or more steam paths connecting two steam headers, this introduces additional degrees of freedom for internal flow distribution allowing to optimise the power generation. Usually, large letdown flows indicate missed opportunity for cogeneration. However, letdown stations play an important role in maintaining the steam at each header at the desired (usually superheated) condition by expanding steam from higher-pressure headers to lower-pressure ones nearly adiabatically. Also, let-down flows might allow a constraint in a steam turbine to be bypassed. For example, if all steam turbines between two headers are all at their maximum flowrates, expanding steam through a letdown station may allow lower-pressure steam turbines or process users to receive more steam.

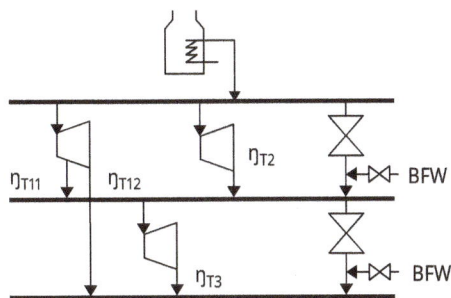

Fig. 4.23: Typical paths and trade-offs in the steam distribution system.

Condensing steam turbines provide utility systems with additional degrees of freedom, generating extra power from higher steam flows (Fig. 4.24). However, the cogeneration efficiency is lower compared with the combination of backpressure turbines and process steam usage. As with condensing steam turbines, venting steam from LP headers also provides an additional degree of freedom to increase power generation. This option also results in lower cogeneration efficiency, even lower than that of the condensing turbine option.

4.6.3 Utility system: an illustrative example

The described models and reasoning can be used for establishing the steam balances of a utility system, be it an existing one or during process design. This section provides an example based on a real site, adapted from Varbanov et al. (2004a).

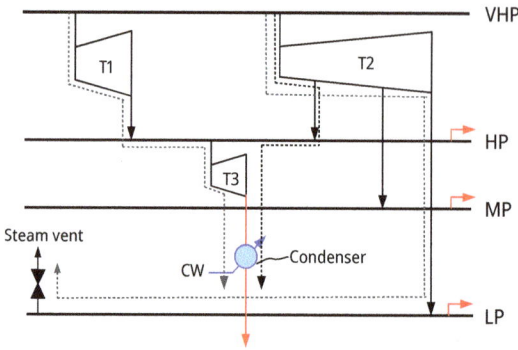

Fig. 4.24: Condensing steam turbines and vents.

Consider the utility system in Fig. 4.25. This includes a gas turbine with a Heat Recovery steam generator, two steam boilers, four steam turbines for electricity generation, and two direct-drive steam turbines. Three fuels are available on the site, specified in Tab. 4.18. The site can import up to 50 MW of power at 0.045 $/kWh, and it can export up to 10 MW of power at the price of 0.060 $/kWh. The site power

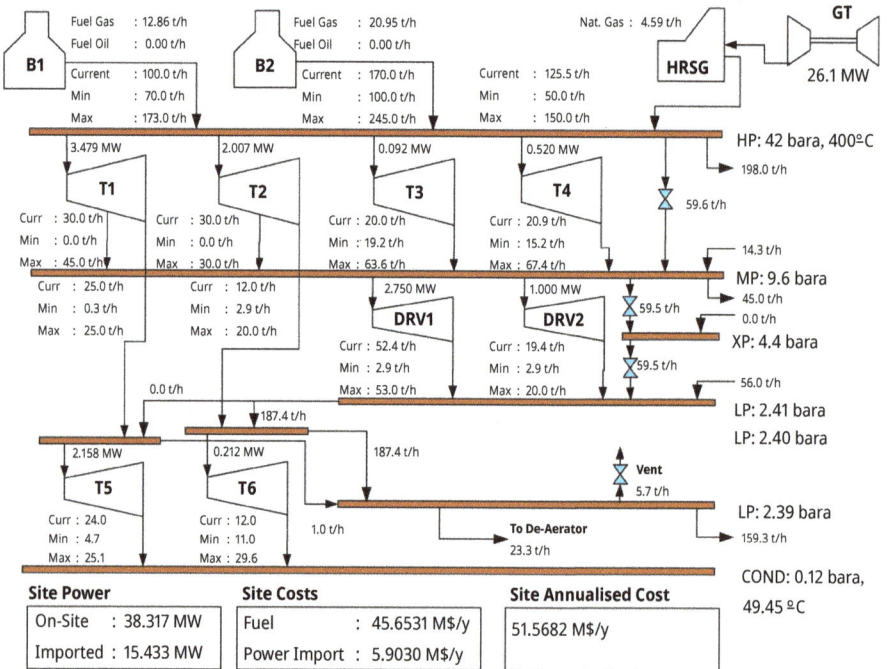

Fig. 4.25: Utility system example: initial operating point.

Tab. 4.18: Fuel data.

	Fuel 1	Fuel 2	Fuel 3
	Fuel gas	**Fuel oil**	**Natural gas**
NHV (kJ/kg)	32502.8	40245.0	46151.8
Price ($/t)	103.41	70.82	159.96

Tab. 4.19: Site configuration data.

Ambient temperature	°C	25.00
Minimum stack temperature	°C	150.00
De-aerator pressure	bara	1.01325
Boiler feed-water temperature	°C	80.00
Condensate return ratio	(1)	0.5557

Tab. 4.20: Steam turbine performance parameters for the illustrative example.

Turbine: stage	L	A	B
	(–)	MW	(–)
T1: HP-MP	0.228	0	1.96
T1: MP-LP-t1	0.010	0	3.15
T2: HP-MP	2.802	0	1.82508
T2: MP-LP-t2	0.193	0	3.15156
T3	0.429	0	1.43
T4	0.289	0	1.47
T5	0.229	0	1.46
T6	0.588	0	1.0445
DRV1	0.100	0	1.5
DRV2	0.040	0	1.53

demand, excluding the drivers, is 50 MW. Additional site configuration data are given in Tab. 4.19. Table 4.20 lists the values of the steam turbine performance parameters.

Figure 4.25 shows the initial operating point, as identified from the existing system. This has been subjected to optimisation, and the result is shown in Fig. 4.26. As a result of the optimisation, turbines T2, T5, and T6 have been turned off. On the other hand, turbines T3 and T4 have been loaded to the maximum. The driver steam turbines remain in operation. Boiler B2 load is reduced to the minimum while the HRSG flow is maximised. As a whole, the on-site power production is increased by 2.447 MW. The power import cost is reduced by 0.9359 M$/y and the fuel cost by 6.4145 M$/y. The total operating cost of the site is reduced by 7.3508 M$/y, a 14% reduction with respect to the initial operation.

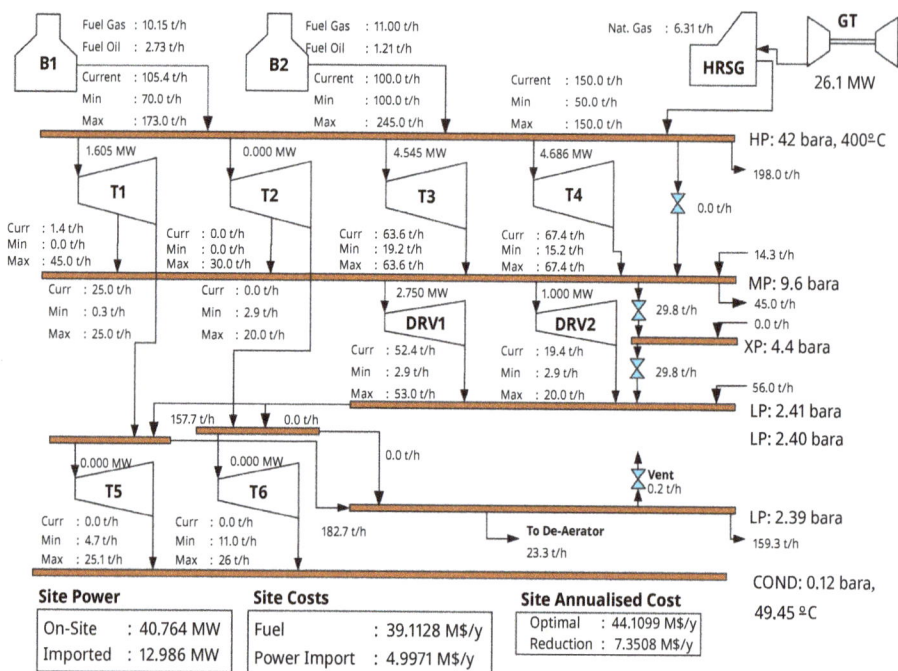

Fig. 4.26: Utility system example: optimal operation.

The reason for the cost trend shown in the optimisation is that power import is cheaper than the on-site power generation on its own. However, the much higher efficiency of the combined power and heat generation makes part of the on-site generated power even cheaper. As a result, on-site power cogeneration, alongside heat required by the site processes, is maximised, but any steam turbine load above the one satisfying the steam requirement is not profitable.

The illustrated example has been calculated using STAR (2013) – a legacy tool. A contemporary replacement of that software is provided by Process Integration Ltd, UK (i-Steam, 2022). An alternative option is to use the PetroSIM software by KBC (2022). Both suites have sufficient modelling components and frameworks for the simulation and operational optimisation of existing utility systems.

4.7 Targeting of Combined Heat and Power generation (CHP, cogeneration) during process design

It can be seen from the SUGCC in Fig. 4.18 that between the HP steam level and the LP steam level, there is an enclosed area. This represents the cogeneration potential if higher-pressure steam is expanded in a turbine to a lower-pressure level. This insight was documented first by Raissi (1994) and further developed into a complete method-ology by Klemeš et al. (1997). Klemeš et al. (1997) defined a simple proportionality co-efficient, whose value is usually evaluated for each industrial site separately. This cogeneration targeting model is referred to as "the T-H model" because it is based on heat flows through the steam system.

Using SUGCCs allowed Klemeš et al. (1997) to set thermodynamic targets for co-generation along with targets for site-scope Heat Recovery minimising the cost of util-ities. Satisfying the goal of maximum Heat Recovery leads to a minimum boiler Very High Pressure (VHP) steam requirement, which in turn can be achieved by maximis-ing steam recovery. At such conditions, the power generation by steam turbines is also minimal, which has the effect of maximising imported power. This scenario can be represented by the site CCs that are shifted to a position of maximum overlap (i.e., pinched). This target represents the thermodynamic limitation on system efficiency, but this is not a specification that has to be achieved. The case of minimising the cost of utilities is handled by exploring the trade-off between steam recovery and cogene-ration potential by steam turbines. If design guidelines are thus based on minimising cost, then the resulting utility network design is usually different from that produced when aiming for minimum fuel consumption.

In the example in Fig. 4.18, from the total 510 MW HP steam generated by the boil-ers, 330 MW is sent for process use, and 180 MW is expanded to the LP steam level. If this amount of steam is expanded in a steam turbine, this can then produce power that can be used on the site.

4.7.1 Targeting CHP using the SUGCC

The simple model from the previous section can be used for targeting the cogenera-tion potential of Total Site Utility Systems. This is performed by representing the Total Site by the corresponding SUGCC. In each expansion zone, where some non-zero area is enclosed by the SUGCC, a single simple steam turbine is assumed; the heat flow at the entrance of the expansion zone is estimated from the net cascaded heat flow and the steam level parameters. The model from Section 4.6.1 is further applied, and the power cogeneration target is estimated.

For Total Site Example 2, the steam turbine parameters to use are listed in Tab. 4.21. Consider the SUGCC of Example 2 again from Fig. 4.18. It reflects a situation where the Heat Recovery through the steam system is maximised, corresponding to the mini-mum amount of HP steam that is required from the site boilers. In producing the Total Site Composite Curves, representing the situation of maximum Heat Recovery and minimal HP steam supply, the starting point was from the positions of the Total Site Profiles, which represents a situation of no Heat Recovery and maximum HP sup-ply from the site boilers. Between these two extreme states, there can be a variation

Tab. 4.21: Steam turbine regression coefficients for Total Site Example 2.

Parameter	Unit	Value
b_0	MW	0.5
b_1	MW/°C	0.008
b_2	–	1.18
b_3	1/°C	0.03
L	–	0.15
η_{mech}	–	0.95

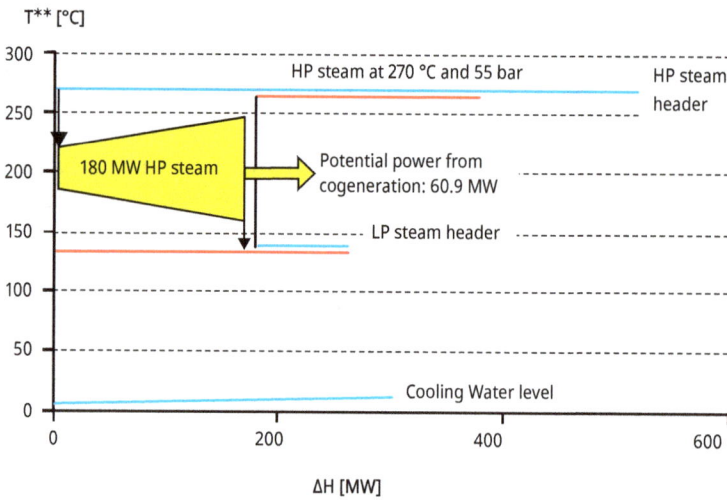

Fig. 4.27: The cogeneration target for Total Site Example 2.

in the HP steam supply from the boilers. At maximum Heat Recovery at the site level, the cogeneration target for Total Site Example 2 is shown in Fig. 4.27.

An even further increase of the HP steam supply from the boilers is possible. Based only on the consideration of the process steam demands and the Total Site Pinch, such an increase makes no sense in terms of Heat Recovery and efficiency. However, if cogeneration is considered with total energy cost as a criterion, this involves cost items for power import from the grid or potential revenue from power export to the grid, in addition to the cost of fuel. Experience has shown (Smith and Varbanov, 2005) that there are certain price ratios at which such an increased steam generation, accompanied by on-site power generation by condensing steam turbines, may be profitable. Such a potential situation for Total Site Example 2 would be reflected by a SUGCC like the one in Fig. 4.28, where about 60 MW extra steam is generated by the HP boiler and run through a condensing steam turbine. For more details

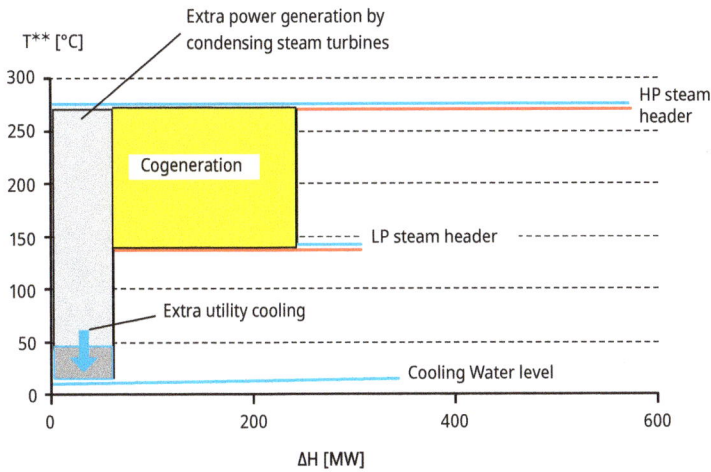

Fig. 4.28: Additional power cogeneration by condensing steam turbines for Total Site Example 2.

on exploring such options, refer to the article on advanced Total Site Analysis by Varbanov et al. (2004b).

4.7.2 Choice of optimal steam pressure levels

Until this point, the steam pressure levels have been considered fixed. However, broadening the view a bit offers new opportunities at the stage of new system design and use of Total Site targeting. If the pressure levels of the intermediate utilities are allowed to vary, it is possible to achieve better Heat Recovery through the utility system (Klemeš et al., 1997). The fundamental trends of the Total Site Profiles and the derived Site Composite Curves can be exploited for this purpose. In this case, the steam pressure level becomes a degree of freedom to be exploited in the optimisation. The main trade-off in a Total Site Utility System with a single intermediate steam header is formed as follows (Fig. 4.29).

(a) If the pressure (and consequently also the temperature) level of the header is lowered, it generally increases the potential amount of utility generation from Site Heat Sources, which would be reflected by a wider horizontal segment of the Site Utility Generation Composite Curve.

(b) On the other hand, the trend regarding the Site Heat Sinks is the opposite: a higher pressure level means larger demand to be covered by the current utility; a lower level means a smaller amount.

Exploiting such a trade-off, it is possible to optimise the exact saturation temperature level and, from there, also the pressure of the particular utility. A defining property of

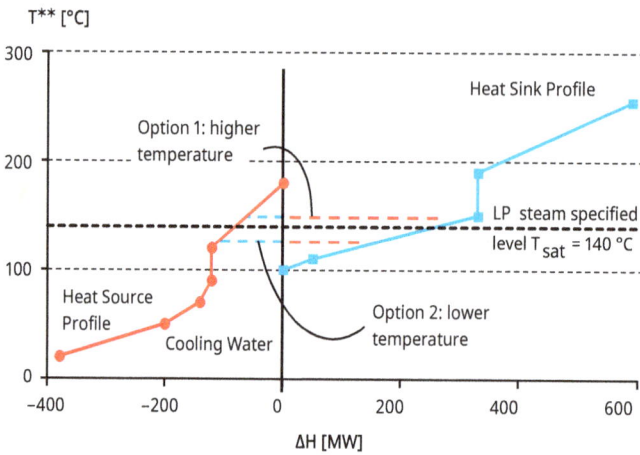

Fig. 4.29: Typical utility generation and use trends for Total Site Profiles in the case of Total Site Example 2.

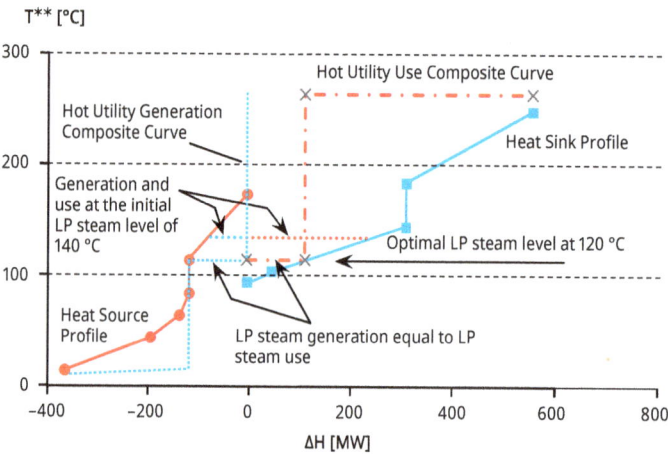

Fig. 4.30: Optimal saturation temperature of the LP steam header for Total Site Example 2.

this optimal level is that the steam generation by Site Heat Sources is equal to the steam used by Site Heat Sinks, which is logical: any extra generated utility also has to be used in order to effect Heat Recovery. Any unused generated utility would anyway be directed to the cooling water.

For the case of Total Site Example 2, the optimal LP steam saturation temperature with minimisation of the boiler steam use is at 120 °C, as illustrated by the diagram in Fig. 4.30. The extent of the improvement is demonstrated in Tab. 4.22.

Tab. 4.22: Results of the LP steam level optimisation for Total Site Example 2 (all quantities are in MW).

	LP steam at 140 °C	LP steam at 120 °C	Change = new – old
LP steam generation	80	120	+40
LP steam use	260	120	−140
HP steam to processes	330	470	+140
HP steam target	510	470	−40
Cooling water target	300	260	−40

In the case when more than one intermediate utility is allowed on a site, the model is more complex, and the optimisation would require using a mathematical optimisation solver. It is also important to mention that level optimisation can also be performed by accounting for cogeneration via the utility system, as has been described by Klemeš et al. (1997) and later elaborated by Shang and Kokossis (2004).

4.8 Advanced Total Site developments

4.8.1 Introduction of the process-specific minimum allowed temperature differences

In the basic examples and considerations in this chapter, all processes on a site, as well as all heat transfer operations between processes and utilities, are assumed to have only one single ΔT_{min} specification. Such an assumption may be too simplistic and lead to inadequate results due to imprecise estimation of the overall Total Site Heat Recovery targets.

When using a uniform ΔT_{min} specification, problems can arise in two cases:
- If the value of the common ΔT_{min} specification is overestimated for some parts of the Total Site Profiles, the achievable Heat Recovery targets for the process would be underestimated, resulting in higher energy cost estimates. This can have economic implications in the form of oversizing the utility system components, most notably the steam boilers, and lead to inefficient use of capital for the utility system.
- In the opposite case, if ΔT_{min} is set smaller than the optimal for the corresponding heat transfer type, the Heat Recovery targets would be overestimated, leading to striving for more Heat Recovery site-wide and excessive capital costs for the heat transfer area. In addition, the underestimated hot and cold utility demands may result in undersized utility facilities such as boilers and cooling towers. This, in turn, can potentially lead to either infeasible operational situations or, at the very least, reduced reliability of the utility system by having to operate near or at its maximum capacity.

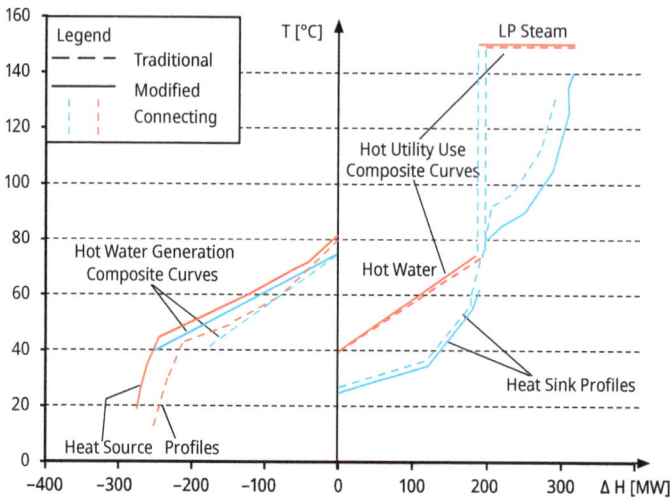

Fig. 4.31: Difference in the Total Site targets when allowing multiple ΔT_{min} specifications.

This issue has been investigated by Varbanov et al. (2012), where a minimum allowed temperature difference specification is defined for each significant Heat Exchange case: inside each site process, plus the cases of generating or using each site utility on the interfaces with each process. The article provides a thorough formulation of the necessary additional Total Site concepts and a modified targeting procedure to suit the new degrees of freedom. These have been illustrated with a case study showing that the difference in the Total Site targets between the cases of single and multiple ΔT_{min} specifications can be as large as 30%; see Fig. 4.31.

4.8.2 Retrofit of industrial energy systems at the site level

Retrofit projects for improving industrial energy efficiency are typically performed by evaluating and maximising the Heat Recovery potentials within the individual process units using Pinch Analysis. Once the potential improvements from the individual process units have been assessed, the Total Site (TS) Heat Integration analysis is performed. Such an approach may steer designers away from promising retrofit opportunities and lead towards suboptimal Heat Recovery from the viewpoint of the overall sites. This gap has been addressed elegantly by a retrofit framework for Heat Recovery on Total Sites (Liew et al., 2014). The paper presents an effective retrofit framework to determine the most cost-effective retrofit options and maximise potential savings. Instead of performing the typical unit-wise process retrofit, the strategy is to start at the site level, determine the baseline Total Site utility consumption and benchmark targets, and identify retrofit options from the Total Site context. The retrofit framework has been tested on a case study involving a petrochemical plant comprising multiple process sections; see Fig. 4.32 for an example of guiding steam turbine placement. The results of the analysis show that significant energy savings can be realised when both direct and indirect Heat Recovery retrofit options are evaluated. Further energy savings can be achieved via the Plus-Minus Principle, which helps pinpoint the correct locations of heat surpluses and deficits and lead to the appropriate TS retrofit solution.

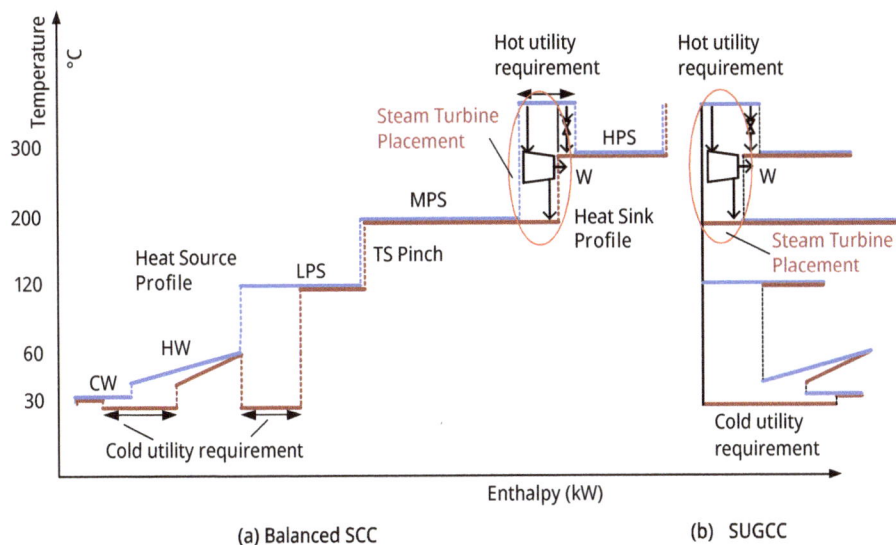

Fig. 4.32: Total Site retrofit guidance: steam turbine placement; after Liew et al. (2014).

4.8.3 Numerical tools for Total Site Heat Integration

This section describes the numerical methodology for Total Site Heat Integration (TSHI) by Liew et al. (2012), which consists of four main steps:
1. Perform Problem Table Algorithm (PTA) for all individual processes;
2. Construct a multiple utility cascade for each individual process;
3. Perform Total Site Problem Table Algorithm (TS-PTA);
4. Construct Total Site Utility Distribution (TSUD) Table.

To illustrate the methodology, Case Study 1, which consists of two plants, Plant A and B, with the data shown in Tabs. 4.23 and 4.24 is used.

Tab. 4.23: Stream data for Plant A of Case Study 1 with ΔT_{min} = 20 °C, a modified example from Canmet ENERGY (2003).

Stream	T_s (°C)	T_t (°C)	ΔH (MW)	mC_p (kW/°C)	T_s' (°C)	T_t' (°C)
A1 Hot	200	100	2.00	20	190	90
A2 Hot	150	60	3.60	40	140	50
A3 Cold	50	120	4.90	70	60	130
A4 Cold	50	220	2.55	15	60	230

Tab. 4.24: Stream data for Plant B of Case Study 1 with ΔT_{min} = 10 °C, a modified example from (Kemp and Lim, 2020).

Stream	T_s (°C)	T_t (°C)	ΔH (MW)	mC_p (kW/°C)	T_s' (°C)	T_t' (°C)
B1 Hot	200	50	0.450	3.0	195	45
B2 Hot	240	100	0.210	1.5	235	95
B3 Hot	200	119	1.863	23.0	195	114
B4 Cold	30	200	0.680	4.0	35	205
B5 Cold	50	250	0.400	2.0	55	255

Step 1: Perform Problem Table Algorithm (PTA) for all individual processes

In the first step, the Pinch Point location for all individual processes needs to be determined. The $\Delta T_{min,pp}$ for Plant A is assumed to be 20 °C, while for Plant B, it is assumed to be 10 °C. The Problem Table Algorithm is constructed as described in Section 2.2.3.4. It was determined that the minimum target for Plant A hot utility requirement is 2,250 kW, and for the cold utility, the requirement is 400 kW with shifted Pinch Temperature of 60 °C. For Plant B, the minimum target for hot utility requirement is 100 kW and for cold utility is 1,543 kW with shifted Pinch Temperature of 195 °C.

Tab. 4.25: Site utility data for Case Study 1.

Utility	Temperature (°C)
High-pressure steam (HPS)	270
Medium-pressure steam (MPS)	179.93
Low-pressure steam (LPS)	133.59
Cooling water (CW)	15–20

Step 2: Construct a multiple utility cascade for each individual process

In the second step, a multiple utility cascade table (MU-PTA) for each individual process needs to be constructed. The MU-PTAs function as Grand Composite Curves (GCC), which are used to determine the number of multiple utility levels to be generated Below the Pinch region, and the number of multiple utility levels to be consumed

Above the Pinch region. In the case of Total Site Profiles, the GCC is composed by omitting the pockets before the multiple utilities' placement for Site Source Profile, and Site Sink Profile is performed. For the numerical version, the multiple utilities' placements for the sources and sinks are done at individual plants instead of the Total Site. Table 4.25 shows the site utility data to be used for illustrating the placement of multiple utilities by using MU-PTA for Case Study 1. It is assumed that the minimum temperature difference between the utility and the process ($\Delta T_{min,up}$) for Case Study 1 is 10 °C.

The steps to construct the MU-PTA are described next by using Plant A as an illustration. The MU-PTA for Plant A is shown in Tab. 4.26.

Tab. 4.26: Multiple Utility Problem Table Algorithm for Plant A of Case Study 1.

1	2	3	4	5	6	7	8	9	
T' (°C)	T'' (°C)	ΔT (°C)	mC_p (kW/°C)	$\Sigma\, mC_p$ (kW/°C)	ΔH (kW)	Multiple Utility utility heat cascade (kW)	Utility consumed/ generated (kW)	Heat sink/source	
			20 40 70 15						
	270					0		HPS	600
		40		0	0				
230	230					0			
		40		−15	−600				
190	190					0			
		10.07		5	50.35				
	179.93					50.35		MPS	0
		39.93		5	199.65				
140	140					250			
		6.41		45	288.45				
	133.59					538.45		LPS	1,650
		3.59		45	161.55				
130	130					700			
		40		−25	−1,000		300		
90	90					0			
		30		−45	−1,350		1,350		
60	60					0		Pinch	
	60					0			
		10		40	400	−400			
50	50					0			
		35		0	0				
	15					0		CW	400

1. In Column 1, the shifted temperatures are arranged in decreasing temperature order, similar to PTA. The Pinch Temperature row, which is at 60 °C, is labelled.
2. Next, all the shifted temperatures (T') in the region Above the Pinch are deducted by $\Delta T_{min,pp}/2$ to return them to normal temperatures. Then, the $\Delta T_{min,pp}$ is added, resulting in a double shifted temperature, which is labelled T''. In the note, all the shifted temperatures (T') in the region Below the Pinch are shifted by adding $\Delta T_{min,pp}/2$ and then $\Delta T_{min,pp}$ subtracted to obtain T''. The resulting T'' values are shown in Column 2. The utility temperatures listed in Tab. 4.25 are also added to Column 2.
3. Column 3 is the temperature difference between each T'' interval.
4. Column 4 represents the hot and cold streams based on their supply and target temperature, similar to PTA. The heat capacity flowrate for each stream is also labelled.
5. Columns 5 and 6 are similar to PTA to determine the net enthalpy change for each temperature interval.
6. For Column 7 onwards, the cascading will be performed based on the Above and Below Pinch regions.
7. The Above Pinch Region represents the Heat Sink Profile. The heat from Column 6 is cascaded starting from the highest temperature segment to the Pinch Temperature in Column 7. The cascading is performed interval-by-interval. If a negative value is encountered in Column 7 while cascading one of the temperatures, it signifies there is an energy deficit. The external utility needs to be added to that row. The external utility added is listed in Column 8, equal to the negative value. The cascade then becomes zero at that temperature. For example, at a shifted temperature of 190 °C, the cascade initially gives a value of −600 kW at Column 7. Consequently, 600 kW of external utility is added at this interval, as listed in Column 8. The cascade now becomes zero here, as shown in Column 7. For positive values in Column 7, this signifies a pocket in GCC and process-to-process recovery. Hence, no external utilities are needed. Once the multiple utility heat cascades are completed, the amount of each type of utility consumed in the process is obtained by adding the utility consumed below the utility temperature from Column 8 to before the next utility temperature. For example, between 270 °C and 179.93 °C, 600 kW of high-pressure steam (HPS) external utility at a temperature of 270 °C is needed. Between 133.59 °C and 60 °C, 1,650 kW of low-pressure steam (LPS) is required. There is no requirement for medium-pressure steam (MPS).
8. Below the Pinch Region represents the Heat Source Profile. The net heat requirement in Column 7 is cascaded from the lowest temperature to the Pinch temperature. A negative external cooling utility is added in Column 8 when a positive value occurs in Column 7 of the cascade, making the cascade become zero at those locations. The positive value in Column 7 signifies energy excess, and hence external utility can be generated. The negative values in Column 7 signify the GCC pockets and process-to-process heat recovery. The amount of utility that can be generated

can be determined by adding the amounts of excess heat from above the utility temperature to the next utility temperature level. For Plant A, as the temperature is low, no utility can be generated. Instead, 400 kW of cooling water (CW) is needed to remove the excess process heat between 50 °C and 10 °C.

The same procedure is repeated for Plant B. The MU-PTA for Plant B is shown in Tab. 4.27. Plant B requires 100 kW of HPS and can generate 216.50 kW of MPS and 996.31 LPS. In addition, it needs 330.19 kW of CW to cool down the lower-temperature excess energy sources.

Tab. 4.27: Multiple Utility Problem Table Algorithm for Plant B of Case Study 1.

T' (°C)	T'' (°C)	ΔT (°C)	mC_p (kW/°C) 3 1.5 23 4 2	ΣmC_p (kW/°C)	ΔH (kW)	Multiple utility heat cascade	Utility consumed/generated (kW)	Heat sink/source	
270						0		HPS	100
		10		0	0		0		
255	260					0			
		20		−2	−40		40		
235	240					0			
		30		−0.5	−15		15		
205	210					0			
		10		−4.5	−45		45		
195	200					45			
	190					0		Pinch	
		10.07		21.5	216.51		−216.50		
	179.93					0		MPS	216.50
		46.34		21.5	996.31		−996.31		
	133.59					0		LPS	996.31
		24.59		21.5	528.69		−330.19		
114	109					−198.5			
		19		−1.5	−28.5		0		
95	90					170			
		40		−3	−120		0		
55	50					−50			
		10		−1	−10		0		
45	40					−40			
		10		−4	−40		0		
35	30					0			
		15		0	0		0		
	15					0		CW	330.19

Step 3: Perform Total Site Problem Table Algorithm (TS-PTA)

The Total Site Problem Table Algorithm (TS-PTA) represents the Site Source and Sink Composite Curves. Table 4.28 shows the TS-PTA for Case Study 1. In Columns 1 and 2, the utilities available are arranged from highest to lowest temperature. In Column 3, the total numbers of external utilities generated Below the Pinch temperature for all sites are added together to represent the net heat source (obtained from Column 8 MU-PTA in Step 2). Similarly, in Column 4, the total amount of external utilities consumed Above the Pinch Temperature for all sites are added together to represent the net heat sink (obtained from Column 8 MU-PTA in Step 2). The net heat requirement in Column 5 is obtained by subtracting the net heat sink from the net heat source. The locations with negative amounts of net heat indicate heat deficits, whereas the locations with positive values indicate heat surpluses. In Column 6, as heat can only be transferred from a higher temperature to a lower temperature, the net heat requirement is cascaded from top to bottom, starting with an initial value of zero. The most negative value in Column 6, which is −1137.19 kW located between the LPS and CW utility rows, represents the amount of external heating needed to be supplied to the whole site after considering TSHI. It also represents the location of the Total Site Pinch Point. This most negative value is made positive, added to the top row of Column 7 and cascaded with the net heat requirement in Column 5 again to give the feasible heat cascade considering single hot and cold utilities giving 1137.19 kW of hot utility and 730.19 kW of cold utility needed after TSHI. Next, the multiple utility heat cascade is performed. For the Above Pinch region, the external utility is consumed (listed in Column 9) when there is a negative value obtained during the cascading from the highest temperature utility to the Total Site Pinch location in Column 8. For the Below Pinch region, the external utility is generated (listed in Column 9) when there is a positive value obtained during the cascading from lower temperature utility

Tab. 4.28: Total Site Problem Table Algorithm (TS-PTA) for Case Study 1.

1	2	3	4	5	6	7	8	9
Utility	Utility temp. (°C)	Net heat source (kW)	Net heat sink (kW)	Net heat requirement (kW)	Initial heat cascade	Final single heat cascade	Multiple utility heat cascade	External utility requirement (kW)
					0	1137.19	0	
HPS	270	0	700	−700				**700**
					−700	437.19	0	
MPS	179.93	216.50	0	216.50				0
					−483.50	653.69	216.50	
LPS	133.59	996.31	1,650	−653.69				**437.19**
					−1137.19	**0 (Pinch)**	0	
CW	15–20	730.19	0	730.19				**−730.19**
					−407	730.19	0	

to the Total Site Pinch location in Column 8. The concept is similar to MU-PTA. From Column 9, it can be identified that 700 kW of HPS and 437.19 kW of LPS need to be generated from the centralised steam system to satisfy the Total Site energy requirement after considering TSHI. In addition, 730.19 kW of CW is needed to dispose of the excess energy at a lower temperature, which cannot be utilized in any sites.

Step 4: Construct Total Site Utility Distribution (TSUD) table

The amounts of utility distribution for each site and the on-site utility systems can be visualised by using the Total Site Utility Distribution (TSUD) table (see Tab. 4.29). All the Heat Sources and heat sinks from the sites are listed separately according to utility type, as shown in Columns 3 and 4. The external utilities calculated from Step 3 are also listed in Tab. 4.29. Arrows within the table show that Heat Sources can be transferred to heat sinks for the same type of utility. If there are extra heat sources, heat can be transferred to the lower utility levels.

Tab. 4.29: Total Site Utility Distribution (TSUD) table for Case Study 1.

1	2	3			4		
		Heat source (kW)			Heat sink (kW)		
Utility	Utility temp. (°C)	Plant A	Plant B	Site utility	Plant A	Plant B	Site utility
HPS	270			700	600	100	
MPS	140		216.5				
LPS	90		996.31	437.19	1,650		
CW	10	400	330.19				730.19

4.8.4 Power Integration

Pinch Analysis is a well-established methodology of Process Integration that has been used in designing optimal networks for the recovery and conservation of resources such as heat, mass, water, carbon, gas, properties, and solid materials for more than four decades. However, its application to power (electricity) systems analysis is still under development.

The work of Wan Alwi et al. (2012) extends the Pinch Analysis concept used in Process Integration to determine the minimum power (electricity) targets for systems comprising hybrid renewable energy sources. Power Pinch Analysis (PoPA) tools described in that paper include graphical techniques to determine the minimum target for outsourced power and the amount of excess power for storage during start-up and normal operations. The work also has a numerical version approach introduced by Rozali et al. (2013a). The PoPA tools can be used by energy managers, electrical and power engineers and decision-makers involved in the design of hybrid power systems. Graphical and numerical PoPA tools are systematic and simple to implement in the optimisation of power systems.

4.8.4.1 Methodology

This section describes the methodology for Power Pinch Analysis (PoPA), which consists of two main steps:

1. Data Extraction,
2. Targeting the *minimum outsourced electricity supply (MOES)* and the *available excess electricity for next day (AEEND)*.

Step 1: Data Extraction

The first step of a Pinch Analysis application for resource conservation typically involves Data Extraction of a system's sources (resource availability) and demands (resource requirements). In the case of hybrid power systems, power sources are the instantaneous on-site electricity generation from the available renewable energy sources such as solar photovoltaic, wind, or biomass. The power demands represent equipment electricity consumption that can be determined from equipment power ratings. The power sources and power demands are recorded at the time they are available. Tables 4.30 and 4.31 show the power sources and demands for the illustrative Case Study 1, which represents the case of an off-grid hybrid power system. In this case, excess power will be stored in a battery system. Power generated and consumed is calculated using eqs. (4.14) and (4.15). Initially, the demands consumed 268 kW of outsourced electricity supply daily. The maximum power demand of 19 kW occurs between 8 to 18 h (see Tab. 4.31).

$$\text{Power generation (kWh)} = \text{Power generated (kW)} \times \text{Time interval (h)} \qquad (4.14)$$

$$\text{Power consumed (kWh)} = \text{Power rating (kW)} \times \text{Time interval (h)} \qquad (4.15)$$

Tab. 4.30: Power sources for Illustrative Case Study 1.

Power source	Time, h		Time interval, h	Power rating generated, kW	Power generation, kWh
	From	To			
Solar	8	18	10	5	50
Wind	2	10	8	5	40
Biomass	0	24	24	7	168

Tab. 4.31: Power demands for Illustrative Case Study 1.

Power demand appliances	Time, h		Time interval, h	Power rating, kW	Power consumption, kWh
	From	To			
Appliance 1	0	24	24	3	72
Appliance 2	8	18	10	5	50
Appliance 3	0	24	24	2	48
Appliance 4	8	18	10	5	50
Appliance 5	8	20	12	4	48

To better illustrate the PoPA method using the illustrative case study, the average value of the generated renewable energy is considered for the specified time range. For real cases, RE supply fluctuates at each time interval, and the designer should use the average amount of RE generated hourly to determine the minimum power targets using the PoPA method.

Step 2: Targeting the Minimum Outsourced Electricity Supply and Available Excess Electricity for the Next Day

This section introduces a new technique known as the Power Composite Curves (PCC) to determine the maximum power transfer from power sources to power demands, the MOES, and AEEND for storage. The PCC is a graphical approach that is based on Pinch Analysis Composite Curves (Linnhoff et al., 1982, latest edition 1994).

Power Composite Curves

The PCC can be constructed as follows:
1. Y-axis represents the time scale from 0 to 24 h while the x-axis represents the power generation or consumption.
2. An individual power source or power demand line is plotted as shown in Fig. 4.33. Note that the gradient of the line can be computed from the inverse of the power rating for the corresponding power demand or power source.
3. The Composite Power Source (or power demand) line is obtained from the sum of power sources (or power demands) within a given time interval. The same procedure is repeated for the rest of the system time intervals in order to yield the Source and Demand Composite Curves for the entire system.
4. The Pinch Point can be determined by shifting the Source Composite Curve to the right-hand side until it touches the Demand Composite Curve. Note that the Source Composite Curve has to be on the right-hand side of the Demand Composite Curve since power can only be transferred from a current to a later time interval. A complete PCC is shown in Fig. 4.34.

5. The excess demand line Below the Pinch Point gives the MOES needed to be purchased during a system start-up, and the excess power source Above the Pinch gives the AEEND. MOES and AEEND targets are 18 kW and 8 kW in illustrative Case Study 1.
6. Now it is possible to determine the amount of MOES needed during daily operation. This can be done by merging the PCC from Day 1 with the next day's PCC by linking the end of the SCC from the current-day PCC to the beginning of the source composite of the next day's PCC. Merging the current-day PCC with the next day's PCC gives the Continuous PCC (CPCC); see Fig. 4.35. There are two possible scenarios for daily operations assumed:

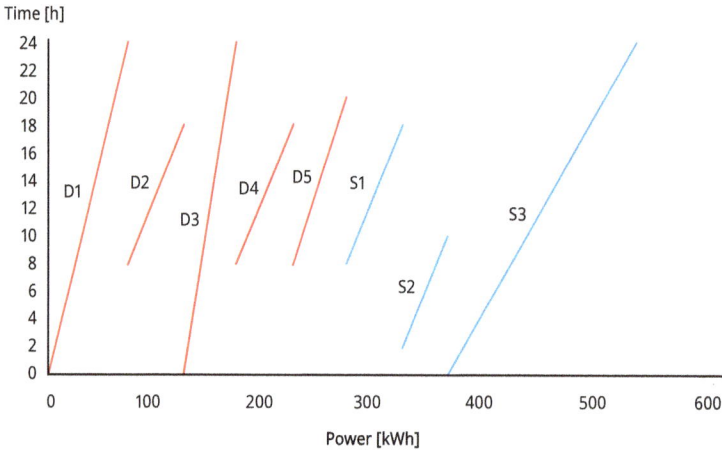

Fig. 4.33: Individual power source and demand lines for Illustrative Case Study 1.

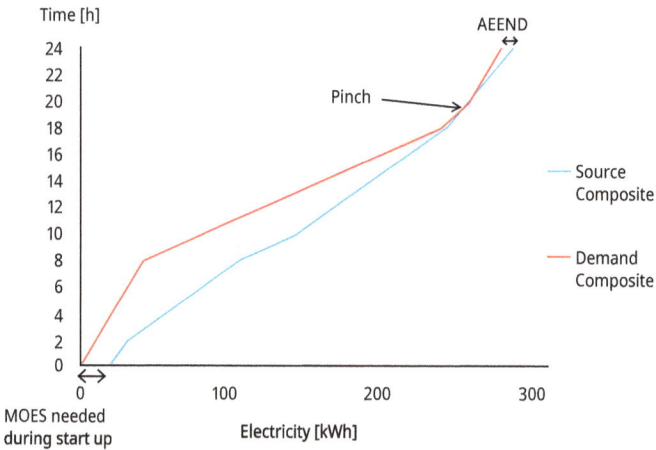

Fig. 4.34: Power Composite Curves (PCC) for a 24 h operation (Illustrative Case Study 1).

(a) Scenario 1 is when AEEND (18 kW) is less than or equal to MOES (8 kW) for Day 1. It indicates that the amount of excess power from the previous day can be used to reduce the amount of outsourced electricity needed for the next day. Illustrative Case Study 1 is classified as Scenario 1. The excess demand line from the second PCC onwards gives the daily outsourced electricity supply needed to be purchased during normal operations. The CPCC reduces the MOES to 10 kW. The hybrid power system results in a savings of 96.3% from MOES cost as compared to a conventional power system without RE.

(b) Scenario 2 is a case where AEEND is more than MOES. It indicates that the amount of excess power sources from the previous day is larger than the amount of outsourced electricity needed for the next day. For Scenario 2, if AEEND is continuously cascaded to the next day, it will cause an accumulation of power inside the storage system. Scenario 2 is explained using Illustrative Case Study 2. In this case, the wind power source rating between times 2 to 10 h in Illustrative Case Study 1 is increased from 5 kW to 6.5 kW. Figure 4.36 shows the CPCC for Scenario 2. The excess source line after integration with the power demand curve for the next day represents the excess wasted power source.

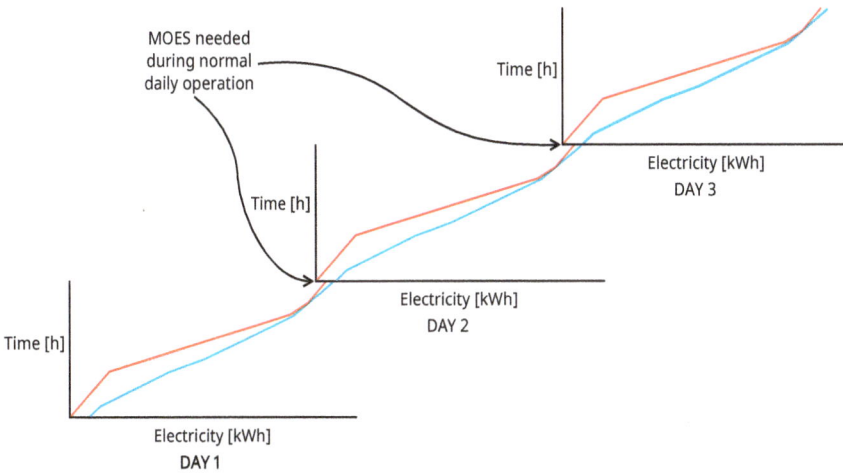

Fig. 4.35: Continuous Power Composite Curves (CPCC) for Illustrative Case Study 1.

However, the power losses incurred in the systems have been considered in the following work: Rozali et al. (2013b). This contribution extends the PoPA method by considering the power losses that occur in the power system conversion, transfer and storage. The losses' effect on the minimum outsourced electrical energy targets and storage capacity are evaluated.

This is a very recent part of the Total Site methodology, and more information can be found in the Sources of Further Information Section.

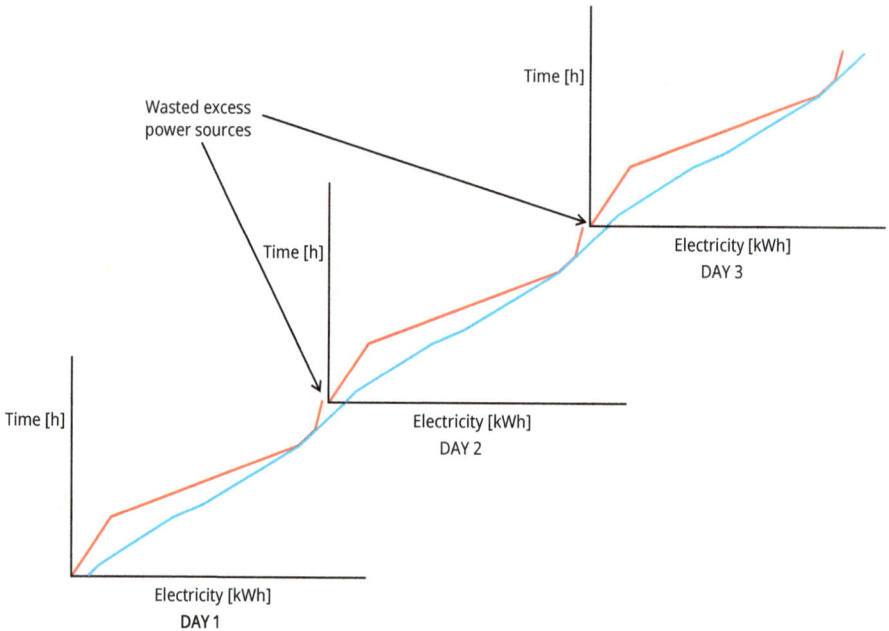

Fig. 4.36: Continuous Power Composite Curves (CPCC) for Illustrative Case Study 2.

Power Cascade Analysis (PoCA)

The Power Cascade Analysis (PoCA) approach was introduced by Rozali et al. (2013a). It can be used to target both the off-grid and on-grid systems' minimum outsourced electricity requirements. The tool consists of a Power Cascade Table (PCT) and Storage Cascade Table (SCT). The PCT represents the same information as the PCC and CPCC. The SCT determines the power storage system capacity and maximum power demand after Power Integration. Case Study 1 data illustrated in Tabs. 4.30 and 4.31 are used to illustrate the PoCA methodology.

(i) Power Cascade Table (PCT)

Table 4.32 shows the PCT for Case Study 1. The following are the steps to construct the PCT:

1. In the first column, the beginning and end times for all power sources and power demands are arranged in ascending order from 0 h to 24 h, with any duplicate time removed.

2. The second column gives the duration between two adjacent time intervals.

Tab. 4.32: Power Cascade Table for Illustrative Case Study 1.

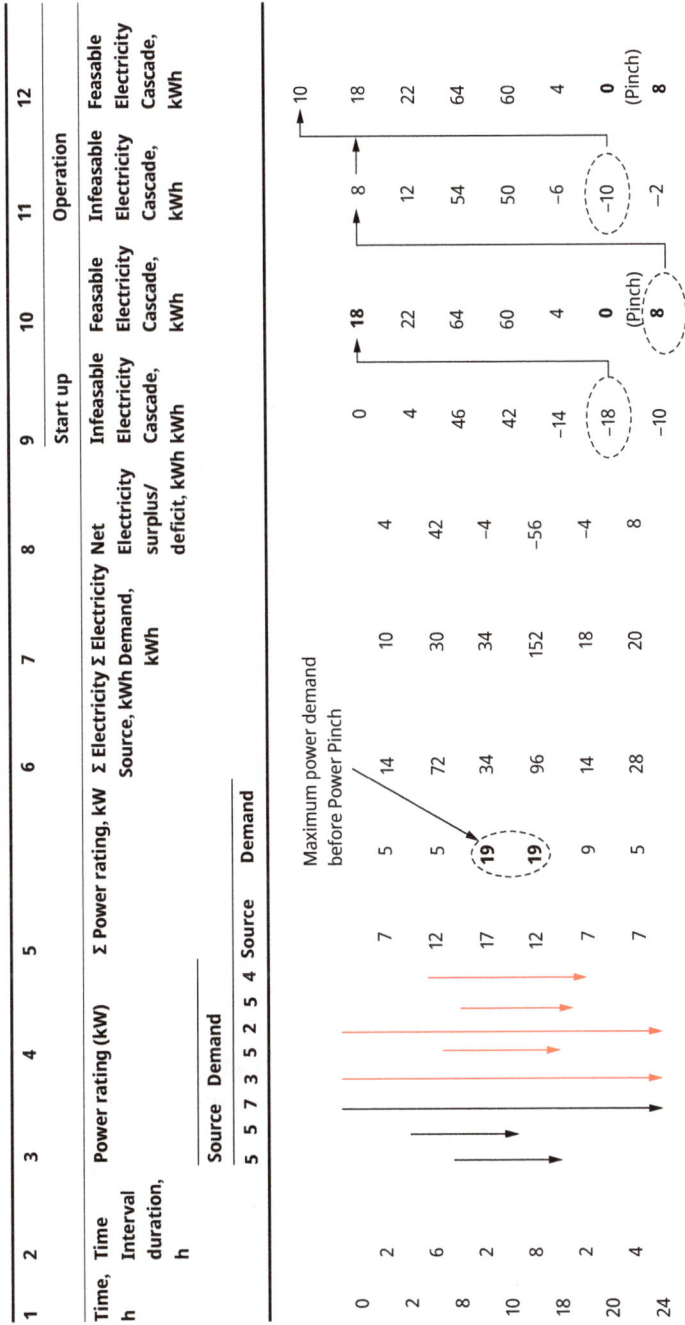

Time, h	Time Interval duration, h	Σ Power rating, kW	Σ Electricity Source, kWh	Σ Electricity Demand, kWh	Net Electricity surplus/deficit, kWh	Infeasable Electricity Cascade, kWh (Start up)	Feasable Electricity Cascade, kWh (Start up)	Infeasable Electricity Cascade, kWh (Operation)	Feasable Electricity Cascade, kWh (Operation)
0						0	18 ↑	8	18 (↑10)
	2	7	14	10	4				
2						4	22	12	22
	6	12	72	30	42				
8						46	64	54	64
	2	17	34	34	−4				
10						42	60	50	60
	8	12	96	152	−56				
18						−14	4	−6	4
	2	7	14	18	−4				
20						−18 (Pinch)	0 (Pinch)	−10	0 (Pinch)
	4	7	28	20	8				
24						−10	8	−2	8

Power rating (kW) — Source / Demand: Source 5 5 7 3   Demand 5 2 5 4

Legend (arrow diagram):
- Source Demand

Maximum power demand before Power Pinch → (19) (19)

3. Columns 3 and 4 show the arrows that represent the power sources and power demands according to the time interval where they exist and their corresponding power ratings.

4. The power ratings for the power sources and power demands are summed together for each time interval and listed in Columns 5 and 6. The maximum power demand before power integration is given by the highest sum of the demand power rating. For this example, it can be seen the maximum power demand is 19 kW, and it occurs between 8 h and 18 h.

5. The total electricity sources and electricity demands in kWh for each time interval shown in Columns 6 and 7 are obtained using eq. (4.16).

$$\sum \text{Electricity Source or Demand} = \sum \text{Power Rating Source or Demand} \quad (4.16)$$

$$\times \text{Time interval duration.}$$

6. The net electricity surplus (+)/deficit (−) between time intervals (Column 8) is obtained from eq. (4.17). A positive net electricity output indicates that electricity sources exceed electricity demands at that time interval. Therefore, the surplus electricity can be stored in a power storage system or sold to the grid. A negative value implies that electricity demands exceed electricity sources. To make up for this electricity deficit, additional electricity needs to be supplied from a power storage system or from an outsourced electricity supply, which can come from the grid or power generator.

$$\text{Net electricity surplus/deficit} = \sum \text{Electricity Source} - \sum \text{Electricity Demand.} \quad (4.17)$$

7. For the off-grid system, the net electricity is cumulatively cascaded in Column 9 from time 0 h to 24 h starting from no outsourced electricity supply. This gives an impossible electricity cascade if there are negative electricity flows. To generate a feasible cascade, the absolute value of the largest negative electricity flow in Column 9, i.e., 18 kW, is cascaded cumulatively again with the net electricity surpluses/deficits (Column 8) in Column 10. The top value of Column 10 gives the Minimum Outsourced Electricity Supply (MOES) of the system (10 kW), while the last value represents the Available Electricity Energy for the Next Day (AEEND) (8 kW) during start-up. This is the same result as obtained using the PCC. Because AEEND ≤ MOES, this can be identified as Scenario 1 (as described in PCC scenarios previously). For Scenario 1, in Column 11, the 8 kW of excess electricity stored at the end of the first day is cumulatively cascaded to the next day from time 0 h to 24 h with the net electricity surpluses/deficits. Because there again are negative electricity flows in Column 11, which signifies that the cascade is not feasible, the absolute value of the most negative electricity flow (−10 kW) in Column 11 is cumulatively cascaded with the net electricity surpluses/deficits in Column 12 to get the MOES of 10 kW and the AEEND of 8 kW during daily operation. For a Scenario 2 case, where AEEND is more than MOES, if AEEND is continuously cascaded to the next day, it

Tab. 4.33: Power Cascade Table for Illustrative Case Study 3.

would cause an accumulation of electricity inside the storage system. Hence, only the amount equal to MOES during start-up is cascaded to the next day in Columns 11 and 12. The difference between MOES and AEEND for the first day represents the amount of wasted electricity sources. Using Illustrative Case Study 2 from Wan Alwi et al. (2012), Tab. 4.33 shows the PCT for Scenario 2. The result in Column 12 of PCT represents the same solution as the CPCC.

8. For the on-grid system, the electricity surplus can be sold to the grid, and a deficit can be satisfied by purchasing electricity from the grid. The sum of the net electricity surplus (54 kWh) gives the total amount of electricity to be sold to the grid (positive values in Column 8, Tab. 4.32). The sum of the net electricity deficit (64 kWh) gives the total amount of electricity to be purchased from the grid (negative values in Column 8, Tab. 4.32).

(ii) Storage Cascade Table (SCT)

The SCT can be used to determine the following:
1. Amount of electricity that can be transferred by each source;
2. An outsourced electricity supply is required at each time interval;
3. Amount of electricity available for storage in real-time;
4. Storage capacity.

Tables 4.34 and 4.35 show the SCT for Scenario 1 and Scenario 2. Following are the steps to construct the SCT:

$$\text{Outsourced power rating} = \frac{\text{Outsourced electricity needed}}{\text{Time interval}} \qquad (4.18)$$

1. The values for Columns 1 to 4, 6 and 10 are extracted from the PCT.

2. Column 5 shows the amount of electricity transfer by the SCT. The maximum amount of electricity transfer is the lower value between the sum of electricity sources (Column 3) and the sum of electricity demands (Column 4) for each time interval.

3. For start-up periods, it is necessary to determine the maximum electricity storage capacity for a battery and the outsourced electricity requirement at various ranges of time intervals. These can be done by cascading the net electricity surplus or deficit cumulatively down the time interval beginning from 0 h. Referring to Tab. 4.34, net cascaded electricity surpluses (positive values) are recorded in the "Storage capacity" column (Column 7), and net electricity deficits are recorded in Column 8 to indicate the instantaneous amount of external electricity supply required for the relevant range of time intervals. Note that when an electricity deficit occurs, the electricity surplus cascade resumes at zero storage capacity value in Column 7. The absolute sum of external electricity requirements at various time interval ranges in Column 8 gives the total system's outsourced electricity requirement of 18 kW. The largest cumulative

Tab. 4.34: Storage Cascade Table for Illustrative Case Study 2.

1	2	3	4	5	6 Start up	7 Start up	8	9	10 Operation	11 Operation	12	13
Time, h	Time interval, h	Σ Electricity Source, kWh	Σ Electricity Demand, kWh	Amount of Net electricity transfer, kWh	Net Electricity surplus/deficit, kWh	Storage capacity, kWh	Outsourced electricity needed, kWh	Outsourced Power rating, kW	Net Electricity surplus/deficit, kWh	Storage capacity, kWh	Outsourced electricity needed, kWh	Outsourced Power rating, kW
0										8		
2	2	14	10	10	4	4			4	12		
8	6	72	30	30	42	46			42	54		
10	2	34	38	34	−4	42			−4	50		
18	8	96	152	96	−56		−14	−1.75	−56		−6	−0.75
20	2	14	18	14	−4		−4	−2	−4		−4	−2
24	4	28	20	20	8	8			8	8		
					Total external electricity needed during start up		−18	Maximum power demand	Total external electricity needed during operation		−10	Maximum power demand

Maximum storage capacity (54)

Maximum power demand (−2)

Tab. 4.35: Storage Cascade Table for Illustrative Case Study 3.

1	2	3	4	5	6	7	8	9	10	11	12	13
					Start up				Operation			
Time, h	Time interval, h	Σ Electricity Source, kWh	Σ Electricity Demand, kWh	Amount of Electricity transfer, kWh	Net Electricity surplus/deficit, kWh	Storage capacity, kWh	Outsourced Electricity needed, kWh	Outsourced Power rating, kW	Net Electricity surplus/deficit, kWh	Storage capacity before setting limit, kWh	Storage capacity with limit, kWh	Excess power source wasted, kWh
0										8	8	
2	2	14	10	10	4	4			4	12	12	
8	6	81	30	30	51	55			51	63	61	2
10	2	37	38	37	-1	54			-1	62	60	
18	8	96	152	96	-56		-2	-0.25	-56	6	4	
20	2	14	18	14	-4		-4	-2	-4	2	0	
24	4	28	20	20	8	8			8	10	8	

Maximum power demand -6

Total outsourced electricity needed during start up -6

Storage capacity limit is set

Express power sources cannot be stored due to battery limit

electricity surplus in Column 7 gives the upper limit of the battery's electricity storage capacity required. For Case Study 1, it can be observed the battery capacity needed is 46 kW during start-up. The kW instantaneous external power demand (Column 9) can be obtained as follows:

For this illustrative case study, the maximum power demand is 2 kW, occurring between time intervals of 18 h to 20 h.

4. For Scenario 1 normal operation periods, the same procedure as for start-up is repeated, except the net electricity surplus or deficit is cumulatively cascaded down the time interval beginning from 0 h starting from the amount of electricity stored from the previous day (in this case 8 kW as determined in PCT). The largest cumulative electricity surplus of 54 kW shown in Column 11 gives the upper limit of the electricity storage capacity of the power storage system. The sum of external electricity requirements at various time interval ranges in Column 12 gives the total system's external electricity requirement of 10 kW. The maximum power demand is 2 kW occurring between time intervals 18 h to 20 h, as can be observed from the biggest negative value in Column 13.

5. For Scenario 2, there is excess energy which cannot be utilised during normal operation periods. Table 4.35 shows the SCT for Case Study 2. To determine the maximum storage capacity, excess electricity from Day 1 is first cascaded to the net electricity surpluses/deficits to give the "available accumulated electricity" (Column 11). The maximum storage capacity of 63 kWh in Column 11 is subtracted from the difference between AEEND and MOES for Day 1 (this value can be obtained from PCT, 8 kWh – 6 kWh = 2 kWh) to yield the maximum storage capacity limit of 61 kWh. Because of this storage capacity limit, part of the power source between time 2 h and 8 h cannot be stored, as shown in Column 13, Tab. 4.35. Column 12 shows the storage capacity for each time interval. Note that 61 kWh is set as the maximum amount of net accumulated power surplus.

The PCC, CPCC, PCT, and SCT methods described in this section can provide a useful guide for hybrid power system designers to estimate the additional external outsource electricity requirement, maximum power demand, and power storage capacity. However, power losses are assumed to be negligible. If the power losses during power system conversion, transfer and storage are considered, it will affect the targeting result. This issue has been addressed in Rozali et al. (2013b) with the introduction of modified PCT and SCT, which consider power losses.

This is a very recent part of the Total Site methodology, and more information can be found in the Sources of Further Information Section.

4.8.5 Targeting for low CO_2 emissions with CO_2 emission Pinch Analysis

Although previous Process Integration applications using Pinch Analysis have contributed to carbon emissions footprint reductions via energy efficiency improvements, they do not emphasize the waste-to-resources approach involving CO_2 exchange. The concept of carbon emissions capture, utilisation and sequestration (CCUS) involves capturing CO_2 from stationary point sources such as power plant boiler flue gas, purifying it to a certain level and either utilising it in processes that can accept impure CO_2 or sequestrating it. Munir et al. (2012) introduced the Carbon Emission Pinch Analysis (CEPA) technique for targeting the minimum fresh CO_2 required and CO_2 emissions for a planned industrial park after considering the maximum CO_2 potential to be exchanged and the Carbon Management Hierarchy (CMH).

There are, again, graphical and numerical targeting methods for CEPA. The graphical method involves constructing the Sources and Demands Curves (SDC). The full methodology can be read in Munir et al. (2012). Another alternative is the use of a Carbon Cascade Table (CCT) by Manan et al. (2014). Both methods yield the same CO_2 targets.

Step 1: Extracting the CO_2 Sources and Demands

The CO_2 sources are typically from the flue gas sources in industrial sites such as furnace stacks, boiler stacks, drying towers, gas turbines, and cogeneration plants. The flue gas composition mainly consists of N_2, O_2, CO_2, CO, NO_x, and SO_x, depending on the type of fuel used and the extent of combustion. The limiting data to be extracted from each source stream are the flue gas flow rate (F_T) and the composition of CO_2 in the flue gas. Not all CO_2 sources should be considered. Those that contain hazardous substances, are remotely located in terms of geography, or have a low concentration of CO_2 should not be considered because the cost of capturing and transporting it will not be economical. The remaining emission sources are treated according to standard treatment procedures before emitting into the atmosphere.

The CO_2 demands are the processes that can accept gases with a certain purity of CO_2. The required data include the flow rate of gases that can be accepted by the demand processes and the limiting CO_2 composition in the gas. Examples of processes that require CO_2 are syngas manufacturing from CO_2, microalgae cultivation, enhanced oil recovery, methanol production, and beverage manufacturing.

Tables 4.36 and 4.37 show the CO_2 sources and demand streams extracted from a refinery industrial park for Case Study 1, which will be used to illustrate the Carbon Pinch Analysis methodology. There are four CO_2 demands and six CO_2 point sources. Initially, there were 20 CO_2 point sources from the refinery. However, only the six-point sources that contribute to the highest CO_2 emissions are selected. The refinery operates 7,920 h/y. As opposed to the concept of Water or Hydrogen Pinch, where the contaminants are measured in ppm, in flue gases, the other gases, such as N_2, O_2, CO, NO_x, and SO_x, may exist at high concentrations. The flowrates of CO_2 and other gases that are required for the carbon Pinch targeting purposes need to be determined using eqs. (4.19) and (4.20):

$$F_{CO_2} = F_T \cdot CO_2 \text{ Composition } (\%) \tag{4.19}$$

$$F_{OG} = F_T - (100\% - CO_2 \text{ Composition } (\%)) \tag{4.20}$$

Where

F_T is the flue gas flowrate,

F_{CO_2} is the flowrate of CO_2 in the flue gas,

F_{OG} is the flowrate of other gases aside from CO_2 in the flue gas, and

CO_2 Composition (%) is the composition of CO_2 in the flue gas.

Tab. 4.36: Data for CO_2 sources for the refinery case study (Munir et al., 2012).

Sources	F_T, t/h	CO_2 Composition (%)	F_{CO_2}, t/h	F_{OG}, t/h	F_{OG}/F_{CO_2}	Cum F_{CO_2}, t/h	Cum F_{OG}, t/h
Gasifier (S_1)	300	75	225	75	0.3333	225	75
FCC (S_2)	55	20	11	44	4.0000	236	119
Stack B (S_3)	70	15	10.5	59.5	5.6667	246.5	178.5
Stack A (S_4)	130	12	15.6	114.4	7.3333	262.1	292.9
Co-generation Plant (S_5)	200	10	20	180	9.0000	282.1	472.9
Gas Turbine (S_6)	90	8	7.2	82.8	11.5000	289.3	555.7

Tab. 4.37: Data for CO_2 demands for the refinery case study (Munir et al., 2012).

Demands	F_T, t/h	CO_2 Composition (%)	F_{CO_2}, t/h	F_{OG}, t/h	F_{OG}/F_{CO_2}	Cum F_{CO_2}, t/h	Cum F_{OG}, t/h
Beverage Plant (D_1)	50	99	49.5	0.5	0.0101	49.5	0.5
Enhanced Oil Recovery (EOR) (D_2)	208.3	80	166.64	41.66	0.2500	216.14	42.16
Methanol Production (D_3)	83.3	50	41.65	41.65	1.0000	257.79	83.81
Microalgae Cultivation (D_4)	220	10	22	198	9.0000	279.79	281.81

Step 2: Targeting the Maximum CO$_2$ Recovery using Carbon Emission Pinch Analysis

As mentioned previously, there are graphical and numerical techniques for CEPA. The steps to perform both methods are described next.

(i) Sources and Demands Curves (SDC)

SDC is the graphical targeting technique for CEPA introduced by Munir et al. (2012). Typically, in previous Pinch Analysis studies of water, mass, gas and property, the mass flow rate of a main/primary carrier fluid was plotted against the mass load/concentration of the contaminants or minor components in a Process Stream (El-Halwagi et al., 2003) and later (Wan Alwi et al., 2009). For CEPA, this is not applicable because CO$_2$ may not be the main carrier fluid in the gaseous stream. The main CO$_2$ sources are typically flue gas from furnaces or boilers resulting from the combustion of fossil fuels or biomass. The flue gas consists not only of CO$_2$, but a larger percentage is constituted by various other gases, such as N$_2$, O$_2$, CO, NO$_x$, and SO$_x$, which can constitute smog/haze (Fan et al., 2018). If the conventional plots of the primary gas flow rate vs the mass load of contaminants are plotted, it will yield the minimum flue gas

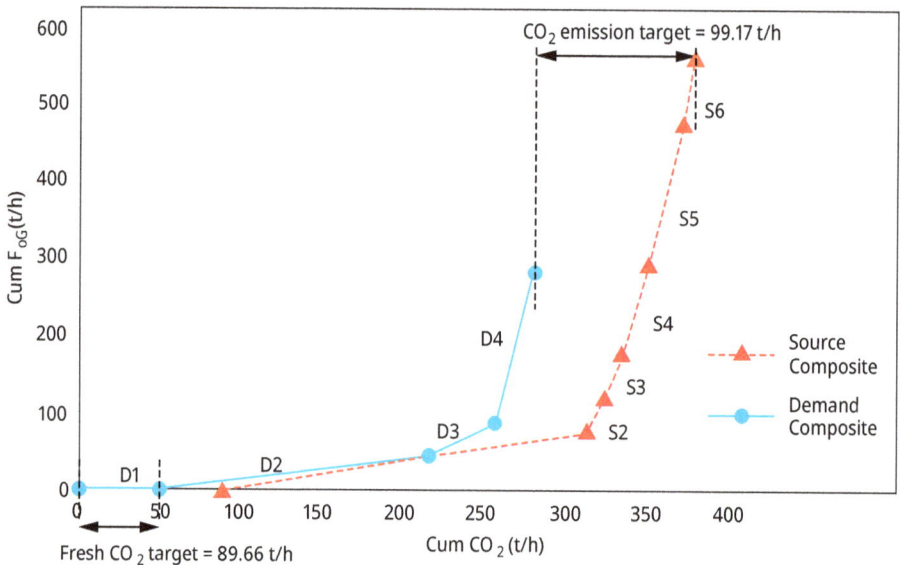

Fig. 4.37: Targeting the Maximum Carbon Emissions Exchange by using Sources and Demands Curves (Munir et al., 2012).

emissions target instead of the minimum carbon emissions target. Hence, the SDC was constructed by plotting the cumulative flow rate of gases other than CO_2, F_{OG}, vs the cumulative flow rate of CO_2, F_{CO_2}.

In demonstrating the construction of the SDC, Case Study 1 data from Step 1 are used. F_{CO_2}, F_{OG} and the ratio F_{OG}/F_{CO_2} are calculated and listed in Columns 4 to 6 of Tabs. 4.36 and 4.37. The sources and demands are arranged in ascending F_{OG}/F_{CO_2} order. The cumulative F_{OG} and cumulative F_{CO_2} values are then calculated as shown in Columns 7 and 8 of Tabs. 4.36 and 4.37. The cumulative F_{OG} vs cumulative F_{CO_2} for the source streams are first plotted from point zero to give the Source Curve. The cumulative F_{OG} versus cumulative F_{CO_2} for the demand streams are plotted from point zero to give the Demand Curve. The whole Source Curve is then moved horizontally at zero F_{OG} until it touches the Demand Curve (Fig. 4.37). The Source Curve should be located on or at the right-hand side of the Demand Curve. The point where the SDC touch is the Pinch Point and represents the bottleneck of CO_2 recovery for the system. Note that the sources and demands are sorted according to increasing F_{OG}/F_{CO_2} to allow the sources with the highest CO_2 content to be matched with the demands that require high CO_2 content. This is similar to the "cleanest to cleanest" rule proposed by Polley and Polley (1998).

The horizontal gap between the demand and source curves at zero F_{OG} gives the minimum fresh CO_2 flow rate, and the overshoot of the source curve gives the minimum CO_2 emission flow rate. For Case study 1, the minimum fresh CO_2 flow rate is 89.66 t/h, and the minimum CO_2 emission flow rate is 99.17 t/h. The Pinch Point is located at F_{CO_2} of 216.14 t/h and F_{OG} of 42.16 t/h. Prior to maximising carbon emissions exchange, 280 t/h of CO_2 was emitted into the atmosphere. This represents a 65% CO_2 emissions reduction.

The left-hand side of the Pinch Point is known as "the region Below the Pinch", and the right-hand side is known as "the region Above the Pinch". The well-established Pinch rule for resource exchange also applies to the case of CO_2 recovery, i.e. sources Above the Pinch should not be used to satisfy demands Below the Pinch, and vice versa. Violation of this rule will result in penalties for increased fresh CO_2 demand as well as CO_2 emissions.

(ii) Carbon Emissions Cascade Table

A Carbon Emissions Cascade Table (CCT) is a numerical technique to maximise CO_2 recovery introduced by Manan et al. (2014). The method can target the minimum fresh CO_2 requirement and the minimum CO_2 emissions from an integrated industrial site. Table 4.38 shows the CCT for Case Study 1. Following are the steps to construct the CCT:

Tab. 4.38: The Carbon Cascade Table for the refinery case study.

1	2	3	4	5	6	7	8	9	10
F_{OG}/F_{CO_2}	$\Delta(F_{OG}/F_{CO_2})$, t/h	ΣF_S, t/h	ΣF_D, t/h	$\Sigma F_S + \Sigma F_D$, t/h	F_C, t/h	F_{OG}', t/h	$F_{OG,cum}'$, t/h	$F_{FC,cum}$, t/h	F_C', t/h
					0.0				F_{FC} = 89.66
0.0101			−49.5000	−49.5			0.0		
	0.2399				−49.5	−11.8750			40.16
0.250			−166.6400	−166.64			−11.8750	−47.50	
	0.0833				−216.14	−18.0117			−126.480
0.3333		225.0		225.0			−29.8867	−89.660	(PINCH)
	0.6667				8.86	5.9067			98.52
1.0			−41.6500	−41.65			−23.9800	−23.980	
	3.0000				−32.79	−98.3700			56.87
4.0		11.0		11.0			−122.350	−30.5875	
	1.6667				−21.79	−36.3167			67.87
5.6667		10.5		10.5			−158.6667	−28.0	
	1.6667				−11.29	−18.8167			78.37
7.3333		15.6		15.6			−177.4833	−24.2023	
	1.6667				4.3100	7.1833			93.97
9.0		20.0	−22.0	−2.0			−170.3	−18.9222	
	2.5				2.31	5.7750			91.97
11.5		7.2		7.2			−164.525	−14.3065	
					9.51				F_{CE} = 99.17

1. The first column of the CCT contains the ratio of other gases flowrate to the CO_2 gas flowrate (F_{OG}/F_{CO_2}), arranged in ascending order of magnitude. A smaller value of F_{OG}/F_{CO_2} indicates higher CO_2 concentration in the flue gas. A bigger value of F_{OG}/F_{CO_2} indicates lower CO_2 concentration in the flue gas.
2. The second column calculates the difference between F_{OG}/F_{CO_2} level at intervals n and $n+1$, as follows:

$$\Delta(F_{OG}/F_{CO_2}) = (F_{OG}/F_{CO_2})_{n+1} - (F_{OG}/F_{CO_2})_n \qquad (4.21)$$

3. The total flowrates for the CO_2 demands ($\sum_j F_{D,j}$) and CO_2 sources ($\sum_i F_{S,i}$) at their corresponding F_{OG}/F_{CO_2} intervals are listed in Columns 3 and 4. The flowrate of CO_2 demand is labelled as negative, while the CO_2 source as positive.
4. In Column 5, the CO_2 demands and sources flowrates are summed up at each *interval* to give the *net interval* carbon dioxide flowrate. A positive value represents a *net carbon source*, and a negative value represents a *net carbon dioxide demand*.
5. The CO_2 flowrate balance (F_C) in Column 6 is obtained by cascading downwards the net interval CO_2 flowrate from Column 5, starting from F_{CO2} = 0 t/h.
6. In Column 7, the value in Column 6 is multiplied by $\Delta F_{OG}/F_{CO_2}$ to yield the F_{OG} surplus or deficit (F_{OG}').

7. $F'_{OG, cum}$ in Column 8 is calculated by cascading the F'_{OG} down the F_{OG}/F_{CO_2} intervals yields.
8. In Column 9, the amount in Column 8 is divided by the F_{OG}/F_{CO_2} interval to give the pure carbon dioxide cascade ($F_{FC,cum}$) in Column 9.
9. The deficits on the pure carbon cascade in Column 9 indicate that the pure carbon cascade is not feasible. A fresh carbon flowrate of exactly the same magnitude as the absolute value of the largest negative $F_{FC,k}$, i.e. 89.66 t/h, should be supplied at the highest F_{OG}/F_{CO_2} level of a feasible carbon cascade (Column 10) to ensure that there is sufficient fresh carbon dioxide at all points in the network. This represents the minimum fresh carbon target (F_{FC}). The location with the largest negative $F_{FC,k}$ also represents the Pinch Point of the CCT. F_{FC} is then cascaded again with the net interval carbon flowrate (Column 5) to yield the feasible carbon cascade (F'_C). The last value of the F'_C gives the minimum carbon emission flowrate target (F_{CE}) of 99.17 t/h.

Step 3: Holistic minimum carbon emissions targets with Carbon Management Hierarchy (CMH)

In Step 2, fresh CO_2 requirement and CO_2 emissions are only reduced by maximising CO_2 reuse. In this step, the target can be further reduced by considering not only reuse but other strategies to reduce CO_2. The CMH shown in Fig. 4.38 was introduced by Munir et al. (2012) to guide designers and planners in the screening and to prioritise CO_2 minimisation strategies to achieve holistic minimum carbon targets in an industrial park. The hierarchy consists of four levels, with direct reuse as the preferred level, followed by source and demand manipulation, regeneration reuse, and lastly, CO_2 sequestration as the least preferred level.

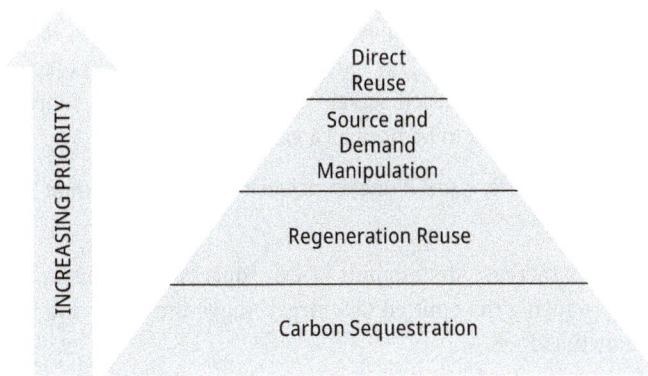

Fig. 4.38: The Carbon Management Hierarchy (CMH) (Munir et al., 2012).

Level 1: Direct reuse

Direct reuse represents the use of the CO_2 point sources for various CO_2 demands without any treatment. It is prioritised first before source and demand manipulation because if CO_2 sources can be consumed by existing CO_2 demands, there may not be a need to perform process changes to manipulate the sources or demand flowrates. Hence, it is recommended to evaluate the potential for maximising the CO_2 exchange in an industrial site by using tools such as SDC or CCT before evaluating other CO_2 reduction strategies (Step 2 previously described).

Level 2: Sources and demands manipulation

Once direct reuse has been evaluated, the next strategy is to evaluate the impact of sources and demand manipulation to reduce fresh CO_2 and CO_2 emissions further. Source manipulation can involve source elimination and reduction. It can be achieved through process parameter changes, equipment modifications, the substitution of re-action chemistry and fuel switching. Demand manipulation can include options such as demand elimination, reduction and generation. This may involve changing the demand flowrates or adding new CO_2 demands.

In manipulating the sources and demands, the concept of Plus-Minus Principle is applied through the observation of SDC and the Pinch region. Three main heuristics were proposed by Munir et al. (2012) to reduce fresh CO_2 and CO_2 emission targets:

H1: *Above the Pinch, apply process changes to eliminate or reduce CO_2 source streams.*
H2: *Increase the limiting flow rate, or add new CO_2 demand processes Above the Pinch.*
H3: *Reduce or eliminate CO_2 demand processes Below the Pinch.*

Level 3: Regeneration reuse

Regeneration reuse involves treating and regenerating the emission sources to increase their CO_2 purity level prior to reuse. This can be done through technologies such as chemical absorption using amine-based solvents. In selecting the CO_2 sources to be regenerated, heuristic 4 is proposed to be used as a guide:

H4: *Regenerate the source stream across the Pinch to achieve purity higher than the Pinch purity.*

Regeneration reuse can further increase the amount of CO_2 sources to be consumed by CO_2 demands, ultimately reducing the emitted CO_2 target Above the Pinch and the fresh CO_2 requirement Below the Pinch.

Level 4: CO_2 sequestration

The final level is CO_2 sequestration. If there is still remaining CO_2 emission after performing the minimum carbon emissions target considering Levels 1 to 3, the CO_2

emission can be sequestrated or stored to achieve temporary zero CO_2 emission. This may involve CO_2 capture and storage in oil wells for use in enhanced oil recovery or CO_2 sequestration by planting more trees.

4.9 Summary

This chapter provides a guide on the fundamentals of Total Site Integration. It discusses the techniques available for analysing Total Sites from the point of view of process Heat Sources and Heat Sinks, whose energy demands can be met to a large extent by Heat Recovery through the generation and use of intermediate utilities (mainly steam), thereby minimising the amount of steam that has to be supplied by central site utility boilers.

The extraction of Heat Sources and Sinks, after maximising process-level Heat Recovery, can be accomplished with the aid of the individual process Grand Composite Curves and the results of the Problem Table Algorithm. The sources of heat from the individual site processes can be integrated to produce the Source Site Profile. Similarly, the sinks of heat from the individual site processes can be integrated to produce the Sink Site Profile. Plotted together, they make up the Total Site Profiles showing the total amount of heat available from all the processes on the site in relation to temperature and the amount of heat needed by all the site processes.

The Heat Sources can be related to the generation of steam from the individual processes at various selected pressures, and the heat sinks can be related to the use of steam at various pressures. The use and generation of steam at these predetermined pressures in relation to the Heat Sources and sinks can be shown in the Total Site Profiles. Without any Heat Recovery between the generation and use of steam, the Total Site Profiles can show the amount of steam that has to be supplied by central site utility boilers. However, the generation of steam from the process Heat Sources can be used to supply the process heat sinks and consequently reduce the steam supply from the site utility boilers. The maximum amount of Heat Recovery from the site processes through the steam distribution system can be shown in the Total Site Composite Curves.

The generation and use of steam related to the process of Heat Sources and Heat Sinks, accompanied by the supply of steam from the site utility boilers, can be shown on the Site Utility Grand Composite Curve. This plot can also indicate the power generation potential that can accompany steam generation and use at various pressure levels. This chapter also highlights the optimisation of a Hybrid Power System by using the Power Pinch Analysis. After minimising energy in a Total Site from a thermal and electricity perspective, the remaining CO_2 still being emitted can be further reduced by exchanging CO_2 across the Total Site. This can be done via CO_2 Emission Pinch Analysis.

References

Aguilar O, Perry S J, Kim J K, Smith R. (2007). Design and Optimization of Flexible Utility Systems Subject to Variable Conditions, Chemical Engineering Research and Design, 85(A8), 1136–1148.

Bruno J C, Fernandez F, Castells F, Grossmann I E. (1998). A Rigorous MINLP Model for the Optimal Synthesis and Operation of Utility Plants, Transaction IChemE, Chemical Engineering Research and Design, 76(March), 246–258.

Canmet ENERGY. (2003). Pinch Analysis: For the Efficient Use of Energy. Water and Hydrogen. Natural Resource Canada, Varennes.

Chin H H, Jia X, Varbanov P S, Klemeš J J, Liu Z-Y. (2021a). Internal and Total Site Water Network Design with Water Mains using Pinch-based and Optimization Approaches, ACS Sustainable Chemistry and Engineering, 9, 6639–6658, doi:10.1021/acssuschemeng.1c00183.

Chin H H, Varbanov P S, Klemeš J J, Alwi S R W. (2021b). Total Site Material Recycling Network Design and Headers Targeting Framework with Minimal Cross-Plant Source Transfer, Computers & Chemical Engineering, 151, 107364, doi:https://doi.org/10.1016/j.compchemeng.2021.107364.

Dhole V R, Linnhoff B. (1993). Total Site Targets for Fuel, Co-generation, Emissions, and Cooling, Computers & Chemical Engineering, 17(Supplement), S101–S109.

El-Halwagi M M, Gabriel F, Harell D. (2003). Rigorous Graphical Targeting for Resource Conservation via Material Recycle/Reuse Networks, Industrial & Engineering Chemistry Research, 42(19), 4319–4328.

Fadzil A F A, Wan Alwi S R, Manan Z A, Klemeš J J. (2017). Total Site Centralised Water Integration for Efficient Industrial Site Water Minimisation, Chemical Engineering Transaction, 61, doi:10.3303/CET1761188.

Fan Y V, Varbanov P S, Klemeš J J. (2018). Andreja Nemet, Process Efficiency Optimisation and Integration for Cleaner Production, Journal of Cleaner Production, 174, 177–183.

Gai L, Varbanov P S, Fan Y V, Klemeš J J, Nižetić S. (2022). Total Site Hydrogen Integration with Fresh Hydrogen of Multiple Quality and Waste Hydrogen Recovery in Refineries, International Journal of Hydrogen Energy, 47(24), 12159–12178, doi:10.1016/j.ijhydene.2021.06.154. i-Steam (2022). i-Steam™. Process Integration Ltd. https://www.processint.com/software/i-steam/. accessed 03/12/2022.

Jamaluddin K, Wan Alwi S R, Manan Z A, Hamzah K, Klemeš J J. (2022). Design of Total Site-integrated Trigenerationsystem using Trigeneration Cascade Analysis Considering Transmission Losses and Sensitivity Analysis, Energy, 252, 123958, doi:10.1016/j.energy.2022.123958.

Jia X, Varbanov P S, Wan Alwi S R, Klemeš J J. (2020). Total Site Water Main Concentration Selection: A Case Study, Chemical Engineering Transactions, 81, 259–264.

Jiménez-Romero J, Azapagic A, Smith R. (2022). STYLE: A New Optimization Model for Synthesis of Utility Systems with Steam Level Placement, Computers & Chemical Engineering, 108060, doi:10.1016/j.compchemeng.2022.108060.

KBC. (2022). Petro-SIM | Process Simulation Software | KBC. <https://www.kbc.global/software/process-simulation-software/>, accessed 03.12.2022

Kemp I C, Lim J S. (2020). Pinch Analysis for Energy and Carbon Footprint Reduction: User Guide to Process Integration for the Efficient Use of Energy, Third ed. Elsevier / Butterworth-Heinemann, Oxford, United Kingdom.

Klemeš J, Dhole V R, Raissi K, Perry S J, Puigjaner L. (1997). Targeting and Design Methodology for Reduction of Fuel, Power and CO_2 on Total Sites, Applied Thermal Engineering, 7, 993–1003.

Klemeš J, Friedler F, Bulatov I, Varbanov P. (2010). Sustainability in the Process Industry: Integration and Optimization. McGraw Hill Companies Inc, New York USA.

Liew P Y, Lim J S, Wan Alwi S R, Manan Z A, Varbanov P S, Klemeš J J. (2014). A Retrofit Framework for Total Site Heat Recovery Systems, Applied Energy, 135, 778–790.

Liew P Y, Wan Alwi S R, Varbanov P S, Manan Z A, Klemeš J J. (2012). A Numerical Technique for Total Site Sensitivity Analysis, Applied Thermal Engineering, 40, 397–408.

Linnhoff B, Townsend D W, Boland D, Hewitt G F, Thomas B E A, Guy A R, Marsland R H. (1982). A User Guide to Process Integration for the Efficient Use of Energy. IChemE, Rugby, UK, latest edition 1994.

Manan Z A, Wan Alwi S R, Sadiq M M, Varbanov P. (2014). Generic Carbon Cascade Analysis Technique for Carbon Emission Management, Applied Thermal Engineering, 70(2), 1141–1147.

Matsuda K, Hirochi Y, Tatsumi H, Shire T. (2009). Applying Heat Integration Total Site-based Pinch Technology to a Large Industrial Area in Japan to Further Improve Performance of Highly Efficient Process Plants, Energy, 34(10), 1687–1692.

Mavromatis S P, Kokossis A C. (1998). Conceptual Optimisation of Utility Networks for Operational Variations – I. Targets and Level Optimisation, Chemical Engineering Science, 53(8), 1585–1608.

Munir S M, Manan Z A, Wan Alwi S R. (2012). Holistic Carbon Planning for Industrial Parks – A Waste-to-Resources Process Integration Approach, Journal of Cleaner Production, 33, 74–85.

Peterson J F, Mann W L. (1985). Steam System Design: How it Evolves, Chemical Engineering, 92(21), 62–74.

Polley G T, Polley H L. (1998). Design Better Water Networks, Chemical Engineering Progress, 96(2), 47–52.

Raissi K. (1994). Total Site Integration, PhD Thesis, UMIST, Manchester, UK.

Rozali N E M, Wan Alwi S R, Manan Z A, Klemeš J J, Hassan M Y. (2013a). Process Integration Techniques for Optimal Design of Hybrid Power Systems, Applied Thermal Engineering, 61(1), 26–35.

Rozali N E M, Wan Alwi S R, Manan Z A, Klemeš J J, Hassan M Y. (2013b). Process Integration of Hybrid Power Systems with Energy Losses Considerations, Energy, 55, 38–45.

Shang Z. (2000). Analysis and Optimisation of total site Utility Systems. PhD Thesis UMIST, Manchester, UK.

Shang Z, Kokossis A. (2004). A Transhipment Model for the Optimisation of Steam Levels of Total Site Utility System for Multiperiod Operation, Computers & Chemical Engineering, 28(9), 1673–1688.

Siddhartha M, Rajkumar N. (1999). Performance Enhancement in Coal Fired Thermal Power Plants. Part II: Steam Turbines, International Journal of Energy Research, 23, 489–515.

Smith R, Varbanov P. (2005). What's the Price of Steam? Chemical Engineering Progress, 101(7), 29–33.

STAR. (2013). Process Integration Software (Centre for Process Integration, School of Chemical Engineering and Analytical Science, University of Manchester, UK). www.ceas.manchester.ac.uk/our-research/themes-challenges/themes/pdi/cpi/software/star/, accessed 14/06/2017.

Sun L, Smith R. (2015). Performance Modeling of New and Existing Steam Turbines, Industrial & Engineering Chemistry Research, 54, 1908–1915.

Varbanov P S, Doyle S, Smith R. (2004a). Modelling and Optimisation of Utility Systems, Transaction IChemE, Chemical Engineering Research and Design, 82(A5), 561–578.

Varbanov P, Perry S, Makwana Y, Zhu X X, Smith R. (2004b). Top-level Analysis of Site Utility Systems, Transaction IChemE, Chemical Engineering Research and Design, 82(A6), 784–795.

Varbanov P, Perry S, Klemeš J, Smith R. (2005). Synthesis of Industrial Utility Systems: Cost-effective De-carbonization, Applied Thermal Engineering, 25(7), 985–1001.

Varbanov P S, Fodor Z, Klemeš J J. (2012). Total Site Targeting with Process Specific Minimum Temperature Difference (ΔT_{min}), Energy, 44(1), 20–28, doi:10.1016/j.energy.2011.12.025.

Varbanov P S, Škorpík J, Pospíšil J, Klemeš J J. (2020). Sustainable Utility Systems: Modelling and Optimisation, Sustainable Utility Systems. De Gruyter, Berlin, Germany, ISBNL 978-3-11-063009-1, doi:10.1515/9783110630091.

Wahab A S A, Liew P Y, Rozali N E M, Wan Alwi S R, Klemeš J J. (2022). Spatial Total Site Heat Integration Targeting Using Cascade Pinch Analysis, Chemical Engineering Transaction, doi:10.3303/CET2294107.

Wang B, Klemeš J J, Varbanov P S, Shahzad K, Kabli M R. (2021). Total Site Heat Integration Benefiting from Geothermal Energy for Heating and Cooling Implementations, Journal of Environmental Management, 290, 112596, doi:10.1016/j.jenvman.2021.

Wan Alwi S R, Aripin A, Manan Z A. (2009). A Generic Graphical approach for Simultaneous Targeting and Design of a Gas Network, Resources, Conservation and Recycling, 53(10), 588–591.

Wan Alwi S R, Mohammad Rozali N E, Manan Z A, Klemeš J J. (2012). A Process Integration Targeting Method for Hybrid Power Systems, Energy, 44(1), 6–10.

Willans P W. (1888). "Economy Trials of a Non-Condensing Steam-Engine: Simple, Compound and Triple. (Including Tables and Plate at Back of Volume)", Minutes of the Proceedings of the Institution of Civil Engineers, 93(1888), 128–188, doi:https://doi.org/10.1680/imotp.1888.21059.

5 An integrated Pinch Analysis framework for low CO₂ industrial site planning

5.1 Introduction

There is a need to reduce the growing rate of industrial sector CO_2 emissions. Industrial site planners can play a big role in developing a low CO_2 emission industrial site by utilising the concept of symbiosis among industries. For example, an industry with excess heat or electricity can transfer the surplus energy to a nearby plant that has a deficit. A plant that generates CO_2 emissions can treat and then supply CO_2 for an industry that consumes CO_2 as its raw material. An industrial site planner can also enact relevant rules and select the most suitable processes to be located on the sites, which can contribute to a symbiotic mechanism and plan the appropriate size of its centralised utility system among the process units when planning for a low CO_2 emission industrial site. In addition, there are various graphical tools based on Pinch Analysis that were developed to guide industries and site planners in minimising their energy and CO_2 emissions, as described in previous chapters. The main strengths of Pinch Analysis tools, compared with a more comprehensive and complicated mathematical model, are that it is simple to construct, they obey the laws of thermodynamics, provide valuable insights that allow planners to take part in the development of the system, and allow the user to perform targets prior to design. Some of the well-known Pinch Analysis tools related to energy and carbon emission are Heat Integration Pinch Analysis, Total Site Heat Integration (TSHI) – see Chapter 4, Combined Heat and Power (CHP) – Chapter 4, Power Pinch Analysis (PoPA) – Chapter 4, and CO_2 Emission Pinch Analysis (CEPA) – which originated from the work of Tan and Foo (2007) and later updated (Tan et al., 2018). In this chapter, a systematic framework to guide industrial site planners in using these tools in an integrated manner for a systematic low CO_2 emission industrial site plan proposed by Abdul Aziz et al. (2017) is described.

5.2 Framework for low CO₂ emissions industrial site planning

The low CO_2 emission industrial site planning framework consists of four stages, as shown in Fig. 5.1. Detailed explanations are given next.

https://doi.org/10.1515/9783110782981-005

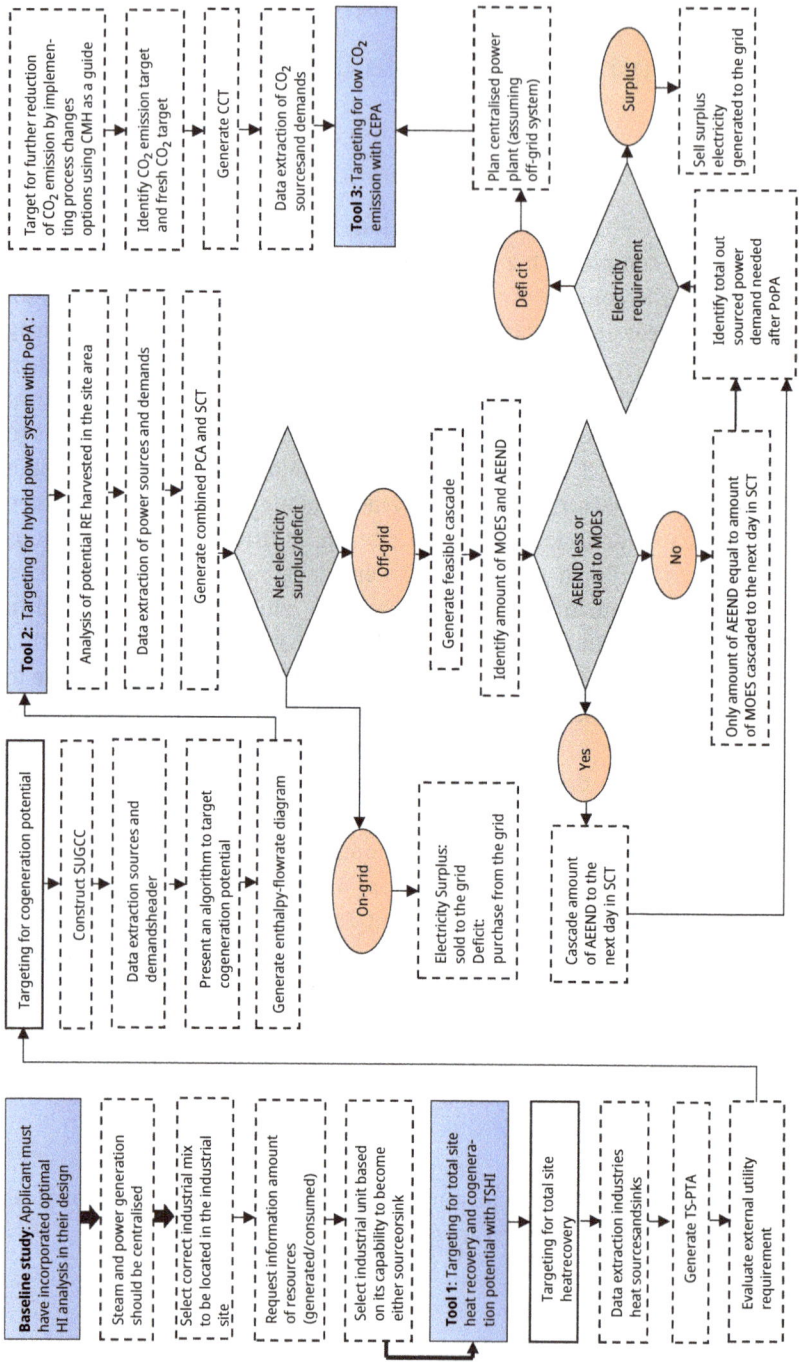

Fig. 5.1: Summary of the developed procedures.

Stage 1: Baseline study

In Stage 1, the baseline information of potential resources to be exchanged is gathered from the industries that have requested to be located at the new industrial site, which includes steam, electricity, and CO$_2$ requirements, as well as the potential amount of steam, electricity, and CO$_2$ the processes can generate to be supplied to others. The required data are:

(i) For steam – the amount of excess steam generated by the units or steam needed, in kW, and the steam level quality (high-pressure steam (HPS)/medium-pressure steam (MPS)/low-pressure steam (LPS)) at the specified pressure or temperature. The cooling water requirement is also required.

(ii) For electricity – the amount of electricity that can be generated from the site either from internal CHP, Rankine system, or Renewable Energy (RE), as well as electricity needed. An estimation of the time the electricity will be produced or consumed is also required.

(iii) For CO$_2$ – the amount of CO$_2$ that can be generated or needed by the units, along with the CO$_2$ purity of the sources or demands.

Other than these, the following data are also obtained:

a) The centralised steam system and cooling water system temperature, pressure, and enthalpy data set by the industrial site.

b) The meteorological and environmental data of the industrial site areas for RE potential calculation.

Note that the industrial site owner can require that all industrial units must have already incorporated Heat Integration analysis (as described in Chapter 2) during the design stage. The remaining requirement for heating and cooling for the total site is the additional utilities (e.g. steam, cooling water) the unit still needs for its processes after considering maximum heat recovery.

Based on these data, the industrial site owner can later decide which industrial units are to be located in its industrial site by correctly selecting the units that can consume and generate the various resources. The owner can also decide the proper location for the units to exchange resources efficiently, by placing them closer to each other.

Stage 2: Tool 1 targeting for Total Site heat recovery with cogeneration using TSHI

In Stage 2, the surplus heat from the industrial sites is analysed using the TSHI method to determine the maximum steam possible to be exchanged among multiple units via the centralised steam system. Among potential methods is the graphical method, Total Site Profiles and Total Site Composite Curves (Klemeš, 2022), explained in Chapter 4 of this book, or the numerical method, Total Site Problem Table Algorithm, explained in Section 4.8.3.

In this stage, the potential to recover further heat by using a CHP system is also analysed from the result of TSHI using the Site Utility Grand Composite Curve (SUGCC), as explained in Section 4.5, the method proposed by Bandyopadhyay et al. (2010), and the follow-up developments accounting for process-specific ΔT_{min} (Varbanov et al., 2012), the numerical algorithm (Liew et al., 2012) which allows multiperiod models, as well as using the seasonal model (Liew et al., 2018). From the two analyses, the requirement for the centralised boiler capacity and potential power generation from CHP can be obtained.

Stage 3: Tool 2 targeting for hybrid power system with PoPA

In Stage 3, the potential RE that can be generated from the industrial site is explored. As the cost of using 100% RE may not be economical yet, the industrial site's power can be supplemented using a conventional power plant, which uses fossil fuels. The system is called a Hybrid Power System (HPS) due to the combination of several RE and fossil fuels for power generation. The system can be an off-grid or an on-grid system.

An off-grid system may involve a power storage system to store surplus energy that cannot be used at the current time, but to be used later. The external electricity requirement can be obtained from generators.

The on-grid system is connected to the grid, and the electricity is assumed to be supplied by the electricity power provider. It is also assumed that electricity generated by the site can be sold to the grid. Users can also consider having a power storage system and generators for the on-grid system to have more flexibility in selling and purchasing electricity, as explained by Liu et al. (2016).

PoPA (Rozali et al., 2013), explained in Section 4.8.4, is a potential tool that can be used to evaluate the potential of a hybrid power system in industrial sites. Using PoPA, the power storage capacity as well as the amount of external electricity still needed, can be determined. The remaining external electricity needed will be supplemented by the power generation from CHP and also the centralised power plant.

Stage 4: Tool 3 targeting low CO_2 emission with CPA

As the centralised power plant and the steam boiler will be burning fossil fuels like coal or natural gas for their combustion process, CO_2 would still be emitted. In Stage 4, CO_2 is captured from the power plant and boiler stacks as well as other processes in the plant that emit CO_2. It is then exchanged among multiple units that can accept CO_2 gas with certain impurities, to be used in its process before finally it is stored or sequestrated. The CEPA tool, explained in Section 4.8.5, can be used to perform this analysis. In this stage, other CO_2 emission minimisation strategies, as described in the Carbon Emission Management Hierarchy (CEMH), are also explored to reduce CO_2 emissions of the Total Site to zero emission.

5.3 Case study

The low CO_2 industrial site planning framework is applied to an industrial site located at Bayan Lepas, Penang, in the region of Peninsular Malaysia. The industrial site owner plans to develop its own centralised utility system that can supply electricity and steam to the industries in the area. Ten industrial units that have applied to be located in the industrial sites were analysed for the potential symbiosis in terms of achieving a low CO_2 emissions industrial site.

5.3.1 Stage 1: baseline study

The baseline information of steam, electricity, and CO_2 generated and required were gathered from the ten industries and listed in Tabs. 5.1(a) and 5.1(b). The amount of resources needed is denoted by a negative sign, while the amount of resources generated is denoted by a positive sign. Table 5.2 shows the temperature, pressure, and enthalpy change for various steam levels and cooling water systems supplied by the centralised steam system. Table 5.3 lists the potential RE available from the multiple sites, which include solar and biomass, for this case.

Tab. 5.1(a): Industrial unit mapping (Industrial Units 1 to 5) for Case Study 1.

1	2			3		4	
Resources	Steam			Power rating		CO_2	
	Amount generated (+)/needed (−) (kW)	Quality	Temp. (°C)	Amount generated from RE (+)/ needed by demands (−) (kW)	Time (h)	Amount generated (+)/ needed (−) (kg/h)	CO_2 purity, (%)
Unit 1	−810	HPS	275	6.5 (solar)	10–16 h	2,054	22.5
	700	MPS	200	−649.5	0–24 h	−12,922	52.6
	−850	LPS	150				
	−366	CW	25				
Unit 2	−250	HPS	275	7.5 (solar)	10–16 h	1,154	22.5
	−625	MPS	200	−620.2	0–24 h	−3,514	25
	196	LPS	150				
	−352	CW	25				
Unit 3	205	HPS	275	6.00 (solar)	10–16 h	1,363	22.5
	−551	MPS	200	−744	0–24 h	−7,743	14.2
	−497	LPS	150				
	370	CW	25				

Tab. 5.1(a) (continued)

1	2			3		4	
Resources	Steam			Power rating		CO_2	
	Amount generated (+)/needed (−) (kW)	Quality	Temp. (°C)	Amount generated from RE (+)/ needed by demands (−) (kW)	Time (h)	Amount generated (+)/ needed (−) (kg/h)	CO_2 purity, (%)
Unit 4	642	HPS	275	225 (biomass)	0–24 h	1,422	22.5
	717	MPS	200	−619.5	0–24 h	−698	13.9
	−105	LPS	150				
	150	CW	25				
Unit 5	−750	HPS	275	−698	0–24 h	1,621	22.5
	−635	MPS	200			−7,508	14.0
	−292	LPS	150				
	431	CW	25				

Tab. 5.1(b): Industrial unit mapping (Industrial Units 6 to 10) for Case Study 1.

1	2			3		4	
Resources	Steam			Power rating		CO_2	
	Amount generated (+)/needed (−) (kW)	Quality	Temp. (°C)	Amount generated from RE (+)/ needed by demands (−) (kW)	Time (h)	Amount generated (+)/needed (−) (kg/h)	CO_2 purity, (%)
Unit 6	676	HPS	275	−696.3	0–24 h	1,410	22.5
	−390	MPS	200				
	−305	LPS	150				
	197	CW	25				
Unit 7	−572	HPS	275	−612	0–24 h	1,388	22.5
	−	MPS	200				
	−850	LPS	150				
	300	CW	25				
Unit 8	−538	HPS	275	−654.1	0–24 h	1,423	22.5
	730	MPS	200				
	163	LPS	150				
	139	CW	25				

Tab. 5.1(b) (continued)

1	2			3		4	
Resources	Steam			Power rating		CO_2	
	Amount generated (+)/needed (−) (kW)	Quality	Temp. (°C)	Amount generated from RE (+)/ needed by demands (−) (kW)	Time (h)	Amount generated (+)/needed (−) (kg/h)	CO_2 purity, (%)
Unit 9	243	HPS	275	−824.7	0–24 h	1,126	22.5
	583	MPS	200				
	−	LPS	150				
	−106	CW	25				
Unit 10	−646	HPS	275	−806.2	0–24 h	1,695	22.5
	−170	MPS	200				
	862	LPS	150				
	296	CW	25				

Tab. 5.2: Properties of utility for Case Study 1.

Steam	Pressure (bar)	Temp. (°C)	Enthalpy (kJ/kg)
VHPS	90	500	1,364
HPS	60	275	1,214
MPS	15	198	845
LPS	5	150	640
CW	0.07	40	163

Tab. 5.3: Data for Renewable Energy (RE) analysis for Case Study 1.

Potential Site	RE resources					
	Solar				Biomass	
	Area PV installations, (m²)	PV efficiency (%)	Daily sun (h)	Mean daily global solar radiation (kWh/m²d)	Power generated (kW)	Time (h)
Unit 1	52	15	10–16	5.0		
Unit 2	60	15	10–16	5.0		
Unit 3	50	15	10–16	5.0		
Unit 4					225	0–24

5.3.2 Stage 2: targeting for Total Site heat recovery with cogeneration using TSHI (Tool 1)

In Stage 2, TSHI is performed to target the maximum heat recovery and cogeneration potential among processes. Tables 5.4 and 5.5 show the data extraction of heat sources and heat demands for the industrial units selected to be located at the industrial site. The initial total steam requirement for HPS was 3,566 kW, MPS was 2,371 kW, and LPS was 2,899 kW. For this case study, the method of Liew et al. (2012) is used for the TSHI analysis. A Total Site Problem Table Algorithm (TS-PTA) is constructed and shown in Tab. 5.6. It can be seen the Total Site Pinch Point is located between LPS and CW utility temperatures. The minimum utility targets are 1,800 kW for HPS, 1,319 kW for LPS and 1,059 kW for CW. The steam demand has been reduced by about 64.7%. The centralised steam boiler would need to generate 3,119 kW of very high-pressure steam (VHPS) to fulfil the requirement for HPS and LPS (assuming no energy losses).

Tab. 5.4: Data extraction of heat sources for Case Study 1.

Utility	Utility temp. (°C)	Heat sources (kW)										Net heat source (kW)
		Unit										
		1	2	3	4	5	6	7	8	9	10	
HPS	275			205	642		676			243		1,766
MPS	200	700			717				730	583		2,730
LPS	150		196						163		862	1,221
CW	25			370	150	431	197	300	139		296	1,883

Tab. 5.5: Data extraction of heat demands for Case Study 1.

Utility	Utility temp. (°C)	Heat demands (kW)										Net heat demand (kW)
		Unit										
		1	2	3	4	5	6	7	8	9	10	
HPS	275	810	250			750		572	538		646	3,566
MPS	200		625	551		635	390				170	2,371
LPS	150	850		497	105	292	305	850				2,899
CW	25	366	352							106		824

Tab. 5.6: Total Site-Problem Table Algorithm (TS-PTA) for Case Study 1.

1	2	3	4	5	6	7	8	9
Utility	Utility temp. (°C)	Net heat source (kW)	Net heat sink (kW)	Net heat requirement (kW)	Initial heat cascade	Final single heat cascade	Multiple utility heat cascade	External utility requirement (kW)
					0	3,119	0	
HPS	275	1,766	3,566	−1,800				1,800
					−1,800	1,319	0	
MPS	200	2,730	2,371	359				0
					−1,441	1,678	359	
LPS	150	1,221	2,899	−1,678				1,319
					−3,119	0	0	(PINCH)
CW	25	1,883	824	1,059				−1,059
					−2,060	1,059	0	

The CHP potential is evaluated next. The SUGCC is constructed based on the data obtained from Columns 3 and 4 of the TS-PTA (Tab. 5.6), as shown in Fig. 5.2. The enclosed area between the VHPS level and HPS, as well as LPS level, represents the amount of potential CHP shaft work. Bandyopadhyay et al. (2010) method is used for the potential CHP shaft work estimation. Tab. 5.7 shows the data extractions of steam sources and demands at the various headers. There is one source header (the VHPS) and two demand headers (HPS and LPS).

VHPS is produced by the site boiler at a pressure P of 90 bar and temperature T of 500 °C. The specific heat load is fixed to 2,022.9 kJ/kg based on the steam table. Because VHPS is used to fulfil the requirement for HPS and LPS, the specific heat loads of HPS and LPS are assumed to be the same as that of the VHPS header. The mass flow rates of the HPS and LPS headers are calculated by dividing the external utility requirement of HPS and LPS (from Column 5, Tab. 5.7) by the VHPS-specific heat load, giving the minimum steam mass flow rate requirement of 0.89 kg/s for HPS and 0.65 kg/s for LPS. Tab. 5.8 presents an algorithm to target the cogeneration potential. The minimum steam generation at the VHPS header is 1.54 kg/s. From Column 5, the Pinch Point (location of the first zero) is located at the LPS header, which shows no heat is passed to the condensate header. In Column 6, the difference between the last two enthalpies is multiplied by the adjusted cumulative steam flows to calculate the cogeneration potential. The cogeneration potential for Case Study 1 is calculated to be 604 kW (= 231 kW + 373 kW) by adding values in Column 6, Tab. 5.8. Fig. 5.3 shows the plot of enthalpies corresponding to saturated water (Column 2, Tab. 5.8) versus cumulative feasible heat flow (Column 5, Tab. 5.8). It can be identified that all turbines are back-pressure type and it is also a topping-cycle cogeneration system.

Tab. 5.7: Data of source and demand headers for Case Study 1.

	Temperature (°C)	Pressure (bar)	Enthalpy (kJ/kg)	Utility requirement (kW)	Mass flow rate (kg/s)
Source header					
VHPS	500	90	1,364	To be determined	To be determined
Demand headers					
HPS	275	60	1,214	1,800	0.89
LPS	150	5	640	1,319	0.65
Condensate return	40	0.07	163	To be determined	To be determined

Tab. 5.8: Data of source and demand headers for Case Study 1.

1	2	3	4	5	6
Pressure (bar)	Enthalpy (kJ/kg)	Net flow (kg/s)	Flow cascade (kg/s)	Feasible flow cascade (kg/s)	Cogeneration potential (kW)
90	1,364	0	0	1.54	0
60	1,214	−0.89	−0.89	0.65	231
5	640	−0.65	−1.54	0	373
0.07	163	0	−1.54	0	0

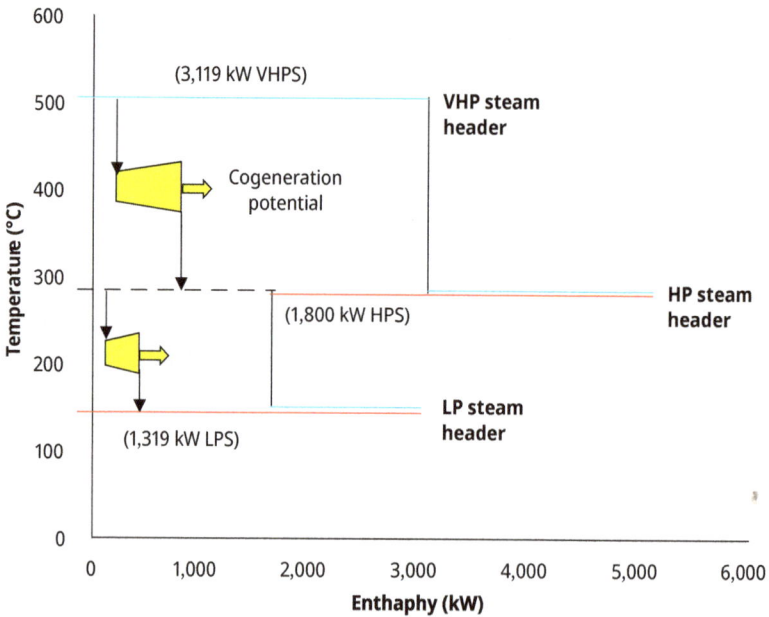

Fig. 5.2: SUGCC for Case Study 1.

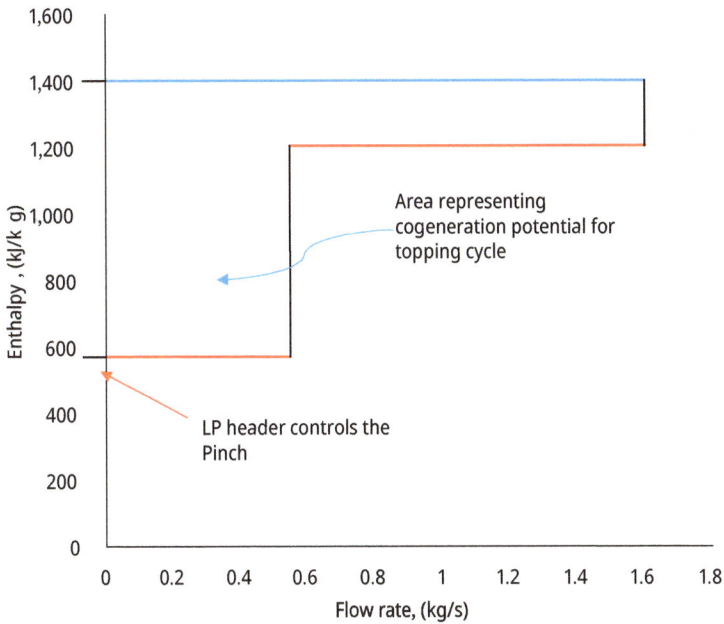

Fig. 5.3: Enthalpy (specific) versus flow rate diagram for Case Study 1.

5.3.3 Stage 3: targeting for hybrid power system, integrating RE resources with PoPA (Tool 2)

In Stage 3, the potential RE sources in the industrial site are evaluated to determine the amount of remaining power plant size that needs to be built to satisfy the industrial site's electricity demand. For Case Study 1, only solar and biomass are considered, as the area has very low wind energy potential (average mean daily wind speed is recorded as only 1.84 m/s). The mean daily global solar radiation for the Bayan Lepas area is 5.11 kWh/(m² d), based on the meteorological and environmental data provided by Engel-Cox et al. (2012). To simplify the case study illustration, 5 kWh/(m² d) of solar radiation is assumed. In addition, 5,400 kWh of biomass energy is also assumed to be available from the site.

Table 5.9 shows the data extraction of power sources and demands. Power source data comprise solar and biomass RE, and also the cogeneration potential obtained from the previous analysis. The power demand data are extracted from each industrial unit's power consumption profile at each time interval. The system is assumed off-grid, as the industrial site can generate its own electricity via the site's centralised power plant. The numerical method of Rozali et al. (2013) is used for the analysis, as explained in Section 4.8.4. The combined Power Cascade Table (PCT) and Storage Cascade Table (SCT) are constructed, as shown in Tabs. 5.10(a) and 5.10(b). A feasible cascade is tabulated in

Tab. 5.9: Data of source and demand headers for Case Study 1 (Rozali et al., 2013).

Time (h)		Sources			Power rating (kW) — Demands — Unit											Total power rating (kW)	
From	To	Biomass	Solar	Cogeneration	1	2	3	4	5	6	7	8	9	10	Σ Power source	Σ Power demand
0	1	225			20	35	22	25	24	24	16	15	24	14	250	219
1	2	225			22	36	25	15	24	34	26	14	25	16	250	237
2	3	225			25	35	34	26	25	36	17	23	27	28	250	276
3	4	225			25	17	33	18	26	36	18	34	27	28	250	262
4	5	225			30	20	35	29	25.5	37	19	37	37	18.5	250	288
5	6	225			28	21	36	39	39	48	20	33	39	20	250	323
6	7	225			22	25	29	35	40	35	30	35	33	50	250	334
7	8	225			21	29	30	30	44	47.3	46	31	24.7	62	250	365
8	9	225			39.5	30	31	32	46.5	27	47	32	33	80	250	398
9	10	225			40	35	48	40	35.7	36	45	38.8	48	45.5	250	412
10	11	225	20		55	50	65	39	35	35	30	47	52	66	270	474
11	12	225	20		46	44	58	41	45	40	39	38	59	66	270	476
12	13	225	20		45	43	52	31	48	40	30	45	88	78	270	500
13	14	225	20		46	56	75	52	41	49	37	43	75	80	270	554
14	15	225	20		40	35	30	29	40	41	30	31	72	30	270	378
15	16	225	20		33	28	33	34	43	35	28	40	55	33	270	362
16	17	225			37	30	35	35	46	26	35	29	20	20	250	313
17	18	225			20	15	22	20	25	24	22	21	20	15	250	204
18	19	225			10	8	12	12	10	7	15	10	15	12	250	111
19	20	225			9	7	9	9	8	7	10	11	10	9	250	89
20	21	225			9	7	9	7	6	6	16	15	18	7	250	100
21	22	225			8.5	5.5	6	8	6	9	13	14	11	10	250	93
22	23	225			9.5	5	7	6	7	7	14	9.8	6	8.7	250	80
23	24	225			9	4	8	7	6	10	9	7	6	9	250	75

Tab. 5.10(a): Combined Power Cascade Table (PCT) and Storage Cascade Table (SCT) for Case Study 1 between times 0 and 12 h (Rozali et al., 2013).

1	2	3	4	5	6	7	8	9	10	11	12	13	14
		Power rating, kW					Start up and Operation						
							PCT			SCT			
Time, h	Time interval, h	∑ Power source	∑ Power demand	∑ Electricity source, kWh	∑ Electricity demand, kWh	Amount of electricity transfer, kWh	Net electricity surplus/deficit, kWh	Infeasible electricity cascade, kWh	Feasible electricity cascade, kWh	Net electricity surplus/deficit, kWh	Storage capacity, kWh	Outsourced electricity needed, kWh	External electricity power rating, kW
1	1	250	219	250	219	219	31		1,783	31	998		
	1	250	237	250	237	237	13	31	1,814	13	1,029		
3	1	250	276	250	276	250	−26	18	1,801	−26	1,042		
4	1	250	262	250	262	250	−12	6	1,789	−12	1,016		
5	1	250	288	250	288	250	−38	−32	1,751	−38	1,004		
6	1	250	323	250	323	250	−73	−105	1,678	−73	966		
7	1	250	334	250	334	250	−84	−189	1,594	−84	893		
8	1	250	365	250	365	250	−115	−304	1,479	−115	809		
9	1	250	398	250	398	250	−148	−452	1,331	−148	694		
10	1	270	412	270	412	270	−142	−594	1,189	−142	546		
11	1	270	474	270	474	270	−204	−798	985	−204	404		
12	1	270	476	270	476	270	−206	−1,004	779	−206	200	−6	−6

Total external electricity needed 998

Maximum storage capacity

Tab. 5.10(b): Combined Power Cascade Table (PCT) and Storage Cascade Table (SCT) for Case Study 1 between times 12 and 24 h.

1	2	3	4	5	6	7	8	9	10	11	12	13	14
		Power rating, kW					Start up and Operation						
							PCT			SCT			
Time, h	Time interval, h	∑ Power source	∑ Power demand	∑ Electricity source, kWh	∑ Electricity demand, kWh	Amount of electricity transfer, kWh	Net electricity surplus/deficit, kWh	Infeasible electricity cascade, kWh	Feasible electricity cascade, kWh	Net electricity surplus plus/deficit, kWh	Storage capacity, kWh	Outsourced electricity needed, kWh	External electricity power rating, kW
12	1	270	500	270	500	270	-230	-1,004	779	-230		-6	-6
13	1	270	554	270	554	270	-284	-1,234	549	-284		-236	-236
14	1	270	378	270	378	270	-108	-1,518	265	-108		-520	-520
15	1	270	362	270	362	270	-92	-1,626	157	-92		-628	-628
16	1	250	315	250	315	250	-65	-1,718	65	-65		-720	-720
17	1	250	204	250	204	204	46	-1,783	0	46		-785	-785
18	1	250	111	250	111	111	139	-1,737	46	139		-739	-739
19	1	250	89	250	89	89	161	-1,598	185	161		-600	-600
20	1	250	100	250	100	100	150	-1,437	-346	150		-439	-439
21	1	250	93	250	93	93	157	-1,287	-496	157		-289	-289
22	1	250	80	250	80	80	170	-1,130	-653	170		-132	-132
23	1	250	75	250	75	75	175	-960	-823	175	38		
24								-785	-998		213		

Pinch (at feasible cascade 0 / infeasible cascade -1,783)

Max power demand (-785)

Column 10, Tabs. 5.10(a) and 5.10(b). The analysis gives a minimum outsourced electricity supply (MOES) required during start up of 1,783 kWh, and there are 998 kWh of available excess electricity for the next day (AEEND). Because AEEND is less than or equal to MOES, the system follows Scenario 1, as described in Rozali et al. (2013). The AEEND can be used for the next day to reduce the amount of outsourced electricity supply. This can be repeated for the following days. The MOES required during continuous operation is 785 kWh (= 1,783 kWh – 998 kWh).

For the power storage system determination, the value of AEEND is cascaded with the net electricity surplus/deficit in Column 12. Note that the storage capacity must always be positive, as it represents the amount of electricity stored in the battery at the respective time. If the battery capacity has been fully utilised, any electricity deficit would instead be supplied by the outsourced electricity. The largest cascaded electricity surplus of 1,042 kWh in Column 12, Tabs. 5.10(a) and 5.10(b) give the maximum storage capacity of the battery. The maximum power demand occurs at 17 h, with a power rating of 785 kW.

As a summary, the MOES needed is 1,783 kW during start up and 785 kWh during normal operation, and a power storage system with a capacity of 1,042 kWh. This utilisation of a hybrid power system has reduced 74.3% of the industrial site power consumption. The centralised power plant (fossil fuel based) that needs to be built should be able to produce 1,783 kWh of power.

5.3.4 Stage 4: targeting for low CO_2 emission with CEPA (Tool 3)

Based on Stages 3 and 4, it has been determined that the centralised steam boiler's capacity is 3,119 kW, and the centralised power plant capacity is 1,783 kW. Both of these systems are assumed to be using natural gas as their fuel, burning of which will still emit CO_2. Taking this into consideration and, in addition, the amount of other CO_2 sources and demands from the other industrial units, Tab. 5.11 lists all the available CO_2 sources and demands for the case study, with their respective flow rates and CO_2 compositions. Source S1 is the CO_2 point sources from furnace stacks, drying towers, or gas turbines. Source S2 represents the CO_2 emissions from initial steam and power requirements at the early stage by all industrial units. The fuels used are assumed to be coming from coal-based fuels. Source S3 represents CO_2 emitted from the site (centralised boiler) from fuel burning for steam and power generation. Examples of CO_2 demands are methanol production, microalgae cultivation, or enhanced oil recovery (Munir et al., 2012).

In this stage, CO_2 is minimised, based on the CMH by Munir et al. (2012). Based on the first level of CMH, the maximum CO_2 exchange is first evaluated. The Carbon Cascade Table (CCT) by Manan et al. (2014), as described in Section 4.8.5, is used to target the maximum CO_2 exchange in the total site. Table 5.12 shows the constructed CCT for the case study. The location of the largest negative value in Column 9 represents the

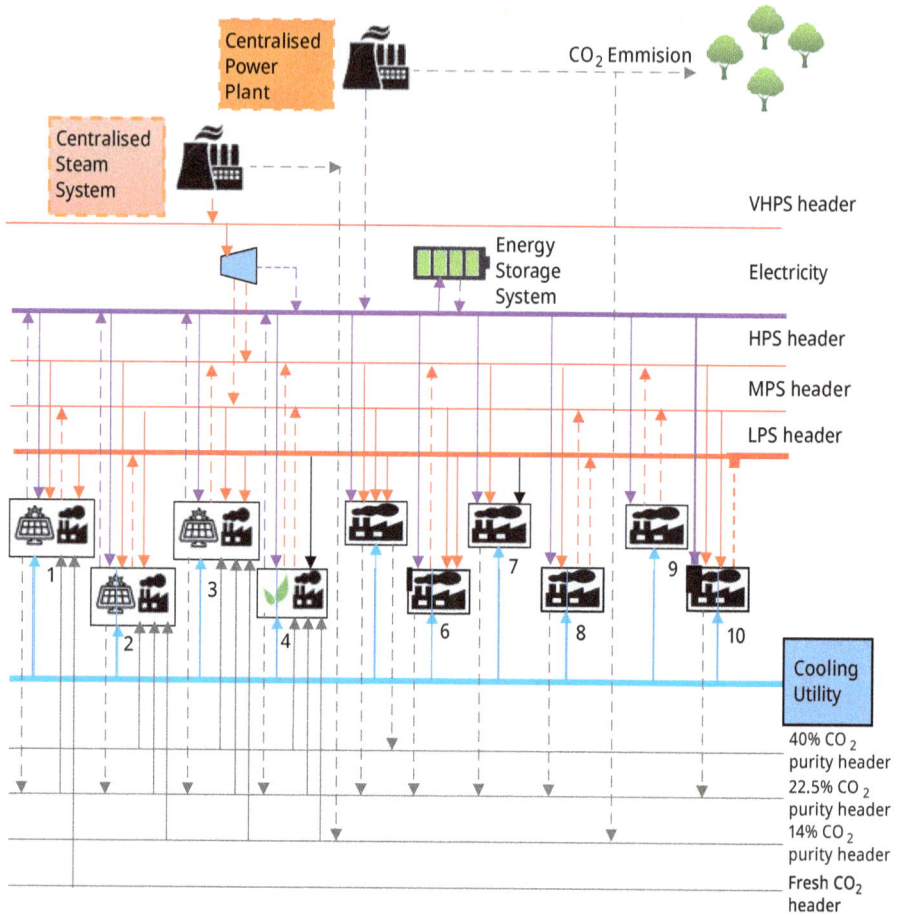

Fig. 5.4: Final low CO_2 emission industrial Total Site network for Case Study 1.

Pinch Point, and is located at F_{OG}/F_{CO_2} of 1.5. From Column 10, the minimum fresh CO_2 target (F_{FC}) is determined to be 5,169 kg/h (first value), and the amount of minimum CO_2 emissions flow rate target (F_{CE}) is 9,226 kg/h (last value).

Next, Level 2 of the CMH, which is Source and Demand Manipulation, is explored. Using Heuristic H1, which states, *"Above the Pinch, apply process changes to eliminate or reduce carbon source streams"*, it can be observed that S_2, S_3, and part of S_1 lie above the Pinch Region. Fuel switching from coal-based fuel to natural gas fuel is considered for source S_2. This reduces the F_{CO_2} from 14,655.75 kg/h to 5,489.34 kg/h. The CCT has been constructed again by using the new flowrate and composition for Source S_2. Table 5.13 shows the CCT after implementing the source reduction process. The new minimum CO_2 emissions flowrate target (F_{CE}) has been reduced from 9,226 kg/h to 59 kg/h. The Pinch location has also changed to F_{OG}/F_{CO_2} of 5.57. Prior to the analysis,

Tab. 5.11: Data Extraction of CO_2 sources and demands for Case Study 1.

1	2	3	4	5	6
	Total gas flowrate, F_T (kg/h)	CO_2 composition (%)	Flowrate CO_2, $F(CO_2)$ (kg/h)	A flowrate of other gases, F_{OG} (kg/h)	Ratio F_{OG}/F_{CO_2}
Sources					
S_1	33,066	40.0	13,227	19,840	1.50
S_2	65,136	22.5	14,656	50,481	3.44
S_3	7,508	14.0	1,051	6,457	6.14
Demands					
D_1	24,552	52.6	12,922	11,630	0.90
D_2	14,054	25.0	3,514	10,541	3.00
D_3	54,588	14.2	7,743	46,845	6.05
D_4	5,027	13.9	698	4,329	6.20

Tab. 5.12: CO_2 Cascade Table (CCT) for Case Study 1.

1	2	3	4	5	6	7	8	9	10
Ratio, F_{OG}/F_{CO_2}	$\Delta(F_{OG}/F_{CO_2})$	Total flowrate CO_2 sources, ΣF_S, kg/h	Total flowrate CO_2 demands, ΣF_D, kg/h	$\Sigma F_S + \Sigma F_D$, kg/h	CO_2 flowrate, F_C, kg/h	F_{OG}', kg/h	Other gases cascade, $F_{OG,cum}'$, kg/h	Pure CO_2 cascade, $F_{FC,cum}'$, kg/h	Feasible CO_2 cascade, F_C', kg/h
					0				5,169
0.9			−12,922	−12,922			0		
	0.6				−12,922	−7,753			−7,753
1.5		13,227		13,226.67			−7,753	−5,169	**PINCH**
	1.5				305	458			5,474
3			−3,514	−3,513.72			−7,296	−2,432	
	0.44				−3,209	−1,412			1,960
3.44		14,656		14,655.75			−8,708	−2,531	
	2.61				11,447	29,877			16,616
6.05			−7,743	−7,743			21,169	3,499	
	0.09				3,704	333			8,873
6.14		1,051		1,051.19			21,502	3,502	
	0.06				4,755	285			9,924
6.2			−698	−698.25			21,788	3,514	
					4,057				9,226

the initial value of CO_2 emission for the total site with the 10 units was recorded as 28,934 kg/h. CO_2 emissions have now been reduced by 99.8%.

Level 3 of the CMH was not considered for this case study. Level 4, which is CO_2 sequestration, is evaluated next. Tree planting is considered the solution for CO_2 sequestration. The amount of minimum CO_2 emissions flow rate target, obtained after

Tab. 5.13: CO_2 Cascade Table (CCT) for Case Study 1 after source reduction.

1	2	3	4	5	6	7	8	9	10
Ratio, F_{OG}/F_{CO_2}	$\Delta(F_{OG}/F_{CO_2})$	Total flowrate CO_2 sources, ΣF_S, kg/h	Total flowrate CO_2 demands, ΣF_D, kg/h	$\Sigma F_S + \Sigma F_D$, kg/h	CO_2 flowrate, ΣF_C, kg/h	F_{OG}', kg/h	Other gases cascade, $F_{OG,cum}'$, kg/h	Pure CO_2 cascade, $F_{FC,cum}'$, kg/h	Feasible CO_2 cascade, F_C', kg/h
						0			5,169
0.9									
	0.6		−12,922	−12,922			0		
					−12,922	−7,753			−7,753
1.5		**13,227**		**13,227**			**−7,753**	**−5,169**	**PINCH**
	1.5				305	458			5,474
3			−3,514	−3,514			−7,296	−2,432	
	2.57				−3,209	−8,247			1,960
5.57		5,489		5,489			−15,543	−2,790	
	0.48				2,280	1,094			7,449
6.05			−7,743	−7,743			−14,448	−2,388	
	0.09				−5,463	−492			−294
6.14		1,051		1,051			14,940	2,433	
	0.06				4,412	−265			757
6.2			−698	−698			−15,205	−2,452	
					−5,110				59

Level 2, is 59 kg/h. This is also equivalent to 517 t CO_2/y. Assuming the industrial site will invest in the replanting of forest with a low sequestration rate of 55 t/ha, a total forest area of 9.4 ha is required to be planted. After implementing all the CMH, this industrial site has achieved the holistic minimum CO_2 targets with zero CO_2 emission.

5.3.5 Summarised network diagram

Fig. 5.4 illustrates the final network design for low CO_2 emissions industrial site planning, which comprises TSHI, CHP system, and hybrid power systems, finally achieving the holistic minimum CO_2 target, with the application of CMH options, including direct reuse, source reduction, and CO_2 sequestration.

5.4 Conclusion

This chapter has presented the framework for low CO_2 emissions industrial site planning using an integrated set of Pinch Analysis techniques, i.e. HPA, TSHI, CHP, PoPA, and CO_2 Emissions Pinch Analysis. A step-by-step guide, with a case study illustration, is presented for use by industrial site planners. Figure 5.5 summarises the detailed

	Framework for Low CO$_2$ Emission Industrial Site Planning
	Baseline Study

<table>
<tr>
<td rowspan="1">STAGE 1</td>
<td>

– Applicant must have incorporated optimal HI analysis design.
– Centralised steam and power generation.
– Request information from applicants on amount of resources (generated/consumed).
</td>
<td>

– Select correct industrial mix to be located in the industrial site.
– Select industrial units based on their capacity to become either source or sink.
</td>
</tr>
</table>

Tool 1: Targeting for total site heat recovery and cogeneration potential with TSHI

<table>
<tr>
<td rowspan="1">STAGE 2</td>
<td>

Targeting for total site heat recovery.

– Data extraction of units' heat sources and sinks.
– Generate TS-PTA.
– Evaluate amount of external heating utility needed.
– Evaluate amount of external cooling utility needed.
</td>
<td>

Targeting for cogeneration potential:

– Construct the SUGCC.
– Data extraction of sources and demands header.
– Generate algorithm for calculation of cogeneration potential.
– Evaluate cogeneration capacity.
– Generate enthalpy versus flowrate diagram.
– Identify type of cogeneration system.
</td>
</tr>
</table>

Tool 2: Targeting for hybrid power system with PoPA

<table>
<tr>
<td rowspan="1">STAGE 3</td>
<td>

– Analyse potential of RE harvested in the site area.
– Data extraction of power sources and demand.
– Generate Combined PCT and SCT.
– Evaluate additional power requirement for power plant capacity after established PoPA (assuming off-grid system).
</td>
<td>

"What if" scenario for construction of PCT and SCT
– PCT construction:
  – Off-grid system
    – Scenario 1
    – Scenario 2
  – Off-grid system
– SCT construction:
  – Scenario 1
  – Scenario 2
</td>
</tr>
</table>

Tool 3: Targeting for low CO emission with CPA 2

<table>
<tr>
<td rowspan="1">STAGE 4</td>
<td>

– Data extraction of CO$_2$ sources and demands.
– Generate CCT.
– Evaluate minimum fresh CO$_2$ and minimum CO emission targets.
</td>
<td>

– Implement CMH for further eduction of fresh CO$_2$ and CO$_2$ emission.
</td>
</tr>
</table>

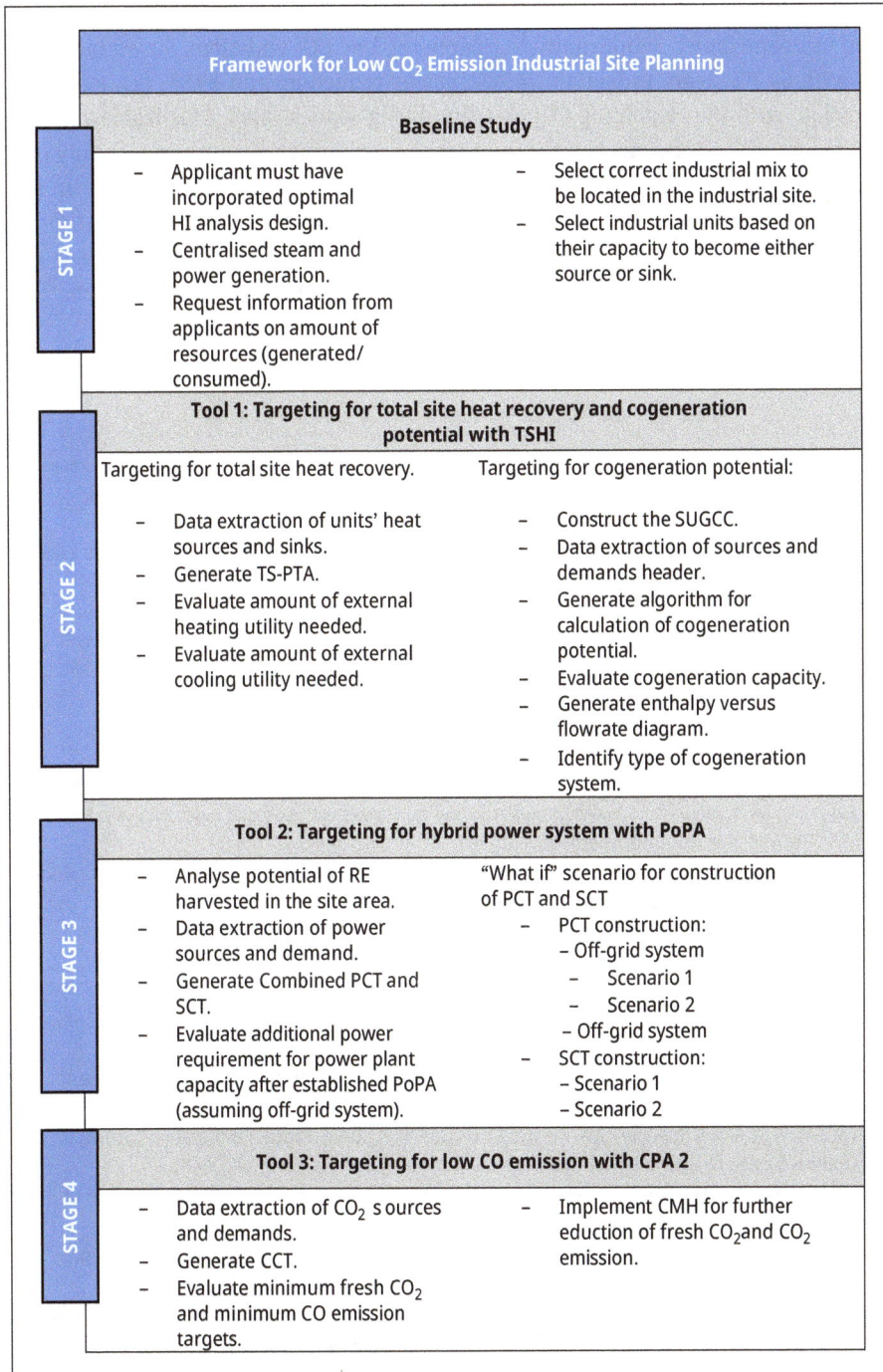

Fig. 5.5: Framework for low CO$_2$ emission industrial site planning.

procedure for the whole framework. Application to the case study shows the framework has contributed significantly to addressing low CO$_2$ emissions for an industrial site, with 64.7% steam reduction, 74.28% power reduction, and 99.8% CO$_2$ emissions reduction, with the remaining CO$_2$ emission being sequestrated. This framework not only reduces CO$_2$ emissions but also reduces the operating cost for overall utility purchase and capital cost, by designing a smaller centralised steam and power plant system. It also provides a holistic guide for a plant to achieve zero CO$_2$ emission by considering maximum energy and CO$_2$ recovery, RE system, increasing efficiency and minimising losses, and process changes, as well as CO$_2$ storage and sequestration.

References

Abdul Aziz E, Wan Alwi S R, Lim J S, Manan Z A, Klemeš J J. (2017). An Integrated Pinch Analysis Framework for Low Carbon Dioxide Emissions Industrial Site Planning, Journ of Cleaner Production, 146, 125–138.

Bandyopadhyay S, Varghese J, Bansal V. (2010). Targeting for Cogeneration Potential through Total Site Integration, Applied Thermal Engineering, 30, 6–14.

Engel-Cox J A, Nair N L, Ford J L. (2012). Evaluation of Solar and Meteorological Data Relevant to Solar Energy Technology Performance in Malaysia, Journal of Sustainable Energy and Environment, 3, 115–124.

Klemeš J J. (2022). Handbook of Process Integration (PI). Handbook of Process Integration (PI) Minimisation of Energy and Water Use, Waste and Emissions, 2nd ed. Woodhead Publishing / Elsevier, Cambridge, United Kingdom. ISBN 978-0-12-823850-9.

Liew P Y, Wan Alwi S R, Ho W S, Abdul Manan Z, Varbanov P S, Klemeš J J. (2018). Multi-period Energy Targeting for Total Site and Locally Integrated Energy Sectors with Cascade Pinch Analysis, Energy, 155, 370–380, doi:: 10.1016/j.energy.2018.04.184.

Liew P Y, Wan Alwi S R, Varbanov P S, Manan Z A, Klemeš J J. (2012). A Numerical Technique for Total Site Sensitivity Analysis, Applied Thermal Engineering, 40, 397–408.

Liu W H, Kaliappan K, Wan Alwi S R, Lim J S, Ho W S. (2016). Power Pinch Analysis Supply Side Management: Strategy on Purchasing and Selling of Electricity, Clean Technologies and Environmental Policy, 18, 2401–2418.

Manan Z A, Wan Alwi S R, Sadiq M M, Varbanov P S. (2014). Generic Carbon Cascade Analysis Technique for Carbon Emission Management, Applied Thermal Engineering, 70, 1141–1147.

Munir S M, Manan Z A, Wan Alwi S R. (2012). Holistic Carbon Planning for Industrial Parks: A Waste-to-Resources Process Integration Approach, Journal of Cleaner Production, 33, 74–85.

Rozali N E M, Wan Alwi S R, Manan Z A, Klemeš J J, Hassan M Y. (2013). Process Integration Techniques for Optimal Design of Hybrid Power Systems, Applied Thermal Engineering, 61, 26–35.

Tan R R, Aviso K B, Foo D C Y. (2018). Carbon Emissions Pinch Analysis of Economic Systems, Journ of Cleaner Production, 182, 863–871, doi: 10.1016/j.jclepro.2018.02.082.

Tan R R, Foo D C Y. (2007). Pinch Analysis Approach to Carbon-Constrained Energy Sector Planning, Energy, 32, 1422–1429, doi: 10.1016/j.energy.2006.09.018.

Varbanov P S, Fodor Z, Klemeš J J. (2012). Total Site Targeting with Process Specific Minimum Temperature Difference (ΔTmin), Energy, 44(1), 20–28. doi: 10.1016/j.energy.2011.12.025.

6 Introduction to Water Pinch Analysis

6.1 Water management and minimisation

The industrial sector is one of the major consumers of water. According to UNESCO (2021), industry consumes 19% of world water, while 69% is consumed by the agriculture sector and 12% by the domestic sector. Water is used for cooling and heating, cleaning, product formulation, and transportation, as well as a mass separating agent. Rising water prices and wastewater treatment costs, stricter environmental regulations for wastewater quality discharge, and shortage of clean water resources have spurred efforts by industry to employ effective means to reduce water footprint via water management and minimisation strategies that include:
(i)   Improving the efficiency of water usage;
(ii) Reducing water wastage;
(iii) Applying water-saving technologies;
(iv) Re-using, recycling, and regeneration (treatment) of wastewater;
(v)   Outsourcing of available water sources, e.g. rainwater, river, snow, or seawater.

The term "Water Footprint" was introduced by Hoekstra (2003) and discussed in connection with the other footprints by Čuček et al. (2012). The water footprint of a product is the volume of freshwater used to produce the product, measured over the full supply chain after considering the direct and indirect water uses. There are three types of water footprints:
1. Blue water footprint: Consumption of blue water resources (surface and groundwater) along the supply chain of a product;
2. Green water footprint: Consumption of green water resources (rainwater before it becomes run-off water);
3. Grey water footprint: Volume of freshwater that is required to absorb a load of pollutants, given the natural background concentrations and existing ambient water quality standards.

Table 6.1 shows the global average water footprints for several products.

Tab. 6.1: Global average water footprints for several products (Water Footprint, 2022).

Product	Global average water footprint, L/kg	Green	Blue	Grey
Butter	5,553	85%	8%	7%
Bioethanol (from sugar cane)	2,107	66%	27%	6%
Chocolate	17,196	98%	1%	1%
Rice	2,497	68%	20%	11%
Sugar (from sugar cane)	1,782	66%	27%	6%

https://doi.org/10.1515/9783110782981-006

6.2 History and definition of Water Pinch Analysis

Water re-use and recycling involve using wastewater as a water source for a process or an activity that does not require freshwater. This is possible if the mass load of the wastewater source is equal to or less than the mass load required by the water sink (i.e. the wastewater source is of the same quality or cleaner than the water sink). Water can also be re-used and recycled by mixing wastewater of different concentrations with freshwater to obtain the desired mass load as well as flowrate of the water sink.

The industry typically consists of many processes that use water and produce wastewater. This creates the need for a systematic method to analyse all the water streams and design the maximum water recovery network that utilises the minimum external freshwater and generates the minimum wastewater. Water Pinch Analysis (WPA) was developed in 1994 by Wang and Smith (1994a) to achieve the mentioned goal.

Initially adapted from the concept of mass exchange network synthesis that was established by El-Halwagi and Manousiouthakis (1989), the earlier emphasis of the WPA was on mass transfer-based water-using processes where water is mainly used as a mass separating agent (a lean stream) to remove impurities from a rich stream, e.g. in solvent extraction, gas absorption, and vessel cleaning, as shown in Fig. 6.1. This case is also widely referred to as the fixed load (water minimisation) problem. The main goal of WPA at that time was to target the minimum freshwater required to remove a fixed amount of impurities. This method assumes that there are no water losses or gains and that the amount of freshwater consumed is equal to the amount of wastewater generated.

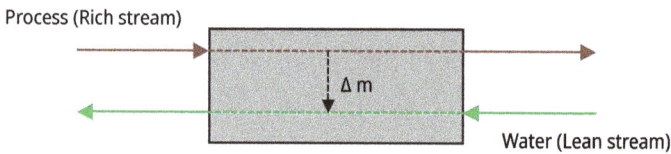

Process (Rich stream)

Δm

Water (Lean stream)

Fig. 6.1: Transfer of species from a rich-to-lean stream in a mass exchanger.

Dhole et al. (1996) later extended this method to a wider application that includes non-mass transfer-based processes. This case is widely known as the fixed flowrate (water minimisation) problem. Non-mass transfer-based processes include water consumed or generated from a reaction, water losses from evaporation, etc. In this new version of WPA, water sources and sinks are considered separate streams during data extraction, as shown in Fig. 6.2.

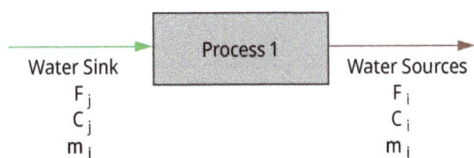

Fig. 6.2: Water source and demand.

WPA can be defined as a systematic technique of implementing water minimisation strategies through the integration of processes to achieve maximum water efficiency (Manan et al., 2004). The term "Pinch" refers to the limit or "bottlenecks" for water recovery. By using WPA, a designer can quickly determine the maximum amount of water recovery possible within a water system. WPA provides a target or "benchmark" for the industry to achieve a "step change", as opposed to incremental water reduction, by designing an optimal water recovery network that implements various water recovery strategies. In a water minimisation project, industry personnel will typically ask the following questions.

– *What could you possibly do (that we have not done)?*
– *How much water could be further saved?*
– *How high are the monetary savings?*
– *How much investment is needed?*
– *Is it going to be worthwhile?*
– *How could the savings be realised?*

By using WPA, all these questions can be answered. The following chapters highlight how WPA answers all these questions.

6.3 Applications of Water Pinch Analysis

WPA has been proven to give beneficial water savings. The potential savings from the WPA application have been widely reported. Tab. 6.2 lists some of the industrial WPA applications reported in the literature.

Tab. 6.2 shows that the range of savings achieved for the reference cases is between 7% to 72% freshwater reduction. The amount of savings depends on the extent of water re-use in the industry as well as on the quantity and quality of water sources and sinks available. The payback periods reported were within two years, which is indeed attractive for most industries. WPA is highly recommended for water-intensive industries, especially in countries where water tariffs and wastewater treatment costs are high.

Tab. 6.2: Applications of Water Pinch Analysis.

Company	Process/industry	Location	Flow reduction, %	Reference
Confidential	Chemical and Fibres	Germany	25	Tainsh and Rudman (1999)
Cerestar	Corn Processing	UK	25	
Gulf Oil	Oil Refining	UK	30	
Monsanto	Chemicals	UK	40	
Parenco	Paper Mill	Netherlands	20	
Sasol	Coal Chemicals	South Africa	50	
Unilever	Polymers (batch)	UK	60	
US Air Force	Military Base	USA	40	
Confidential	Oil Refining	Netherlands	40	
Confidential	Chemicals	USA	40	
Confidential	Chemicals and fibers	USA	25	
Confidential	Semiconductors	Malaysia	72	Wan Alwi and Manan (2008)
Confidential	Chloralkali	Malaysia	7	Handani et al. (2010)
Confidential	Paper Mill	Malaysia	13	Manan et al. (2007)
Confidential	Beet Sugar	Slovenia	69	Zbontar Zver and Glavič (2005)
Confidential	Citrus	Argentina	30	Thevendiraraj et al. (2003)
Confidential	Brick	England	56.4	Skouteris et al. (2018)
Regional	Water scarcity	Czech Republic	Up to 100	Jia et al. (2020)

6.4 Water Pinch Analysis steps

WPA involves five key steps as follows.

Step 1: Analysis of the water network
The existing or the base case water network is analysed during a plant audit. All the water-using operations are identified, and water balances are developed. The scope of analysis is also defined.

Step 2: Data extraction
Water sources and water sinks having potential for re-use and recycling are identified. The limiting water flowrate and limiting concentration data are extracted.

Step 3: Setting the minimum utility targets
The minimum freshwater requirement and wastewater generation, or the maximum water recovery, targets are established using either graphical or numerical targeting methods. This is further elaborated in Chapter 7.

Step 4: Water network design/retrofit
Design a water recovery network to achieve the minimum water targets. This is further elaborated in Chapter 8.

Step 5: Economics and technical evaluations

Water Pinch Analysis (WPA) has been applied for grassroots and retrofit cases. In the case of retrofit, the repiping cost may be needed. Retrofitting may also cause process disruption. All these factors must be considered before undertaking a WPA project. The final step of WPA involves evaluating the economics and the technical constraints of the water network. Mathematical modelling techniques have been widely used to calculate the network cost by considering the repiping needs, the geographical layout and constraints, and the associated fixed as well as variable costs. For preliminary economic calculations, the payback period is widely used as a criterion to assess the feasibility of a proposed network solution. The payback period is calculated using Equation (6.1). The net capital investment is the investment needed for pipe rerouting, pumping, and storage tanks for intermittent processes and control systems. The net annual savings are the potential savings from freshwater reduction:

$$\text{Payback period (y)} = \frac{\text{Net Capital Investment } t(\$)}{\text{Net Annual Savings } (\$/\text{y})} \tag{6.1}$$

Other than the payback period, net present value (NPV) as well as Return on Investment (ROI) are also widely used. In order to evaluate the technical feasibility, the amount of water saving, the technology risk, the implementation period, and the impact on normal operation need to be evaluated.

6.5 Analysis of water networks and data extraction

Industrial implementation of WPA begins with the analysis of the water network and water stream data extraction. These steps, which are the critical path of a WPA project, are described in detail next.

6.5.1 Analysis of water networks

The first step in the implementation of WPA involves analysing the existing water network of a facility by developing an overall water mass balance. This includes identifying all areas to where the freshwater is being distributed, the processes, and the process from which wastewater is being generated. Process Flow Diagrams (PFDs), Process Instrumentation Diagrams (P&IDs), or the DCS system will help, but in most cases, not all water-using processes are recorded in these diagrams. However, line tracing still needs to be performed.

Water is used for various uses in industry, such as cooling and heating, transportation agent, cleaning, product formulation, mass separating agent, general plant services, and potable/sanitary. There are two important parameters that need to be determined for WPA, i.e. flowrate and contaminant concentration of each of the water streams.

6.5.1.1 Water flowrate determination

Extracting water flowrate data is often not so straightforward because the data for water inlets and outlets are not normally monitored or recorded in a plant. Freshwater is taken from the Water Mains and, thereafter, distributed throughout a plant. There are typically no devices installed to measure the flowrate of the distributed water sources. Wastewater from various processes is typically collected in a common drain, and the quantity of the mixed wastewater is only known at the wastewater treatment plant. Mixed wastewater is no longer useful for WPA implementation because the quality of this wastewater is at its lowest, as it has already been degraded as a result of being mixed with all sorts of spent water. For WPA implementation, the flowrate of the segregated individual wastewater streams generated from each process or utility stream needs to be extracted.

As a start, water balance data can be obtained from the existing process material balances, computer monitoring, routine measurements, previous plant studies, and laboratory reports (Liu et al., 2004). If the water data are still unavailable, a plant owner can decide either to install a fixed flow meter or use a portable ultrasonic flow meter to determine the missing flowrates of the water sources and sinks. However, the limitation of using an ultrasonic flow meter is that the pipe coating and insulation may need to be scrapped (this is not preferred by most plant owners). In addition, the readings may not be accurate if a pipe is not full or when there is the possibility of fouling in the pipe. Knowledgeable estimates can be made as a last resort. For water data that fluctuates according to the users' pattern of water usage, e.g. for potable water usage such as toilet flushing, water surveys can also be conducted to estimate the average or the range of water consumption.

Stream data should be extracted during normal operations, and not during plant shutdowns, start ups, or upsets. Water balance should be conducted to account for all water flows. As a general rule, the aim is to complete at least 80% of the site's water balance. The priority is to select streams that have large flowrates and/or contribute significantly to the contaminant loading. Streams that have been verified to have relatively much smaller flowrates can be ignored.

6.5.1.2 Contaminant concentration determination

Contaminant concentration data are also not easily available in a plant. For water sinks (or water streams entering a processing unit), the contaminant concentration limits can be extracted from equipment specification sheets, engineers' estimates, or literature reviews. Other sources of constraints include (Foo et al., 2009):

(i) The manufacturer's design data;
(ii) Physical limitations such as weeping flowrate, flooding flowrate, channelling flowrate, saturation composition, minimum mass transfer driving force, and maximum solubility;

(iii) Technical constraints such as avoiding scaling, precipitation, corrosion, explosion, contaminant build-up, etc.

As an example, according to a maintenance manual, the water quality standard for a cooling tower should be maintained at a pH between 7.2 to 8.5, total hardness less than 200 ppm, total alkalinity less than 100 ppm, chloride ion less than 200 ppm, total ion (Fe) less than 1.0 ppm, silica less than 50 ppm, ammonium ion less than 1.0 ppm, and Total Dissolved Solids (TDS) less than 1,500 ppm. These data can be taken as the contaminant limits for the cooling tower water sink.

For water sources (or wastewater leaving a process), water samples can be taken from a sampling point, and the contaminant concentration can be determined by a lab test or by using a water-quality test kit. Several water samples should be taken from each water source in order to determine the reliable upper and lower bounds of the contaminant concentration. The maximum outlet concentration for water sources may also need to be adjusted to take into account the mass pickup. For example, if contaminants are dissolving, then an increase in the inlet concentration would also be reflected at the outlet concentration to balance the system.

6.5.2 Data extraction

In a water-intensive process plant, specifying the limiting data can be a very tricky, arduous, and time-consuming task. This is typically the bottleneck and, more importantly, the critical success factor for a water minimisation project. Once the water balances have been developed, the next step is to select the potential candidates for water re-use, i.e. to extract the limiting water data. Processes chosen are preferably located close together, and chemically related. This can potentially reduce the piping as well as pumping cost and allow the use of a common water regeneration unit. Almost similar chemical properties will also facilitate water integration.

The system can be modelled as a single contaminant or multiple contaminant problem, based on the water quality requirement of a process plant. However, problems involving multiple contaminants necessitate a complex modelling procedure and have to rely on mathematical modelling and optimisation techniques. Hence, the typical simplifying assumption is the use of aggregated contaminants, such as suspended solids and Total Dissolved Solids (TDS) that allow the consideration of multiple quality factors to be modelled as a single contaminants system (The Institution of Chemical Engineers, 2000). The aggregated contaminants, modelled as a single contaminant system, are known as a pseudo-single contaminant system.

The primary contaminants that prevent direct re-use in the water system are then chosen. As an example, for a semiconductor plant, conductivity is the most important contaminant, which is monitored throughout the system, and is selected for WPA implementation, even though there are also other contaminants in the water system. The proposed network, however, still needs to be reassessed by checking that all other contaminant concentrations not considered are still within allowable limits, before implementing WPA.

The water sources' and sinks' flowrates and quality requirements for each of the selected water-using processes are then extracted. Most processes normally operate between the upper and lower bounds of their process parameters to maintain product quality while preventing the equipment from overloading. For operability reasons, the maximum concentration limit and the minimum flowrate limit are extracted to ensure that the worst-case scenario is considered when designing the water re-use network.

Once the limiting water flowrate and contaminant concentration have been extracted, the contaminant mass load of the stream can be calculated. The mass load for each water-using process can be calculated by using Equation (6.2).

$$m = \frac{F \times C}{1000} \tag{6.2}$$

Where
m = contaminant mass load, kg/h,
F = flowrate, t/h,
C = contaminant concentration, ppm,
C_{in} = inlet contaminant concentration, ppm, and
C_{out} = outlet contaminant concentration, ppm.

6.5.3 Example

6.5.3.1 Example 6.1: Speciality chemical plant

Figure 6.3(a, b) show the utility and process flowsheets of the speciality chemical plant case study taken from Wang and Smith (1995). Suspended solids were chosen as the main contaminant of the system. The original water balance data extracted from the flowsheet is shown in Tab. 6.3.

All the existing inlet concentrations are initially zero, meaning that freshwater is used throughout all existing systems; see Tab. 6.3. However, this constraint needs to be re-evaluated as it does not allow for any water re-use. The maximum inlet concentrations were re-evaluated to include the appropriate re-use constraints, such as those based on solubility limits, corrosion limits, and contamination of products. The maximum outlet concentrations were also adjusted to take into account the mass pick-up due to the increase in the maximum inlet concentration. For example, if contaminants are dissolving, then an increase in the inlet concentration will also be reflected at the outlet concentration to balance the system. Table 6.4 shows the adjusted maximum inlet and outlet concentrations.

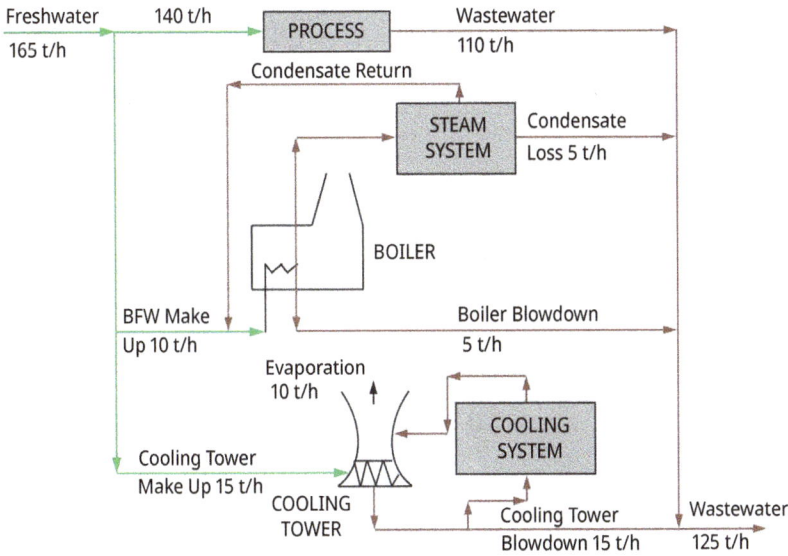

(a) Utility flowsheet

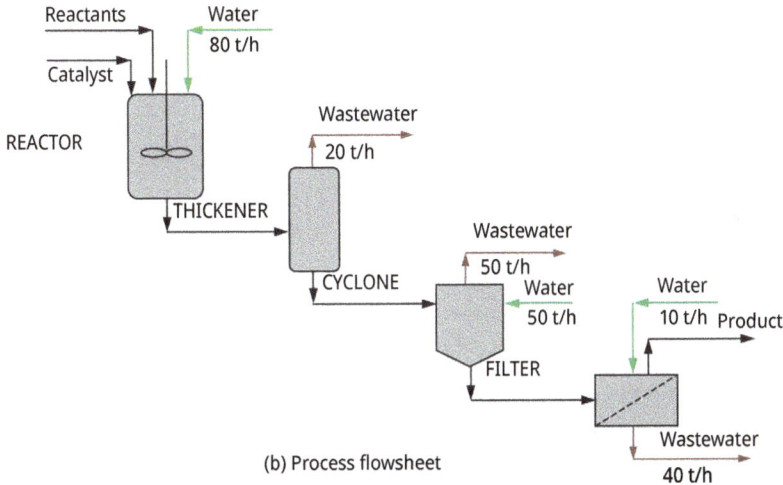

(b) Process flowsheet

Fig. 6.3: Example 6.1: Speciality chemical plant.

Note that the water inlet and outlet are not the same for the reactor (water loss), filtration (water gain), and cooling system (water loss). The outlet and inlet flowrates are extracted as the source and sink data, as shown in Tab. 6.5. The mass load can be calculated by using Equation (6.2), as shown in the last column of Tab. 6.5.

Tab. 6.3: Water balance for Example 6.1.

Operation	F_{in} (t/h)	F_{out} (t/h)	C_{in} (ppm)	C_{out} (ppm)
Reactor	80	20	0	900
Cyclone	50	50	0	500
Filtration	10	40	0	100
Steam System	10	10	0	10
Cooling System	15	5	0	90

Tab. 6.4: Modified stream data for Example 6.1.

Operation	F_{in} (t/h)	F_{out} (t/h)	C_{in} (ppm)	C_{out} (ppm)
Reactor	80	20	100	1,000
Cyclone	50	50	200	700
Filtration	10	40	0	100
Steam System	10	10	0	10
Cooling System	15	5	10	1,00

Tab. 6.5: Source and Sink data extraction.

Operation	Sink	F (t/h)	C (ppm)	m (kg/h)	Source	F (t/h)	C (ppm)	m (kg/h)
Reactor	SK1	80	100	8	SR1	20	1,000	20
Cyclone	SK2	50	200	10	SR2	50	700	35
Filtration	SK3	10	0	0	SR3	40	100	4
Steam System	SK4	10	0	0	SR4	10	10	0.1
Cooling System	SK5	15	10	0.15	SR5	5	100	0.5

6.5.3.2 Example 6.2: Acrylonitrile (AN) production plant

Figure 6.4 shows an Acrylonitrile (AN) production plant flowsheet from El-Halwagi (1997), with its water balances. Ammonia (NH_3) was identified as the main contaminant in the water system that limits water re-use. The plant wished to expand its AN production. However, their biotreatment facility was already operating at full capacity. The plant had to either reduce its wastewater discharge or build an additional wastewater treatment plant. The plant decided to implement WPA in order to reduce wastewater discharge and avoid building a new treatment plant.

From the flowsheet, two water sinks can be observed. The first is the scrubber, which uses freshwater to remove the reactor's remaining off-gas. The second sink is the Boiler Feed Water (BFW), which is used in the boiler to generate steam for the steam jet ejector. Four streams are identified as water sources. These include wastewater generated from the off-gas condensate, an aqueous layer of the decanter, the

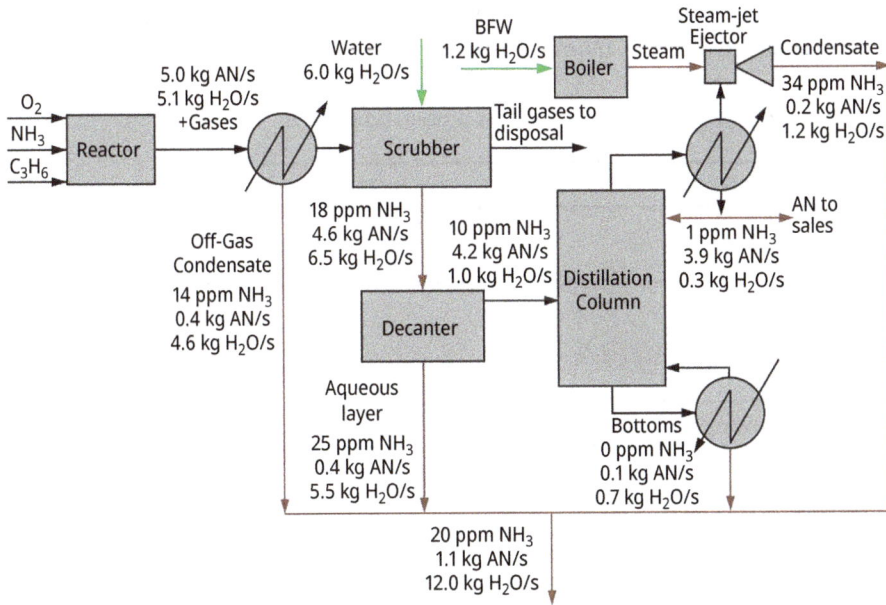

Fig. 6.4: Flowsheet of AN production for Example 6.2.

bottom product of the distillation column, and condensate from the steam ejector. Note that data for the individual wastewater streams were extracted, as opposed to the mixed stream going to the wastewater treatment plant.

Note that water that is further used in subsequent processes, for example, between the decanter and the distillation column, is not extracted as a water source or sink, as it does not require freshwater and does not generate wastewater. Water coming from the distillation top product, which is part of the product to be sold, is also not considered a water source, as it is not sent to the wastewater treatment plant.

The following are the technical constraints for the water sinks.
1. Scrubber
   - $5.8 \leq$ flowrate of wash feed (kg/s) ≤ 6.2
   - $0.0 \leq NH_3$ content of wash feed (ppm) ≤ 10.0
2. Boiler feed water
   - NH_3 content = 0.0 ppm
   - AN content = 0.0 ppm

For the scrubber, upper and lower bounds were given for both the flowrate and contaminant concentration. As explained in the previous section, the minimum flowrate and maximum contaminant concentration are extracted as the limiting water data, i.e. 5.8 kg/s and 10 ppm. Tabs. 6.6 and 6.7 show the extracted limiting water sink and source data for the AN plant. The mass load of each water source and water sink is calculated by using Equation (6.2) and shown in the last column.

Tab. 6.6: Limiting water sinks data for AN production plant (Example 6.2).

Sink	Description	F, kg/s	C, ppm	m, mg/s
SK1	Scrubber	5.8	10	58
SK2	Boiler	1.2	0	0

Tab. 6.7: Limiting water sources data for AN production plant (Example 6.2).

Source	Description	F, kg/s	C, ppm	m, mg/s
SR1	Distillation bottoms	0.8	0	0
SR2	Off-gas condensate	5.0	14	70.0
SR3	Aqueous layer	5.9	25	147.5
SR4	Ejector condensate	1.4	34	47.6

6.6 Summary

This chapter introduced the basics of Water Pinch Analysis. The methodology, even when it was introduced, based on a single-contaminant analysis, made a significant impact in the industry, as shown at the beginning of the chapter. Moreover, while the single-contaminant limitation was preventing Water Pinch Analysis from a proper integration with Mathematical Programming-based methods of water network optimisation and water saving, there have been significant developments in the last few years by Chin et al. (2021b), who introduced multi-contaminant water targeting, Water Mains targeting (Chin et al., 2021a), and other key tools. These methodology extensions are discussed in detail in Chapter 10.

References

Chin H H, Jia X, Varbanov P S, Klemeš J J, Liu Z-Y. (2021). Internal and Total Site Water Network Design with Water Mains Using Pinch-Based and Optimization Approaches, ACS Sustainable Chemistry & Engineering, 9, 6639–6658. doi: 10.1021/acssuschemeng.1c00183.

Chin H H, Varbanov P S, Liew P Y, Klemeš J J. (2021b). Pinch-based Targeting Methodology for Multi-contaminant Material Recycle/Reuse, Chemical Engineering Science, 230, 116129. doi: 10.1016/j.ces.2020.116129.

Čuček L, Klemeš J J, Kravanja Z. (2012). A Review of Footprint Analysis Tools for Monitoring Impacts on Sustainability, Journal of Cleaner Production, 34, 9–20.

Dhole V R, Ramchandani N, Tainsh R A, Wasilewski M. (1996). Make Your Process Water Pay for Itself, Chemical Engineering, 103, 100–103.

El-Halwagi M M, Manousiouthakis V. (1989). Synthesis of Mass-Exchange Networks, AIChE Journal, 35(8), 1233–1244. doi: 10.1002/aic.690350802.

El-Halwagi M M. (1997). Pollution Prevention Through Process Integration: Systematic Design Tools. Academic Press, San Diego, California, USA.

Foo D C Y. (2009). State-of-the-art Review of Pinch Analysis Techniques for Water Network Synthesis, Industrial & Engineering Chemistry Research, 48(11), 5125–5159.

Handani Z B, Wan Alwi S R, Hashim H, Manan Z A. (2010). A Holistic Approach for Design of Minimum Water Networks Using Mixed Integer Linear Programming (MILP) Technique, Industrial & Engineering Chemistry Research, 49, 5742–5751.

Hoekstra A Y. (ed.) (2003). Virtual Water Trade: Proceedings of the International Expert Meeting on Virtual Water Trade, Value of Water Research Report Series No 12, UNESCO-IHE, Delft, Netherlands, www.waterfootprint.org/Reports/Report12.pdf, accessed 16/10/2022.

Jia X, Klemeš J J, Alwi S R W, Varbanov P S. (2020). Regional Water Resources Assessment Using Water Scarcity Pinch Analysis, Resources, Conservation and Recycling, 157, 104749. https://doi.org/10.1016/j.resconrec.2020.104749.

Liu Y A, Lucas B, Mann J. (2004). Up-to-date Tools for Water-system Optimisation, Chemical Engineering Magazine, 111(1), 30–41.

Manan Z A, Tan Y L, Foo D C Y, Tea S Y. (2007). Application of Water Cascade Analysis Technique for Water Minimisation in a Paper Mill Plant, International Journal of Pollution Prevention, 29(1–3), 90–103.

Manan Z A, Tan Y L, Foo D C Y. (2004). Targeting the Minimum Water Flowrate Using Water Cascade Analysis Technique, AIChE Journal, 50(12), 3169–3183. doi: 10.1002/aic.10235.

Noureldin M B, El-Halwagi M M. (1999). Interval-based Targeting for Pollution Prevention via Mass Integration, Computers & Chemical Engineering, 23, 1527–1543.

Skouteris G, Ouki S, Foo D, Saroj D, Altini M, Melidis P, Cowley B, Ells G, Palmer S, O'Dell S. (2018). Water Footprint and Water Pinch Analysis Techniques for Sustainable Water Management in the Brick-manufacturing Industry, Journal of Cleaner Production, 172, 786–794.

Tainsh R A, Rudman A R. (1999). Practical Techniques and Methods to Develop an Efficient Water Management Strategy, Linnhoff March International, Paper Presented at: IQPC Conference "Water Recycling and Effluent Re-Use". In: KBC. https://www.environmental-expert.com/articles/practical-techniques-and-methods-to-develop-an-efficient-water-management-strategy-2956, accessed on 16 October 2022.

The Institution of Chemical Engineers. (2000). Guide to Industrial Water Conservation. Part 1. Water Re-use. Single Contaminant Systems, ESDU 0020, Rugby, UK.

Thevendiraraj S, Klemeš J, Paz D, Aso G, Cardenas J. (2003). Water and Wastewater Minimisation Study of a Citrus Plant, Resources, Conservation and Recycling, 37, 227–250.

UNESCO. (2021). The United Nations World Water Development Report 2021: Valuing Water. World Water Assessment Programme.

Wan Alwi S R, Manan Z A. (2008). A Holistic Framework for Design of Cost-effective Minimum Water Utilisation Network, Journal of Environmental Management, 88, 219–252. doi: 10.1016/j.jenvman.2007.02.011.

Wang Y P, Smith R. (1994b). Design of Distributed Effluent Treatment Systems, Chemical Engineering Science, 49, 3127–3145. doi: 10.1016/0009-2509(94)E0126-B.

Wang Y P, Smith R. (1994a). Wastewater Minimisation, Chemical Engineering Science, 49(7), 981–1006. doi: 10.1016/0009-2509(94)80006-5.

Wang Y P, Smith R. (1995). Wastewater Minimisation with Flowrate Constraints, Chemical Engineering Research and Design, 24, 2093–2113.

Water Footprint, Product Gallery. https://waterfootprint.org/en/resources/interactive-tools/product-gallery/, accessed on 16 Oct 2021.

Zbontar Zver L, Glavic P. (2005). Water Minimisation in Process Industries: Case Study in Beet Sugar Plant, Resources, Conservation and Recycling, 43(2), 133–145.

7 Setting the Maximum Water Recovery targets

7.1 Introduction

After completing the limiting water data extraction, as explained in Chapter 6, the Maximum Water Recovery (MWR) targets can now be determined.

Techniques for MWR targeting include the following.

1. Limiting Composite Curves (Fig. 7.1) by Wang and Smith (1994): Water-using processes are plotted on a concentration versus flowrate diagram. This method is applicable for both fixed flowrate and fixed load problems. However, for fixed flowrate problems, it involves the pairing of sources and sinks to include the water losses and water gains.

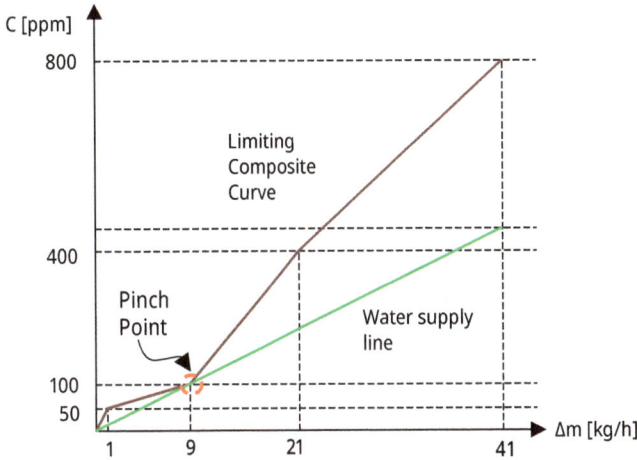

Fig. 7.1: Limiting Composite Curve.

2. Water Surplus Diagram (Fig. 7.2) by Hallale (2002): Water sources and sinks are plotted separately and plotted on a concentration vs flowrate diagram to determine the water surplus and deficit. The water surplus and deficit are then summed to form the Water Surplus Diagram (WSD). The freshwater targets are then predicted by using a trial and error method until all the WSD lie on the right-hand side of the y-axis. This method is applicable for both fixed flowrate and fixed load problems.

https://doi.org/10.1515/9783110782981-007

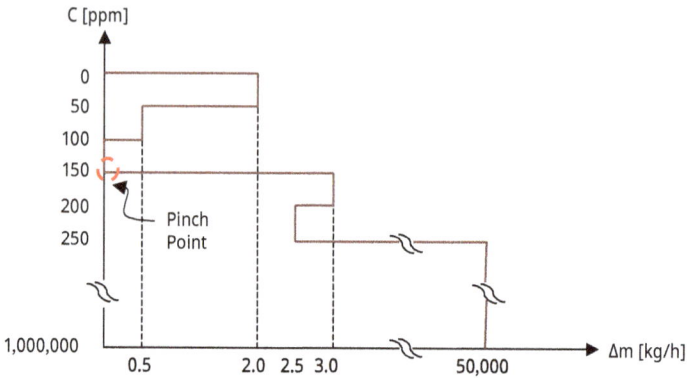

Fig. 7.2: Water Surplus Diagram.

3. Source/Sink Composite Curves (Fig. 7.3) by El-Halwagi et al. (2003) and little later also Prakash and Shenoy (2005): Water sources and sinks are plotted on a contaminant mass load vs flowrate. It was first introduced by El-Halwagi et al. (2003) for material recycle/reuse. However, later Prakash and Shenoy (2005) introduced an almost similar method for water systems. The Source/Sink Composite Curves overcome the limitation of WSD, which requires two graphs and an iterative process. This method is applicable for both fixed flowrate and fixed load problems.

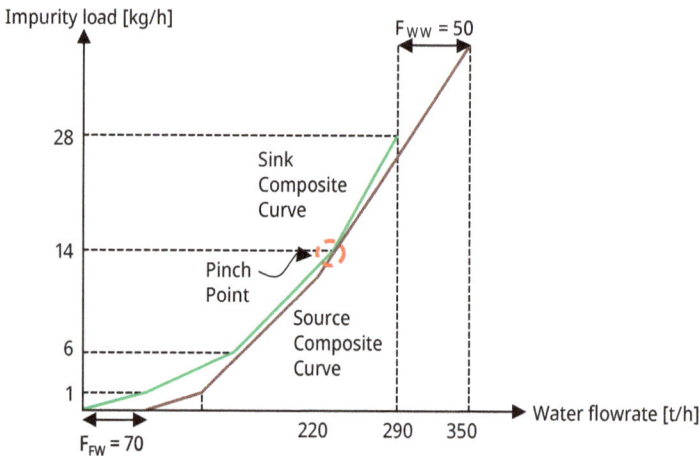

Fig. 7.3: Source/Sink Composite Curve.

4. Water Cascade Analysis (WCA) (Tab. 7.2) by Manan et al. (2004) and later Foo et al. (2006): The original WCA that was based on purity intervals (Manan et al., 2004) was a numerical version of WSD by Hallale (2002). WCA eliminates the iterative process of WSD

and provides more accurate results. Foo et al. (2006) later used concentration intervals in the WCA technique. Since the WCA is based on algebraic calculations, its steps and formulas can be readily programmed into Microsoft Excel. This allows the tabulated values of WCA to be easily duplicated to other rows by using the Excel formula drag function.

5. Algebraic Targeting Approach (Fig. 7.4) by Al-Mutlaq et al. (2005): A numerical version of Source/Sink Composite Curves. It uses the Load-Interval Diagram.

k	C_k	$\Sigma_i F_i - \Sigma_j F_j$	Cum $F_{C,k}$	Δm_k	Cum Δm_k	$F_{WW,k}$
k	C_k	$[\Sigma_i F_i - \Sigma_j F_j]_k$		Δm_k		
			$[\Sigma_i F_i - \Sigma_j F_j]_k$			
k+1	C_{k+1}	$[\Sigma_i F_i - \Sigma_j F_j]_{k+1}$		Δm_{k+1}	Cum Δm_{k+1}	$F_{WW,k+1}$
			Cum $[\Sigma_i F_i - \Sigma_j F_j]_{k+1}$			
⋮	⋮	⋮	⋮	⋮		
					Cum Δm_{n-1}	$F_{WW,n-1}$
n-1	C_{n-1}	$[\Sigma_i F_i - \Sigma_j F_j]_{n-1}$	Cum $[\Sigma_i F_i - \Sigma_j F_j]_{n-1}$	Δm_{n-1}		
					Cum Δm_n	$F_{WW,n}$
n	C_n	$[\Sigma_i F_i - \Sigma_j F_j]_n$				

Fig. 7.4: Algebraic steps for Water Source Diagram.

6. Source Composite Curves (Fig. 7.5) by Bandyopadhyay (2006): A hybrid between numerical and graphical approaches. It uses an almost similar cascading approach as WCA but only requires single cascading instead of double cascading. The results of the numerical step are then used to plot Source Composite Curves, which is a plot of concentration versus mass load. The plot consists of a Source Composite Curve and a wastewater line. The advantage of this method is that it can predict the average outlet wastewater concentration.

All these MWR methods apply to continuous processes. For batch processes, these methods can also be used by employing the use of storage tanks. Since water sources may be generated at different times compared to when water is needed at the water sinks, a storage tank can be used to store the water source until it is needed at the water sinks. The use of storage tanks allows WPA to be applied to continuous or batch processes. However, there are also methods to target the Maximum Water Recovery for batch processes without the use of storage tanks (direct reuse). For more details on batch Maximum Water Recovery targeting, readers can refer to the work of Foo et al. (2005) who presented a time-dependent Water Cascade Analysis (WCA) technique and Majozi et al. (2006), who presented a graphical technique.

Section 7.2 provides detailed descriptions of the graphical Source/Sink Composite Curves (SSCC) and the algebraic Water Cascade Analysis (WCA) techniques that have been widely used to determine the Maximum Water Recovery target for single pure

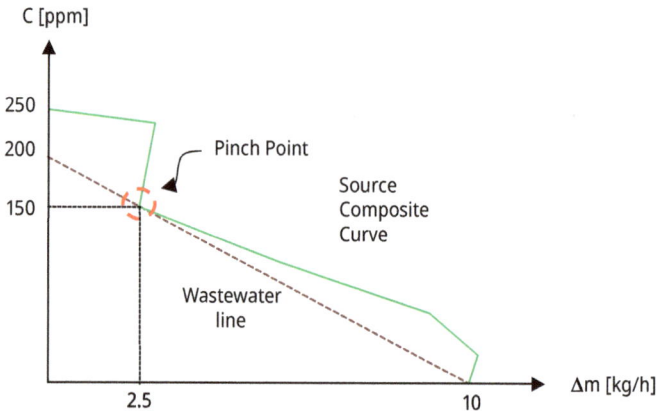

Fig. 7.5: Source Composite Curve.

freshwater as the available utility. The WCA technique is easier to construct and gives more precise and accurate results due to its numerical nature. Its steps and formulas can be readily programmed into Microsoft Excel. This allows the tabulated values from the WCA to be easily duplicated to other rows by using the Excel formula drag function. On the other hand, the SSCC provides useful visualisation insights for the engineers to understand the proposed solution for them to influence the design.

For simplification, in most WPA applications, freshwater is typically assumed as the only water utility available at zero contaminant concentration, even though freshwater may contain some small amount of contaminants in practice. Furthermore, various freshwater sources such as demineralised, deionised, and potable water may also be available as utilities. Apart from that, "outsourced water", i.e. water from the environment, includes rainwater, snow, borehole water, river water, and even "imported" spent water, may also be available for use in the plant area. In Section 7.3, the use of SSCC to determine the water targets for cases involving impure freshwater as well as multiple water sources, are described.

7.2 Maximum Water Recovery target for single pure freshwater

7.2.1 Water Cascade Analysis technique

Water Cascade Analysis (WCA) (Manan et al., 2004) is an algebraic or numerical targeting method that is used to determine the minimum water targets, i.e. the overall freshwater requirement and wastewater generation for a process after looking at the possibility of using the available water sources within a process to satisfy the water sinks. To achieve this objective, the net water flowrate, water surplus and deficit at the different water concentration levels within the process under study have to be established. The WCA was initially developed by using purity levels as the water quality measure instead of concentrations. Foo et al. (2006) later simplified the WCA method by using the concentration levels.

Figure 7.6 is a conceptual illustration of how water cascading can minimise freshwater needs and wastewater generation. In Fig. 7.6a, 100 kg/s of wastewater is produced by Operation 1 water source at the concentration level of 100 ppm and 50 kg/s of water is needed by Operation 2 water sink at the concentration level of 200 ppm. Without considering water reuse, 100 kg/s of wastewater would be generated, while 50 kg/s of freshwater would be required. However, as shown in Fig. 7.6b, by making use of 100 kg/s of the water source at the concentration level of 100 ppm from Operation 1 to satisfy the water sink of 50 kg/s at the concentration level of 200 ppm from Operation 2, it is possible to avoid sending part of the water source directly to the effluent. Doing so not only reduces the wastewater generation but also the freshwater consumption, in both cases, by 50 kg/s.

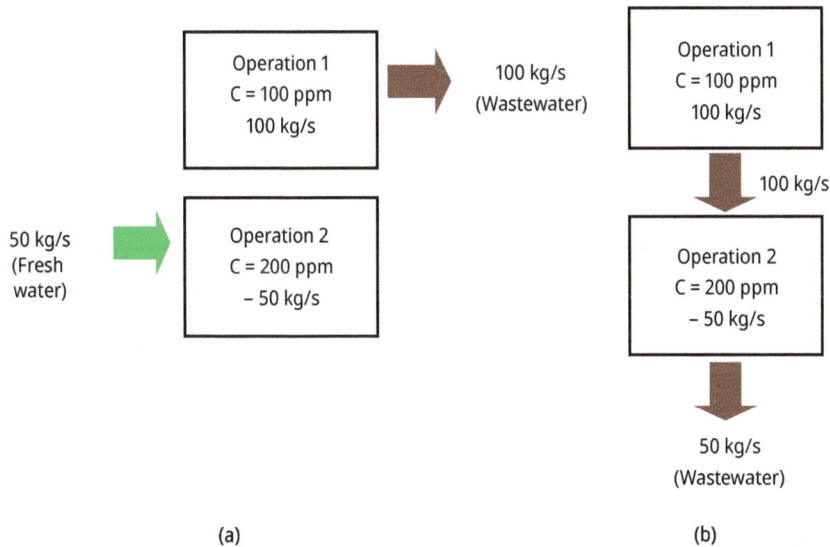

Fig. 7.6: The principle of water cascading.

Example 7.1 describes the construction of the Water Cascade Table (WCT) that is represented by Tab. 7.1.

Tab. 7.1: Limiting water data for Example 7.1 (Polley and Polley, 2000).

D_j	F_j, t/h	C, ppm	m, kg/h
1	50	20	1
2	100	50	5
3	80	100	8
4	70	200	14
S_i	F_i, t/h	C, ppm	m, kg/h
1	50	50	2.5
2	100	100	10
3	70	150	10.5
4	60	250	15

To construct the WCT, the contaminant concentrations (C) of the water streams are listed in ascending order (see Column 1 of Tab. 7.2). Duplicate concentration should be listed only once. Column 2 lists the concentration difference (ΔC) computed using eq. (7.1):

$$\Delta C = C_{n+1} - C_n. \tag{7.1}$$

Tab. 7.2: Water Cascade Table for Example 6.1.

C_k, ppm	ΔC_k, ppm	$\sum F_{SKi}$, t/h	$\sum F_{SRj}$, t/h	$\sum F_{SKi}$ $+ \sum F_{SRj}$, t/h	F_C, t/h	Δm, kg/h	Cum. Δm, kg/h	$F_{FW, cum}$, t/h	F_C, t/h
									$F_{FW} = 70$
0				0			0		
	20				0	0			70
20		−50		−50			0	0	
	30				−50	−1.5			20
50		−100	50	−50			−1.5	−30	
	50				−100	−5.0			−30
100		−80	100	20			−6	−65	
	50				−80	−4.0			−10
150			70	70			−10	**−70**	(PINCH)
	50				−10	−0.5			60
200		−70		−70			−11	−55	
	50				−80	−4.0			−10
250			60	60			−15	−60	
					−20	0			$F_{WW} = 50$
							−15		

The water sinks (F_{SK}) and sources (F_{SR}) flowrates in Columns 3 and 4 are added at each concentration level. For simplicity, the word sum be designated as Σ next. Note that water sinks are assigned with negative values while the water sources are positive. Water sinks and sources are summed in Column 5 ($\Sigma F_{SKi} + \Sigma F_{SRj}$) at each concentration level. A positive value in this column indicates a net surplus of water present at the respective concentration level, while a negative value indicates a net deficit of water. Any water sources at a lower concentration can be used as a source for water sinks at a higher concentration.

In Column 6, a zero freshwater flowrate is first assumed. This freshwater flowrate is then cascaded with Column 5 to give the cumulative flowrate (F_C) for each concentration level. The first row in this column represents the estimated flowrate of freshwater required for the water-using processes (F_{FW}). The total cumulative water flowrate value in the final column represents the total wastewater generated in the process (F_{WW}). However, this is the preliminary infeasible cascade and not the true freshwater and wastewater targets.

In Column 7, the product of the cumulative flowrate and concentration difference ($F_C \times \Delta C$) is calculated at every concentration level to give the mass load (Δm). The mass load is then cumulated down each concentration level ($Cum\ \Delta m$) in Column 8. An interval freshwater flowrate ($F_{FW,K}$) is then determined by using eq. (7.2), where C_{FW} is the freshwater concentration (Column 9). For this Example 6.1, C_{FW} of 0 ppm is assumed:

$$F_{FW,k} = \frac{cum\Delta\ m_k}{C_k - C_{FW}} \tag{7.2}$$

If a negative value exists in this column, this means that there is insufficient water purity in the networks. More freshwater needs to be added until no negative value exists in this column. The largest negative value of $F_{FW,k}$ is taken, and this value is replaced with the earlier assumed zero freshwater flowrate in Column 6 to be re-cascaded to obtain the feasible water cascade (Column 10). Note that the location of the concentration level with the largest negative value of $F_{FW,k}$ is also the location of the Pinch Point, i.e. at 150 ppm. The water source which exists at the Pinch is called the Pinch-causing source. Part of this source is located above the Pinch, and part of it is Below the Pinch.

The new cascade now gives the minimum freshwater target (in the first row) and wastewater target (last row) of 70 t/h and 50 t/h.

7.2.2 Source/Sink Composite Curves (SSCC)

The SSCC is a plot of mass load (m) versus flowrate (F). It is used to target the minimum usage of fresh resources for material recycle/reuse networks (El-Halwagi, 2003). Example 7.1 is used to illustrate the steps to construct the SSCC. The Sink Composite Curve is created by connecting each sink with its corresponding mass load and flowrate cumulatively in ascending concentration order. The same is performed for the Source Composite Curve. This is shown in Fig. 7.7a. The Source Composite Curve is then shifted to the right until it touches the Sink Composite Curve, with the Source Composite Curve located below the Sink Composite Curve in the overlapped region (Fig. 7.7b). The Pinch occurs where the two Composite Curves touch. The Pinch concentration is determined by the concentration of the Pinch-causing source stream, i.e. S3 at 150 ppm. The minimum freshwater target is the flowrate distance difference between the beginning of the Source Composite Curve and the Sink Composite Curve. The minimum wastewater target is the flowrate distance difference between the end of the Source Composite Curve and the Sink Composite Curve. This is illustrated in Fig. 7.7b. The freshwater and wastewater targets are the same as those obtained using the WCT, i.e. 70 t/h and 50 t/h.

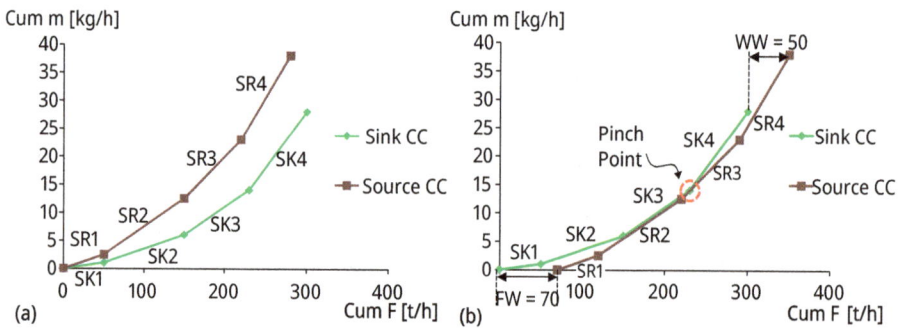

Fig. 7.7: Source/Sink Composite Curve for Example 7.1: (a) Before shifting Source Composite Curve, (b) After shifting Source Composite Curve.

7.2.3 Significance of the Pinch region

As in the case of Heat Pinch Analysis, the Pinch point in WPA is crucial in guiding designers towards the correct network design. Fig. 7.8 shows the definition of the Pinch region for a Multiple Pinch case.

A water source located on one side of the Pinch should only be used to satisfy a water sink located in the same Pinch region, or otherwise a water penalty will occur. An exception is the water source which is the Pinch-causing stream, as part of it is located above the Pinch while the other part is below the Pinch. In the following is the significance of the Pinch regions.

1. Below the Pinch region (lower concentration level):
   The entire water sinks mass load and flowrate should be satisfied either by water sources or freshwater. No wastewater should be generated.

2. Between Pinch region:
   All the water sources should satisfy the water sinks flowrate and mass load. No freshwater should be added or wastewater generated.

3. Above the Pinch region (higher concentration level):
    The water sources which satisfies the water sinks can have an equal or lower mass load amount. No freshwater should be added to this region. Excess water sources is to be discharged as wastewater.

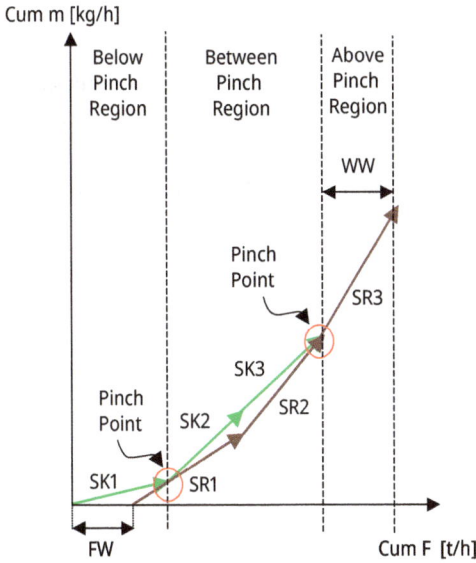

Fig. 7.8: Source/Sink Composite Curve: Pinch region classification.

7.3 Maximum Water Recovery target for a single impure freshwater source

Wan Alwi and Manan (2007) proposed a method to determine the Maximum Water Recovery target for a single impure freshwater source that may exist at a concentration lower or higher than other streams' concentrations. The authors divided the problem into Pinched and threshold problems as described next.

7.3.1 Pinched problems

Prior to determining the minimum new utilities flowrate (F_{MU}), it is important to establish if a utility is suitable for a given process, using the following heuristic:

Heuristic 7.1: Only consider a water source as a utility if its concentration is lower than the concentration of the Pinch.

A utility at a concentration higher than the Pinch point will only increase wastewater. Given that a water source at a concentration lower than the Pinch concentration is available for the limiting data in Example 7.1 (Tab. 7.1). The F_{MU} can be obtained by systematically moving the Source Line Above (SLA) along the utility line and the Source Line Below (SLB) until Utilities/Process Pinches are obtained. Definitions of the Source Lines Above (SLA) and Below (SLB) the U are shown in Fig. 7.9.

Systematic shifting of SLA and SLB to get F_{MU} involve two key steps:

1. Move the Utility Line along with SLB to the right-hand side of the Sink Composite Curve until the line meets either the first Utility or Process Pinch (SLB and utility lines must Pinch the Sink Composite Curve).
2. Shift SLA upwards along the utility line until the entire SLA Pinches the Sink Composite Curve.

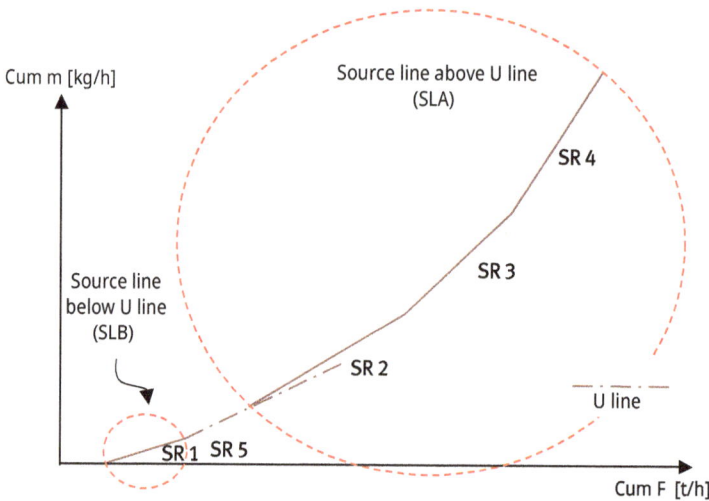

Fig. 7.9: Location of various water sources relative to the Utility Line, SR5.

1. Example A: A utility with no SLB creates a Utility Pinch.

SR5 at a concentration of 10 ppm is a utility added to the limiting water data from Example 6.1. There is no water source below SR5. The SR5 line is drawn first and shifted until a Utility Pinch (C_{Pinch} = 10 ppm) occurs at Cum m = 0 according to Step 1 (Fig. 7.10).

SLA (SR1 to SR4) is moved upwards according to Step 2 along the SR5 line until a process Pinch occurs at C_{Pinch} = 150 ppm (Fig. 7.10). The horizontal gap between the Sink Composite and the intersection of SLA and SR5 gives the minimum utility flowrate (F_{MU}), i.e. 75 t/h.

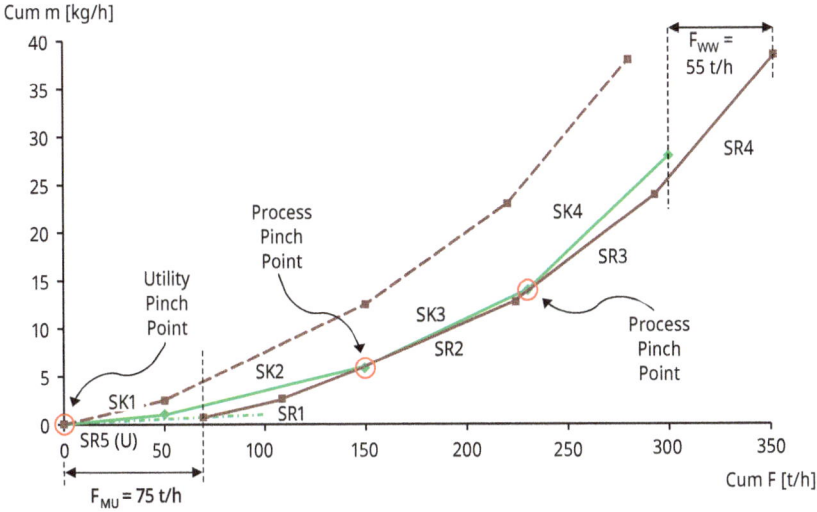

Fig. 7.10: SLA shifted along SR5. Final Composite Curve with minimum utility addition.

2. Example B: A utility with an SLB creates a Utility Pinch.

SR5 was a utility at 80 ppm added to the limiting data from Polley and Polley (2000). Shifting SLB (SR1) and SR5 along zero "Cum m" axis according to Step 1 created a Utility Pinch at C_{Pinch} = 80 ppm (Fig. 7.11). SLA (SR2 to SR4) was next shifted according to Step 2 along

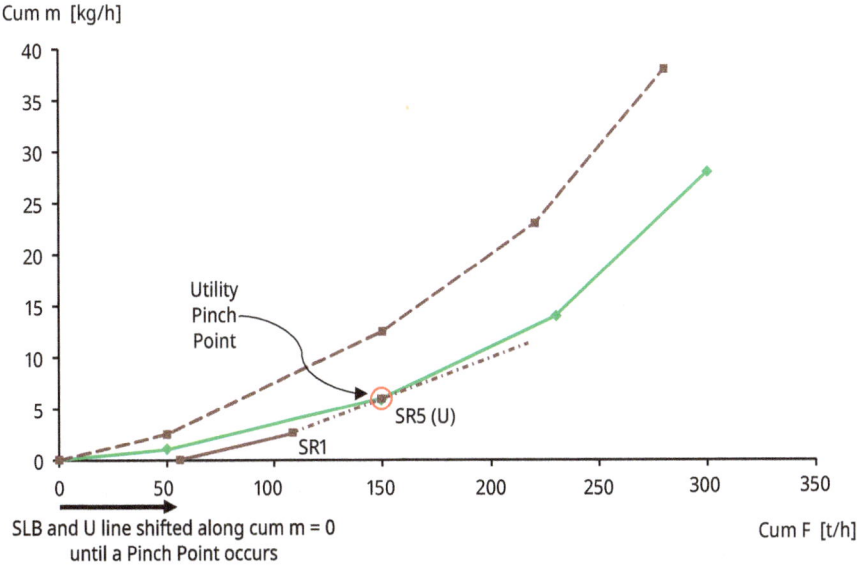

Fig. 7.11: SLB (S1) and SR5 shifted along the Cum m = 0 line.

SR5 from the new Pinch point onwards until a process Pinch point was created at C_{Pinch} = 100 ppm (Fig. 7.12). Fig. 7.13 shows the final Composite Curves with F_{MU} of 43.75 t/h.

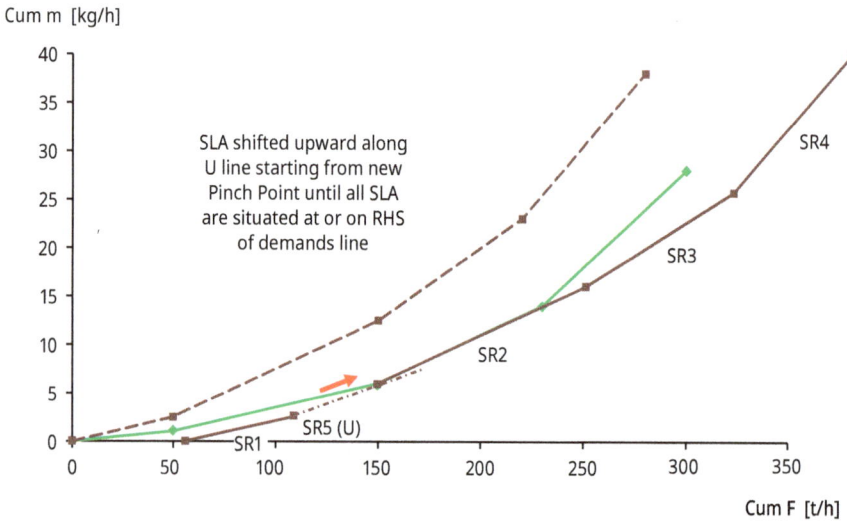

Fig. 7.12: SLA (S2 to S4) shifted upwards along SR5 from the new Pinch Point until SLA created another Pinch Point at C_{pinch} = 100 ppm.

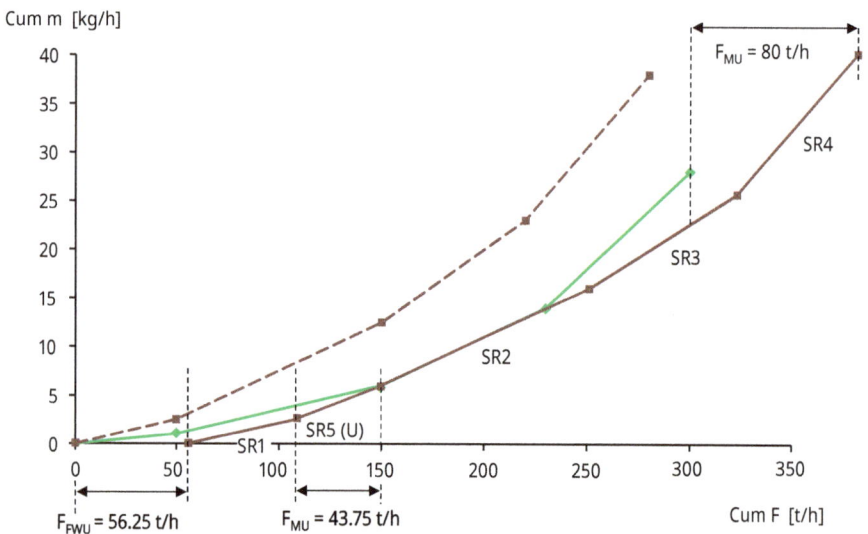

Fig. 7.13: Final Composite Curves with addition of S5.

3. Example C: The utility is a regenerated water source that creates a Utility Pinch. The utility may be a water source regenerated from a concentration above to below the Pinch point. This may create new Utility and Process Pinch points at lower concentrations and reduce the length of the regenerated source line. Tab. 7.3 shows the limiting data from Sorin and Bedard (1999) which resulted in multiple Pinch points at 100 ppm and 180 ppm. To have beneficial water savings, a source above or at concentration of 180 ppm should be regenerated to a concentration below 100 ppm. SR6 is a new Utility created by regenerating SR5 from 250 to 30 ppm. SR6 is then shifted along the freshwater line until a Utility Pinch occurred at C_{Pinch} = 30 ppm (Fig. 7.14). The SLA (SR1 to SR5) is next shifted along SR6 with SR5 original flowrate maintained until a Process Pinch occurs at 100 ppm (Fig. 7.15). For F_{MU} = 114.3 t/h, the amount of SR5 after reduction is calculated at 80.7 t/h. Fig. 7.16 shows the final Composite Curves after SR5 reduction. The minimum freshwater and wastewater flowrates are 120 t/h and 40 t/h.

Tab. 7.3: Limiting data for Example 7.2 from Sorin and Bedard (1999).

Sink	F, t/h	C, ppm	m, kg/h	Cum F, t/h	Cum m, kg/h
SK1	120	0	0	120	0
SK2	80	50	4	200	4
SK3	80	50	4	280	8
SK4	140	140	19.6	420	27.6
SK5	80	170	13.6	500	41.2
SK6	195	240	46.8	695	88
Source					
SR1	120	100	12	120	12
SR2	80	140	11.2	200	23.2
SR3	140	180	25.2	340	48.4
SR4	80	230	18.4	420	66.8
SR5	195	250	48.75	615	115.55

4. Example D: Utility addition does not create a Utility Pinch.

Tab. 7.4 shows the limiting data for Example 7.3 (Wan Alwi and Manan, 2007). SR7 is a utility added at 130 ppm. Note that, unlike in the previous three examples (Examples A to C), shifting SR7 and SLB in this case creates a Process Pinch at C_{Pinch} = 100 ppm instead of a Utility Pinch. Fig. 7.17 shows the final Composite Curves after shifting SR7 and SLB along the zero "Cum m" axis line and after moving SLA upwards from the process Pinch along SR7. The F_{MU} for this case is 52.9 t/h. The Process Pinch Points are at 100 ppm (on line SR3) and 200 ppm (on line SR4). The freshwater and wastewater targets are 188.0 t/h and 140.9 t/h

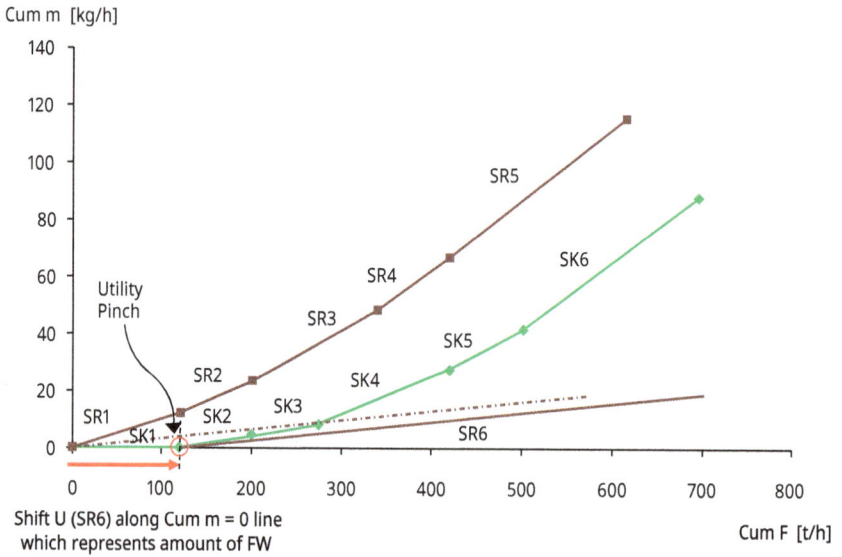

Fig. 7.14: SR6 utility line shifted along Cum m = 0 line.

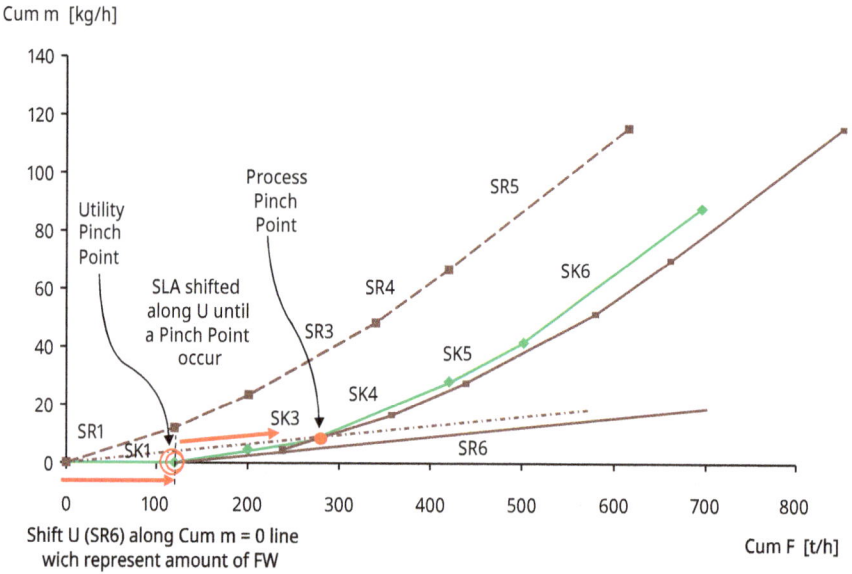

Fig. 7.15: SLA shifted along SR6 until a Pinch Point occurred.

Cum m [kg/h]

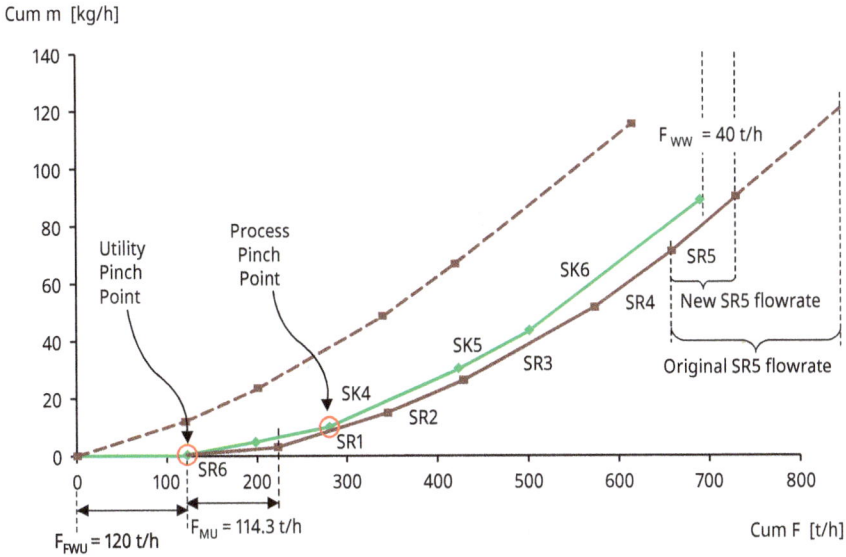

Fig. 7.16: SSCC after SR6 utility addition. SR5 flowrate reduction was exactly the same as the SR6 utility flowrate increment.

Tab. 7.4: Limiting data for Example 7.3 (Wan Alwi and Manan, 2007).

Sink	F, t/h	C, ppm	m, kg/h	Cum F, t/h	Cum m, kg/h
SK1	10	0	0	10	0
SK2	120	5	0.6	130	0.6
SK3	50	30	1.5	180	2.1
SK4	80	40	3.2	260	5.3
SK5	50	50	2.5	310	7.8
SK6	30	100	3	340	10.8
SK7	90	150	13.5	430	24.3
Source					
SR1	10	10	0.1	10	0.1
SR2	70	50	3.5	80	3.6
SR3	80	100	8	160	11.6
SR4	50	200	10	210	21.6
SR5	30	300	9	240	30.6
SR6	90	500	45	330	75.6

Cum m [kg/h]

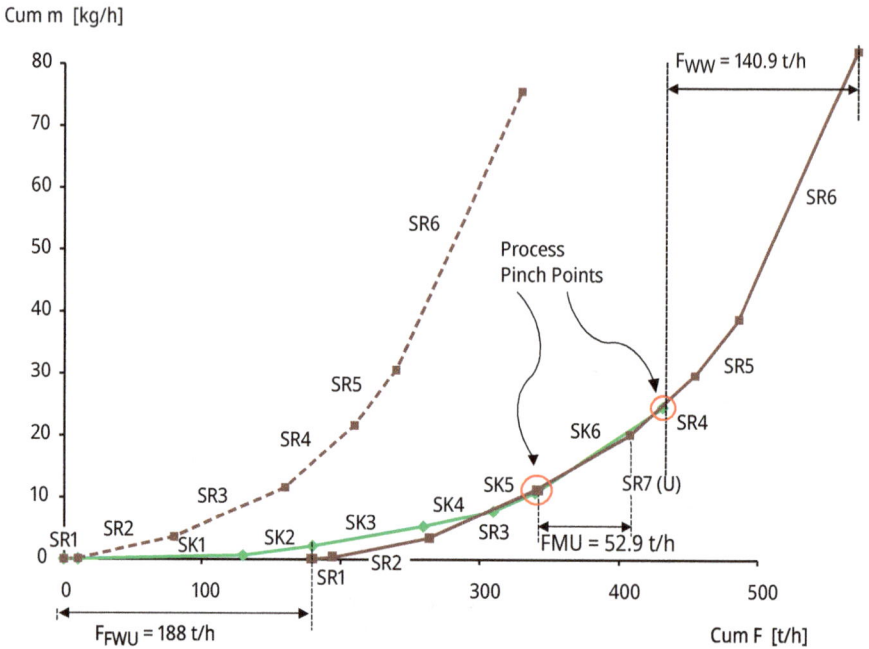

Fig. 7.17: Composite Curves with utility (SR7) addition.

7.3.2 Threshold problems

A process which has either freshwater or wastewater as "utility" is a threshold problem. This example focuses on a threshold problem that does not produce wastewater. Note that since the flowrate of regenerated source equals the flowrate of reduced wastewater source, regeneration would not change the flowrate of freshwater and wastewater for this case. Harvesting external water source provides more room for savings and is therefore the more preferred option. For this type of threshold problem, a Pinch may or may not exist. If a Pinch exists, the utility targeting technique for Pinch problem applies. For threshold problems without a Pinch Point, the following steps should be taken.
1. Shift utility line with SLB until a Utility/Process Pinch Point is created.
2. Shift SLA (SR3) above the Utility/Process Pinch Point upwards along the Utility Line and SLB until all sink flowrates are satisfied.

Tab. 7.5 is the limiting data for Example 7.4 from Wan Alwi and Manan (2007) and Fig. 7.18 is the corresponding initial SSCC for a threshold problem. The initial freshwater target is 269.99 t/h. Utility Line (SR4) at 80 ppm was added and shifted with SLB (SR1 and SR2) to the right of the Sink Composite until a Pinch Point occurred at $C_{Pinch} =$ 80 ppm. SLA (SR3) above the Utility Pinch Point was shifted upwards along the SR4 until the water sink flowrates were fully satisfied (see Fig. 7.19). The F_{MU} for this case is 50 t/h, and the Pinch Point is at 80 ppm. The new freshwater target is 220 t/h.

Tab. 7.5: Example 7.4: limiting data for threshold problem (Wan Alwi and Manan, 2007).

Sink	F, t/h	C, ppm	m, kg/h	Cum F, t/h	Cum m, kg/h
SK1	10	0	0	10	0
SK2	120	5	0.6	130	0.6
SK3	130	10	1.3	260	1.9
SK4	50	50	2.5	310	4.4
SK5	30	100	3	340	7.4
SK6	90	200	18	430	25.4
Source					
SR1	10	10	0.1	10	0.1
SR2	70	50	3.5	80	3.6
SR3	80	100	8	160	11.6

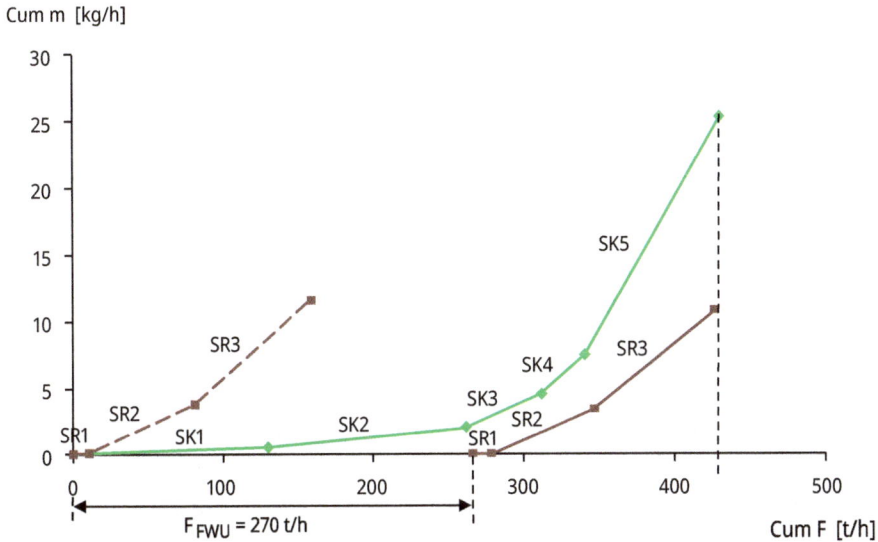

Fig. 7.18: SSCC for Example 6.4: A threshold problem.

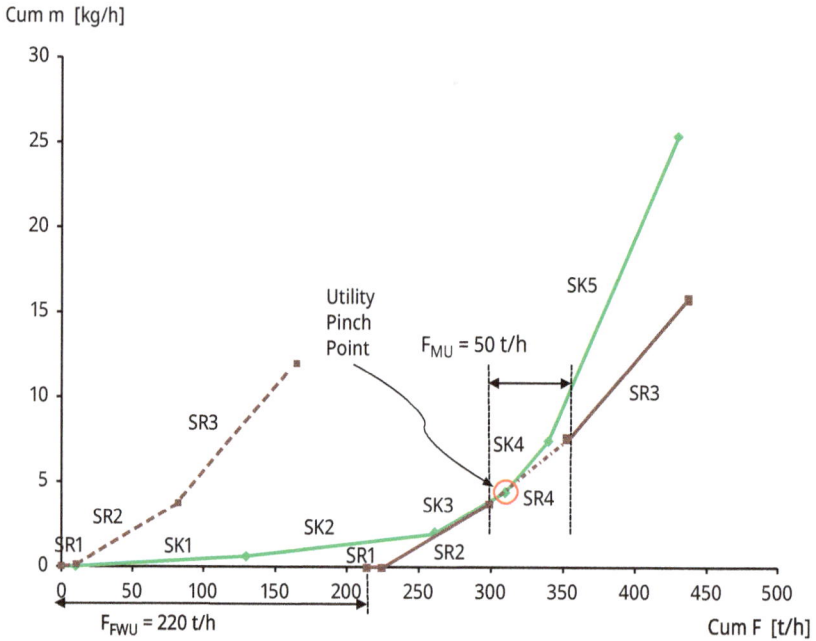

Fig. 7.19: SSCC for threshold problem with the addition of SR4 utility.

7.4 Maximum Water Recovery targets for multiple freshwater sources

A higher quality (cleaner) utility is usually more valuable, particularly for cases involving regeneration. When multiple sources of water and regenerated wastewater are available as utilities, the general rule is to minimise the use of higher quality utility in order to maximise savings. This could be achieved using the following heuristic:

> **Heuristic 7.2:** Using Water Composite Curves, obtain the F_{MU} one by one, starting from the cleanest to the dirtiest water source.

Heuristic 7.2 means that the F_{MU} for the cleanest *new utility* has to be obtained first using the SSCC procedure described earlier. Adding a utility creates new Utility and Process Pinch points. The next utility could only be considered if its concentration is lower than the highest Pinch concentration. Note that the maximum utility freshwater savings had already been reached with addition of the first utility. The addition of a dirtier utility should only reduce the flowrate of the cleaner utility added previously. The same procedure is repeated until all available utilities have been utilised. Example E explains how the technique is implemented.

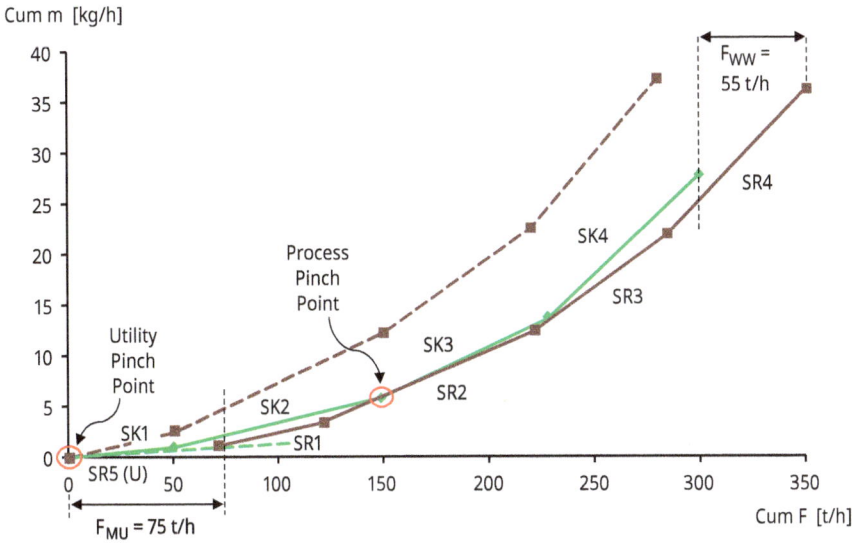

Fig. 7.20: SSCC with addition of U1 (SR5) at C = 10 ppm.

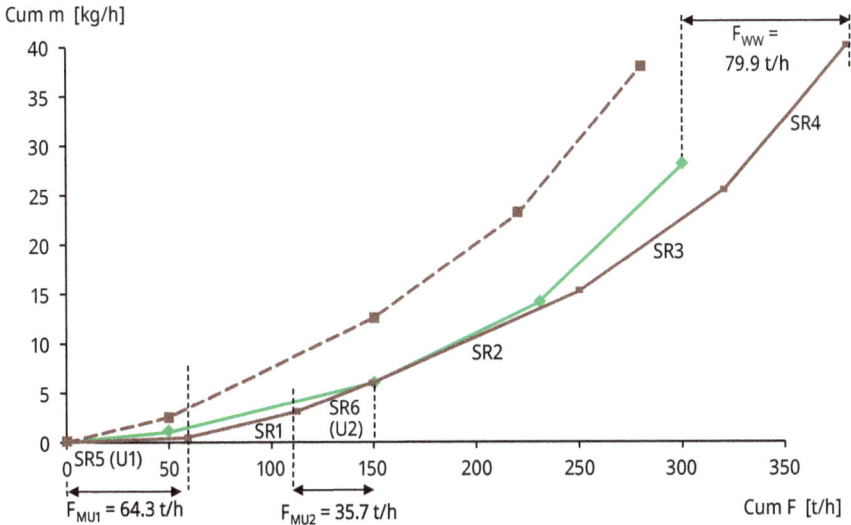

Fig. 7.21: Shifting of SLB and *U2* line (C = 80 ppm) along *U1* (C = 10 ppm) line until a Pinch Point occurred.

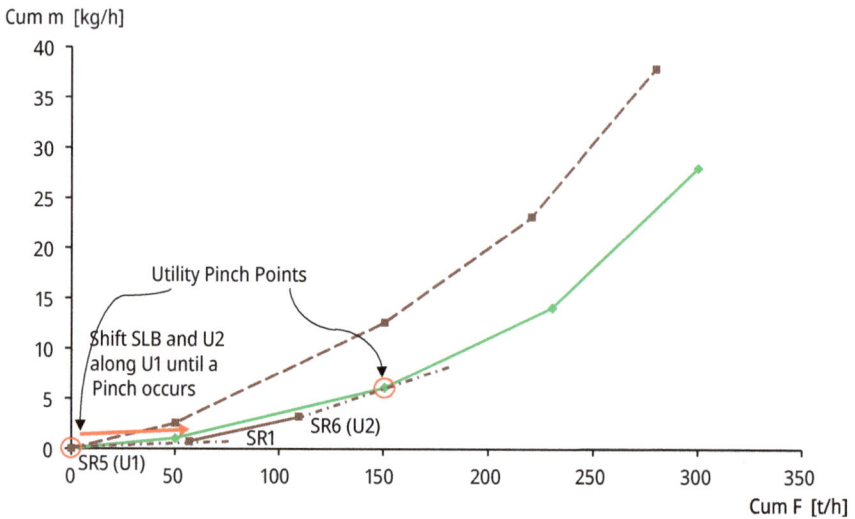

Fig. 7.22: Final SSCC after addition of *U1* and *U2*.

5. Example E: Two new utilities, *U1* at 10 ppm and *U2* at 80 ppm
To illustrate the technique, the limiting data from Example 6.1 are assessed for possible integration with two new utilities, *U1* and *U2* available at 10 ppm and 80 ppm. Note that *U1* is SR5 from Example A. *U1* as the cleaner utility at 10 ppm is considered first according to Heuristic 7.1. Fig. 7.20 shows the F_{MU1} for SR5 is 75 t/h and the utility and Process Pinches are at 0 and 100 ppm. *U2* (SR6) at 80 ppm is a viable utility since it existed below the new Pinch Point at 100 ppm. SR6 is drawn next and shifted with SLB (SR1) downwards along the first utility line (SR5) until another Utility Pinch occurred at 80 ppm (Fig. 7.21). Next, the SLA above SR6, i.e. SR2 to SR4, are drawn at the new Utility Pinch of 80 ppm and shifted until it completely appeared on the right-hand side of the Sink Composite Curve, or until it created a new Pinch Point. This give F_{MU1} and F_{MU2} of 64.3 t/h and 35.7 t/h (Fig. 7.22). The multiple utility targeting procedures ultimately yield freshwater and wastewater targets at 0 t/h and 79.9 t/h. Note that if SLA had created a new Process Pinch when it was shifted along SR6, any new water source at concentration lower than the new process Pinch Point could still be added to further reduce SR6.

7.5 Working session

Assuming pure freshwater with zero contaminant is available as utility, determine the Maximum Water Recovery target for Examples 6.1 and 6.2 by using:
(a) Water Cascade Analysis (WCA) method;
(b) Source/Sink Composite Curve (SSCC) method.

7.6 Solution

(a) Water Cascade Analysis (WCA) method.

The procedure described in Section 7.2.2 is used to construct the WCT. Table 7.6 shows the WCT for Example 6.1. It can be observed that the largest negative cumulative $F_{FW,k}$ is at 700 ppm, indicating the Pinch location. This value also gives the minimum freshwater target of 90.64 t/h. Cascading the flowrate in the last column for the feasible cascade gives the minimum wastewater target of 50.64 t/h.

Tab. 7.6: Water Cascade Table for Example 6.1.

C_k, ppm	ΔC_k, ppm	$\sum F_{SKi}$, t/h	$\sum F_{SRj}$, t/h	$\sum F_{SKi} + \sum F_{SRj}$, t/h	F_C, t/h	Δm, kg/h	Cum. Δm, kg/h	F_{FW}, Cum, t/h	F_C, t/h
					0				$F_{FW} = 90.64$
0		−20		−20			0		
	10				−20	−0.20			70.64
10		−15	10	−5			−0.20	−20.00	
	90				−25	−2.25			65.64
100		−80	45	−35			−2.45	−24.50	
	100				−60	−6.00			30.64
200		−50		−50			−8.45	−42.25	
	500				−110	−55.00			−19.36
700			50	50			−63.45	**−90.64**	**(PINCH)**
	300				−60	−18.00			30.64
1,000			20	20			−81.45	−81.45	
	−1,000				−40	40.00			$F_{ww} = 50.64$

Similarly, Tab. 7.7 shows the WCT for Example 6.2. The largest negative cumulative $F_{FW,k}$ which is also the Pinch location is at a concentration of 14 ppm. The minimum freshwater target is 2.06 t/h and the wastewater target is 8.16 t/h.

Tab. 7.7: Water Cascade Table for Example 6.2.

C_k, ppm	ΔC_k ppm	$\sum F_{SKi}$, t/h	$\sum F_{SRj}$, t/h	$\sum F_{SKi} + \sum F_{SRj}$, t/h	F_C, t/ h	Δm, kg/ h	Cum. Δm, kg/h	$F_{FW,cum}$, t/h	F_C, t/h
					0.00				$F_{FW} = 2.06$
0		−1.20	0.80	−0.40			0.00		
	10				−0.40	−4.00			1.66
10		−5.80		−5.80			−4.00	−0.40	
	4				−6.20	−24.80			−4.14
14			5.00	5.00			−28.80	−2.06	**(PINCH)**
	11				−1.20	−13.20			0.86

Tab. 7.7 (continued)

C_k, ppm	ΔC_k, ppm	$\sum F_{SKi}$, t/h	$\sum F_{SRj}$, t/h	$\sum F_{SKi} + \sum F_{SRj}$, t/h	F_C, t/h	Δm, kg/h	Cum. Δm, kg/h	$F_{FW,cum}$, t/h	F_C, t/h
25		5.90	5.90				−42.00	−1.68	
	9				4.70	42.30			6.76
34		1.40	1.40				0.30	0.01	
					6.10	0.00			$F_{WW} = 8.16$

(b) Source/Sink Composite Curve (SSCC) method.

For Example 6.1, starting from zero, the sources are first plotted cumulatively in ascending concentration order in a plot of cumulative mass load vs. cumulative flowrate diagram. This gives the Source Composite Curve as shown in Fig. 7.23. The sinks are also plotted cumulatively in ascending concentration order starting from zero. This gives the Sink Composite Curves, as shown in Fig. 7.24. The Source Composite Curves are then shifted to the right-hand side of the Sink Composite Curves until a Pinch Point is established as shown in Fig. 7.25. It can be observed that the Pinch Point is located at SR4, which has a concentration of 700 ppm. The minimum freshwater target is the flowrate distance difference between the beginning of the Source Composite Curve and the Sink Composite Curve, i.e. 90.64 t/h. The minimum wastewater target is the flowrate distance difference between the end of the Source Composite Curve and the Sink Composite Curve, i.e. 50.64 t/h.

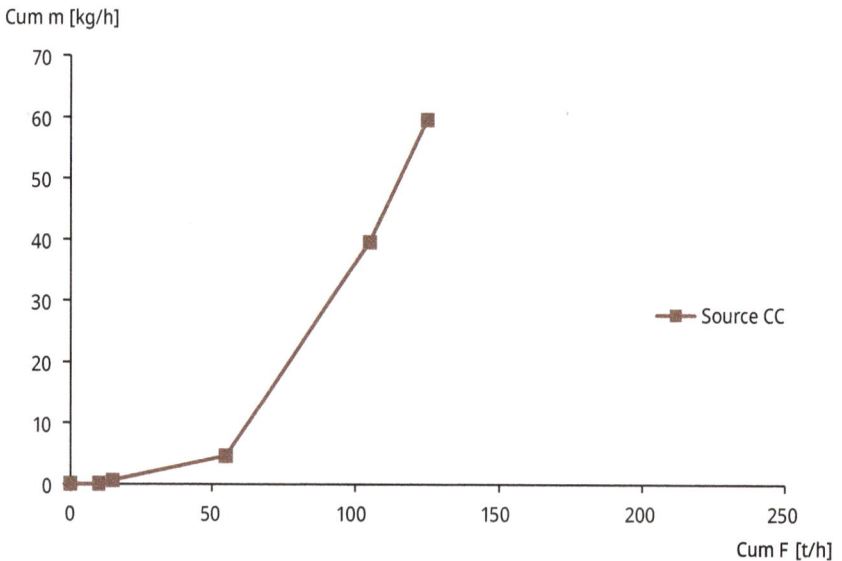

Fig. 7.23: Source Composite Curve for Example 6.1.

Cum m [kg/h]

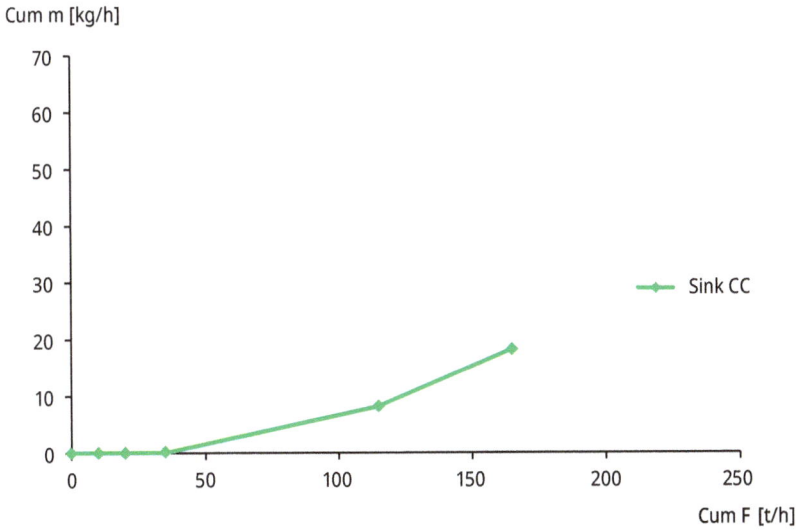

Fig. 7.24: Sink Composite Curve for Example 6.1.

Cum m [kg/h]

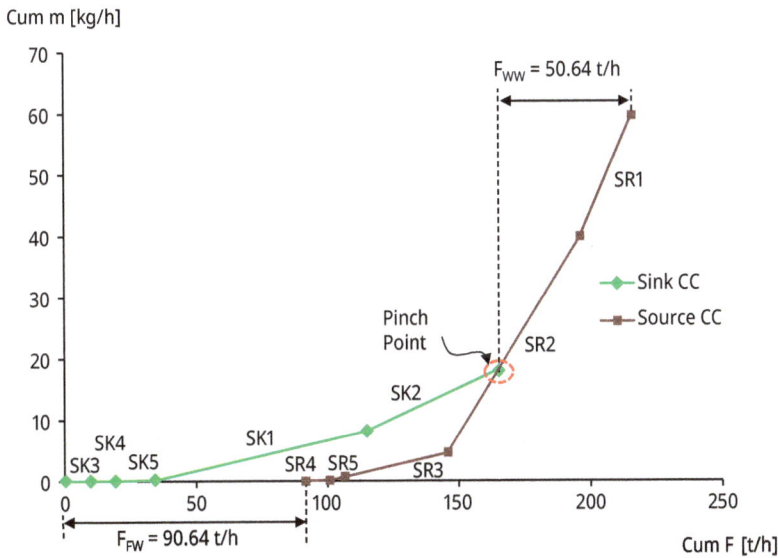

Fig. 7.25: Source/Sink Composite Curve for Example 6.1.

Similar procedures are performed for Example 6.2. The final SSCC is shown in Fig. 7.26. It can be observed that the pinch point is also located at SR2, 14 ppm. The minimum freshwater target is 2.06 t/h and the minimum wastewater target is 8.16 t/h.

Cum m [kg/h]

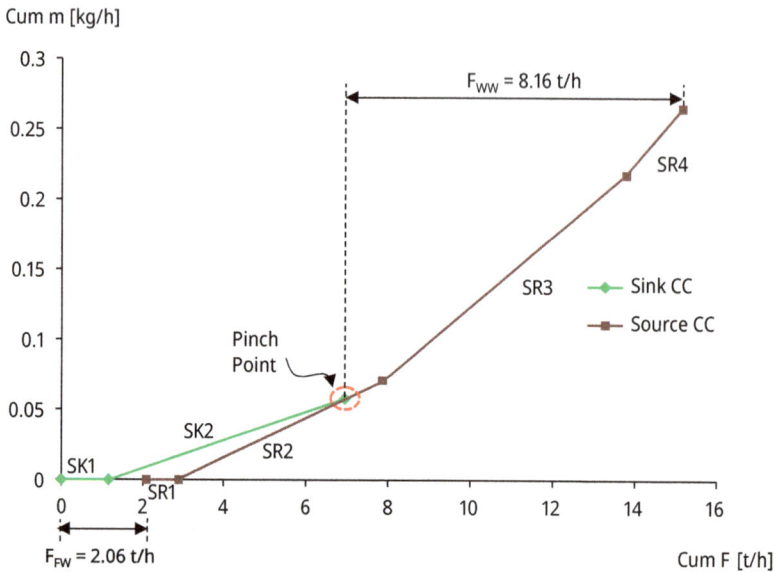

Fig. 7.26: Source/Sink Composite Curve for Example 6.2.

References

Almutlaq A M, Kazantzi V, El-Halwagi M. (2005). An Algebraic Approach to Targeting Waste Discharge and Impure Fresh Usage via Material Recycle/Reuse Networks, Cleaner Technology and Environmental Policy, 7(4), 294–305.

Bandyopadhyay S. (2006). Source Composite Curve for Waste Reduction, Chemical Engineering Journal, 125(2), 99–110.

El-Halwagi M M, Gabriel F, Harell D. (2003). Rigorous Graphical Targeting for Resource Conservation via Material Recycle/ Reuse Networks, Industrial & Engineering Chemistry Research, 42, 4319–4328.

Foo D C, Manan Z A, Tan Y L. (2005). Synthesis of Maximum Water Recovery Network for Batch Process Systems, Journal of Cleaner Production, 13(15), 1381–1394.

Foo D C, Manan Z A, Tan Y L. (2006). Use Cascade Analysis to Optimize Water Networks, Chemical Engineering Progress, 102(7), 45–52.

Hallale N. (2002). A New Graphical Targeting Method for Water Minimization, Advances in Environmental Research, 6(3), 377–390, doi: 10.1016/S1093.

Majozi T, Brouckaert C J, Buckley C A. (2006). A Graphical Technique for Wastewater Minimisation in Batch Processes, Journal of Environmental Management, 78, 317–329.

Manan Z A, Tan Y L, Foo D C Y. (2004). Targeting the Minimum Water Flowrate Using Water Cascade Analysis Technique, AIChE Journal, 50(12), 3169–3183.

Polley G T, Polley H L. (2000). Design Better Water Networks, Chemical Engineering Progress, 96(2), 47–52.

Prakash R, Shenoy U V. (2005). Targeting and Design of Water Networks for Fixed Flowrate and Fixed Contaminant Load Operations, Chemical Engineering Science, 60(1), 255–268. doi: 10.1016/j.ces.2004.08.005.

Sorin M, Bedard S. (1999). The Global Pinch Point in Water Reuse Networks, Process Safety and Environmental Protection, 77, 305–308.

Wan Alwi S R, Manan Z A. (2007). Targeting Multiple Water Sources Using Pinch Analysis, Industrial & Engineering Chemical Research, 46(18), 5968–5976.

Wang Y P, Smith R. (1994). Wastewater Minimisation, Chemical Engineering Science, 49(7), 981–1006.

8 Water network design/retrofit

8.1 Introduction

Once the Maximum Water Recovery target has been determined, the next step is to design a water network which can achieve this target. Two established methods for water network design for a single contaminant are the Source/Sink Mapping Diagram (SSMD) by Polley and Polley (2000) and the Source and Sink Allocation Curves (SSAC) by Wan Alwi and Manan (2008), and will be described further in this chapter. In addition, there are many new advances in water network design, for example, simultaneous mass and energy recovery networks (Wan Alwi et al., 2011), biogas, water regeneration and water reuse network (Misrol et al., 2021), flexible water network design under uncertainty (Moghaddam et al., 2022), and batch water network design (Huang et al., 2023).

8.2 Source/Sink Mapping Diagram (SSMD)

The SSMD was introduced by Polley and Polley (2000). The authors proposed a design principle that involves the systematic mixing of source streams with each other and, where necessary, with freshwater to produce streams that meet the individual sinks in terms of both quantity and quality. The quantity of water to be sent to the contaminated source is determined by calculating the contaminant load associated with the sink. The result takes one of the following forms:
(a) If this quantity is equal to the required value, then no freshwater needs to be supplied;
(b) If the quantity is less than the sink, then some freshwater is required;
(c) If the quantity required from the source exceeds the sink, then part of the sink's contaminant load should be satisfied using a source having a higher contaminant concentration.

Constraints for network design between water sources i and sinks j are given as follows (Hallale, 2002):
(a) Sinks
   1. Flowrate

$$\sum_i F_{i,j} = F_{SK,j} \tag{8.1}$$

   Where F_i is the total flowrate available from source i to Sink j and $F_{SK,j}$ is the flowrate required by Sink j.

https://doi.org/10.1515/9783110782981-008

2. Concentration

$$\frac{\sum_i F_{SRi,SKj}C_i}{\sum_i F_{SRi,SKj}} \leq C_{max,SKj} \tag{8.2}$$

Where C_{SKj} is the contaminant concentration of source i and $C_{max,SKj}$ is the maximum acceptable contaminant concentration of sink j.

(b) Sources
1. Flowrate

$$\sum_i F_{SRi,SKj} \leq FSRi \tag{8.3}$$

This network design method can stand on its own even without the targeting stage. However, for some cases, the network design is not the same as in the targeting stage since the cleanest to cleanest rule does not hold for the mass load deficit case (this will be explained further in Section 8.3). If the designers would like to use this method, it is important to ensure that the resulting freshwater and wastewater flowrates match the minimum water targets.

Example 7.1 from the previous chapter are used to illustrate this method. All the water sinks are aligned horizontally at the top, while all water sources are lined up vertically on the left-hand side. Both are arranged according to increasing contaminant concentration. All the water flowrates and mass loads for each source and sink are listed. Pinch Regions are also labelled on the diagram (see Fig. 8.1).

Start the source-sink matching process by matching the sink with the lowest contaminant concentration with the source with the lowest contaminant concentration. The quantity of water from the contaminated source required is determined by the quality (contaminant load) and quantity (flowrate) of the sink. For example, SK1 requires 50 t/h and can accept a mass load of up to 1 kg/h. SR1 has a flowrate of 50 t/h and a mass load of 2.5 kg/h. Hence, only a flowrate of 20 t/h, which corresponds to 1 kg/h of mass load, can be used from SR1 to satisfy the SK1 quality requirement. The remaining 30 t/h of flowrate required by SK1 needs to be satisfied by using freshwater, which has zero mass load (assuming freshwater does not have any contaminants). This satisfies both the quantity and quality required by the SK1.

Next, SK2 is matched with the remaining SR1. Since SK2 requires a water flowrate of 100 t/h and a mass load of 5 kg/h, all the remaining SR1 (30 t/h of flowrate with 1.5 kg/h of mass load) can be sent to SK2. Since SK2 quantity and quality have still not been satisfied yet (still requires F = 70 t/h and m = 3.5 kg/h), the next source, i.e. SK2, is considered. 35 t/h of SK2 with a 3.5 kg/h mass load added to SK2 to satisfy the quality needed. The remaining flowrate of SK2 is satisfied by using freshwater (35 t/h). This step is repeated for the remaining sinks.

Note that Below the Pinch Region, both the quantity and quality of the sink must be satisfied by either the sources or in combination with freshwater. However, for Above the Pinch Region, only the quality of the sink needs to be satisfied while the quantity can be equal or less. The excess water source, when all water sink have been matched, is sent to the wastewater treatment plant. Since the freshwater and wastewater are the same as those targeted in the previous chapter, this design is correct. Fig. 8.1 shows the completed SSMD.

Fig. 8.1 shows one possible network design generated by the SSMD. Note that this is only one of many possible designs that can achieve the targets. The designers can influence the solution by imposing other constraints, such as forbidden or forced connection for safety or economic reasons. However, these additional constraints may sometimes result in a water penalty.

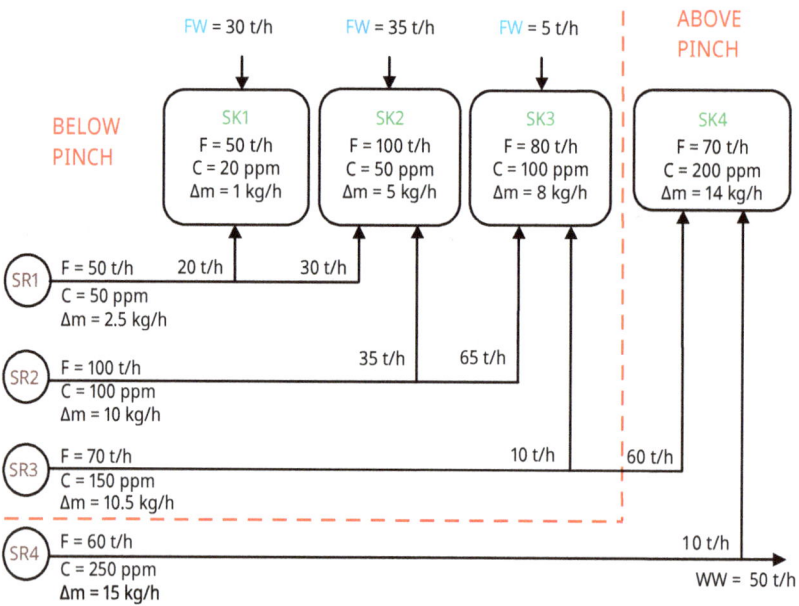

Fig. 8.1: Network design by Source/Sink Mapping Diagram (Polley and Polley, 2000).

8.3 Source and Sink Allocation Curves (SSAC)

The SSAC can be used to show the water network allocation or design between the source and the sink. SSAC was initially proposed by El-Halwagi et al. (2003) for pure freshwater sources and Kazantzi and El-Halwagi (2005) for impure freshwater sources. Both methods are based on the cleanest to cleanest rule, similar to the SSMD method by Polley and Polley (2000). All three methods are, however, limited to the *flowrate deficit case*, i.e. the case where the mass load of a sink is satisfied but not its flowrate, as shown in Fig. 8.2. There is another important case which is the *"mass load deficit case"*, where the source(s) meets only the flowrate of a sink but not the mass load (see Fig. 8.2b). The cumulative water sources Below and Between Pinch Regions have to have the exact same flowrate and mass load as the cumulative sink. Note that using the cleanest source that does not meet the mass load of a sink results in an excess mass load of the source and, ultimately unsatisfied flowrate of subsequent sinks. Wan Alwi and Manan (2008) introduced another rule for the "mass load deficit case", which is to satisfy the mass load and flowrate of a sink using the cleanest as well as the dirtiest sources in the Pinch Region and making sure the Pinch Point and the utility targets are satisfied. The authors also introduced the Network Allocation Diagram (NAD) which translates the Source and Sink Allocation Composite Curves into a network mapping diagram for easy visualisation.

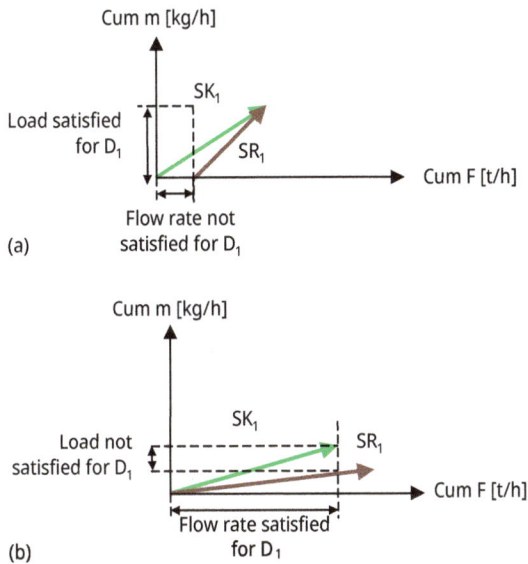

Fig. 8.2: Satisfying the cleanest sink with the cleanest water source using SSAC (a) flowrate deficit case, (b) mass load deficit case (Wan Alwi and Manan, 2008).

The method by Wan Alwi and Manan (2008) comprises three steps:

Step 1: Plot the sources and sinks from the cleanest to the dirtiest cumulatively to form a Source/Sink Composite Curve (SSCC) as described in Chapter 7. Note the minimum freshwater and wastewater flowrate. Note also the Region where the sources and sinks are located.

Step 2: Draw the Source and Sink Allocation Curves (SSAC) for the two separate Pinch Regions based on the following rules.

(a) Matching sources and sinks in the Region Below the Pinch and Between Pinches.
   1. Begin by matching the cleanest source with the cleanest sink, and pick the subsequent matches as follows:
   2. For the case where freshwater or utility purity is superior to *all* other streams (see Fig. 8.3):
      (i) If the cumulative water source(s) satisfies the mass load but not the flowrate as shown in Fig. 8.2a, add water utility until the sink flowrate is satisfied (only valid for Below Pinch Region). Once both the flowrate and the mass load of the sink have been satisfied, match the remaining sources with the next sink.
      (ii) If the cumulative water source(s) satisfies the flowrate but not the mass load as shown in Fig. 8.2b, pick the dirtiest source in this Pinch Region to satisfy the remaining mass load of the first sink. Once both the flowrate and the mass load of the first sink have been satisfied, use the remaining dirtiest source to satisfy the mass load of the next sink.

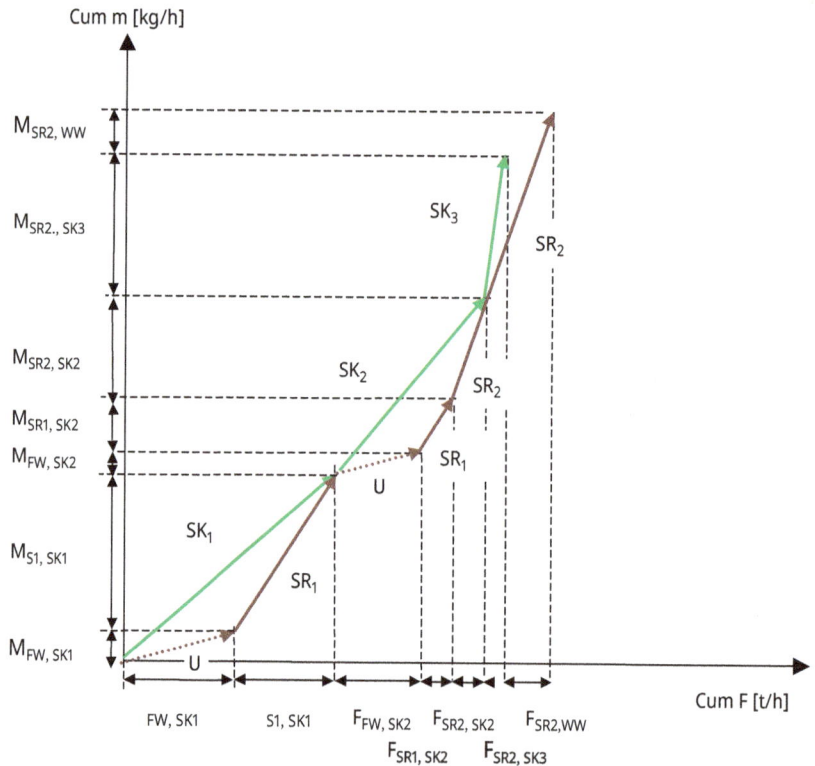

Fig. 8.3: Source and Sink Allocation Curves with utility stream concentration superior to all other streams.

3. For the case where freshwater or utility purity is not superior to *all* other streams (see Fig. 8.4) (this rule can also be applied for the availability of utility superior to other streams):
   (i)   Treat the utility as one of the water sources.
   (ii) Satisfy the mass load and flowrate of a sink by using the cleanest and dirtiest sources in this Pinch Region. Note the minimum amount of utility targets that have already been established in Step 1.
   (iii) Once both the flowrate and the mass load of the sink have been satisfied, match the remaining sources with the next sink.

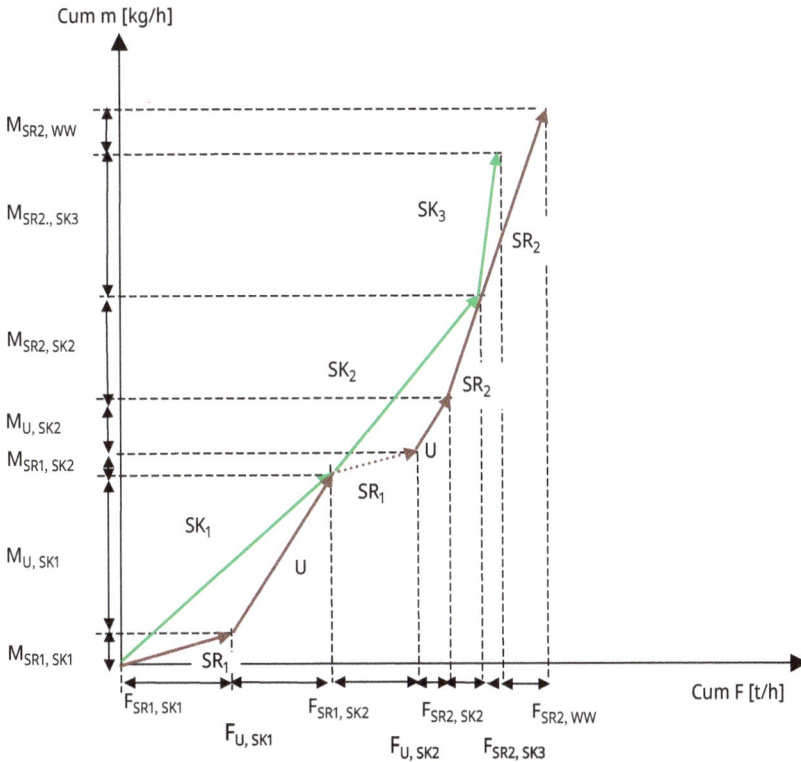

Fig. 8.4: Source and Sink Allocation Curves with utility stream concentration not superior to all other streams.

   (a) Matching sources and sinks in the Region Above the Pinch. Note that Region Above the Pinch refers to the starting point of the last Pinch Point to the point before the water source(s) becomes wastewater.
4. The sources Above the Pinch are used in ascending concentration order to satisfy the sink flowrate requirements. Note that no freshwater or higher concentration source shifting is required Above the Pinch. All sources' mass

loads Above the Pinch are less than the mass loads of sinks. The flowrate provided by the sources is sufficient for the sinks, but the mass load would be less. This means that sinks Above the Pinch will be fed by a higher purity source than required.

5. If there are no more sinks to satisfy, the remaining sources will become wastewater.

Step 3: Draw the Network Allocation Diagram (NAD) based on the SSAC. The exact flowrate of each source allocated to each sink can be obtained directly from the length of the x-axis.

8.3.1 Example of network design using SSCC for utility purity superior to all other streams

Example 7.2 will be used to illustrate the network design step for the case where utility purity is superior to all other streams. In this case, the utility has a concentration of 0 ppm.

Step 1: Plot the sources and sinks from the cleanest to the dirtiest cumulatively to form a Source and Sink Composite Curve (SSCC).
The water sink and source lines are first plotted according to increased concentration. A locus of the utility line is then drawn starting from the origin, with the slope represented by the utility concentration. The water source line is then shifted to the right along the utility line (Kazantzi and El-Halwagi, 2005) until all the water source lines are on the right-hand side of the sink line. The point where the water source line touches the water sink line is the Pinch Point. The new utility flowrate is given by the length of the horizontal utility line after shifting the water source lines to the right. The minimum wastewater target is the overshoot of the source line. The SSCC for Example 7.2 are as shown in Fig. 8.5. The Pinch Points are noted at 100 ppm and 180 ppm. This defines the Region Below, Between, and Above Pinch. The freshwater and wastewater targets are 200 t/h and 120 t/h.

Step 2: Draw the SSAC based on the proposed rules.
1. Region Below the Pinch
   There are three sinks located Below the Pinch Region, i.e. SK1, SK2, and SK3 (see Fig. 8.5). SK1 required 120 t/h of water with zero mass load. Since there was no water source with zero mass load, hence the first sink flowrate was completely satisfied using freshwater. This corresponded to 120 t/h of freshwater usage.

   For SK2, only part of SR1 could satisfy the mass load of SK2, as shown in Fig. 8.6. As stated in *Rule 2(i)*, if the cumulative water source(s) satisfies the mass load but not the flowrate, add water utility until the sink flowrate is satisfied.

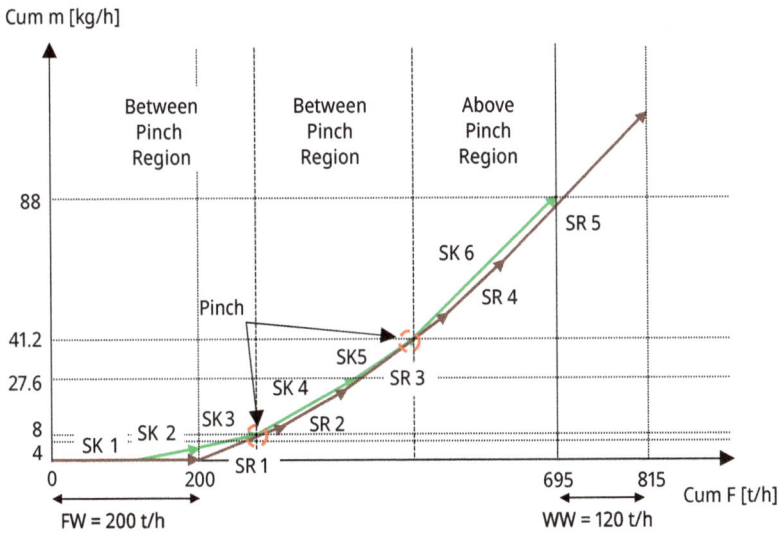

Fig. 8.5: Source/Sink Composite Curve for Example 7.2.

Freshwater was added until the entire sink flowrate was satisfied, and 40 t/h of SR1 and 40 t/h of freshwater were needed to satisfy SK2.

The final sink located Below the Pinch Region was SK3. The remaining SR1 located Below the Pinch Region was used to satisfy the SK3 mass load requirement. Again, *Rule 2(i)* applied to this case. Freshwater was added until the entire sink

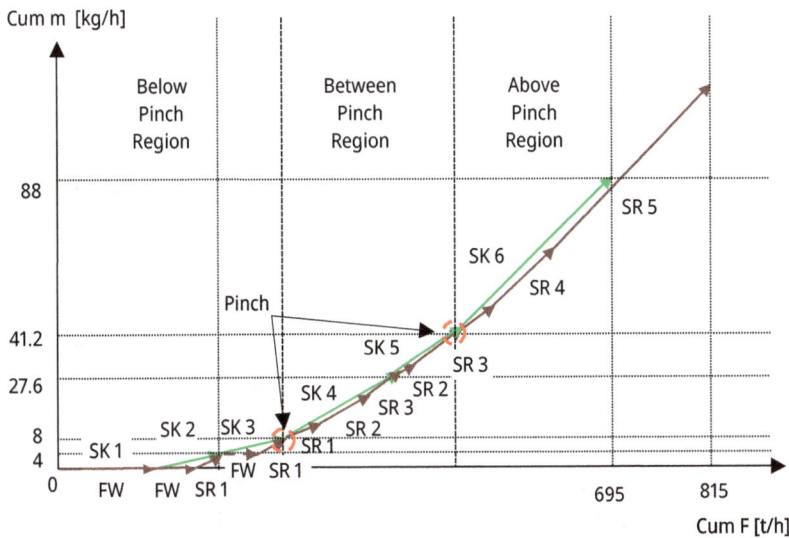

Fig. 8.6: Final SSAC for Example 7.2.

flowrate requirements were satisfied, and 40 t/h of SR1 and 40 t/h of freshwater were fed to SK3. The whole process of the source and sink allocation Below the Pinch Region is shown in Fig. 8.6.

2. Region between Pinches

There were two sinks located Between the Pinch Region, i.e. SK4 and SK5, as illustrated in Fig. 8.5. SK4 flowrate was satisfied using part of SR1, all of SR2 and part of SR3 in ascending order of source concentration based on *Rule 1*. However, as shown in Fig. 8.7, the flowrate of SK4 has already been satisfied but not the mass load requirement. Using *Rule 2(ii)*, SR3, as the dirtiest line in the Region, was shifted downwards along the SR2 line until all the mass load and flowrate requirements of SK4 were satisfied. Hence 40 t/h of S1, 60 t/h of SR2, and 40 t/h of SR3 were used for SK4. The remaining SR2 and SR3, corresponding to 20 t/h and 60 t/h in the Pinch Region, were used to satisfy the remaining sink, i.e. SK5. Fig. 8.6 shows the final source and sink allocation for the Region Between Pinches.

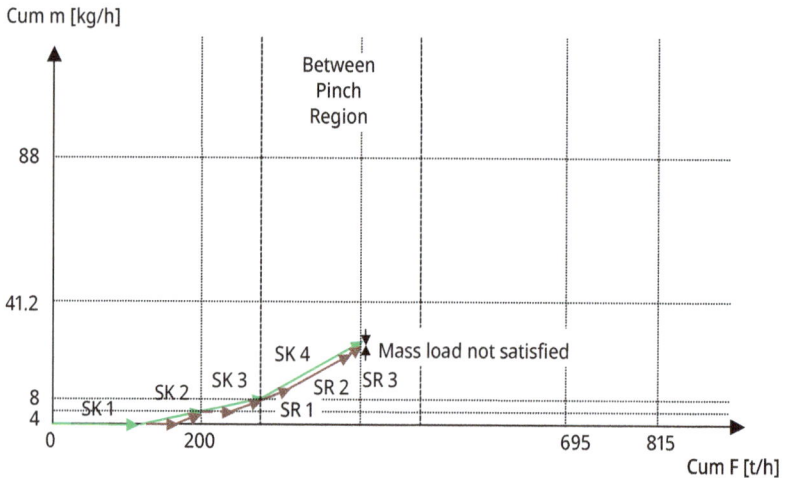

Fig. 8.7: SSAC for Example 6.2 where SK4 mass load was not satisfied for the Region Between Pinches.

3. The Region Above the Pinch

Finally, to satisfy the flowrate for the Above Pinch Region, the Pinch sources in the Region were used in ascending order to satisfy the sink flowrate requirement (*Rule 4*). Hence, 40 t/h of SR3, 80 t/h of SR4, and 75 t/h of SR5 were used to satisfy the SK4 flowrate. The amount of mass load cumulated from these three sources only totalled 44.35 kg/h, but SK4 could actually accept 48.75 kg/h. The remaining SR5 was rejected as wastewater since there was no more sink left to satisfy (*Rule 5*). The Above Pinch source and sink allocation are shown in Fig. 8.6.

Step 3: Draw the Network Allocation Diagram (NAD) using the SSACs.

Now that Steps 1 and 2 are completed, the Network Allocation Diagram (NAD) can be drawn. The x-axis was segmented into sink flowrate intervals by drawing vertical lines. All the water sinks were aligned horizontally at the top while all water sources were lined up vertically on the left-hand side. Both axes were arranged according to increasing contaminant concentration. Note that the lengths of water sinks were drawn to match the sink flowrate obtained from the final Source and Sink Allocation Curves. The x-axis was further segmented into source flowrate intervals by drawing vertical lines. The sources were then drawn horizontally just below the water sinks, according to the

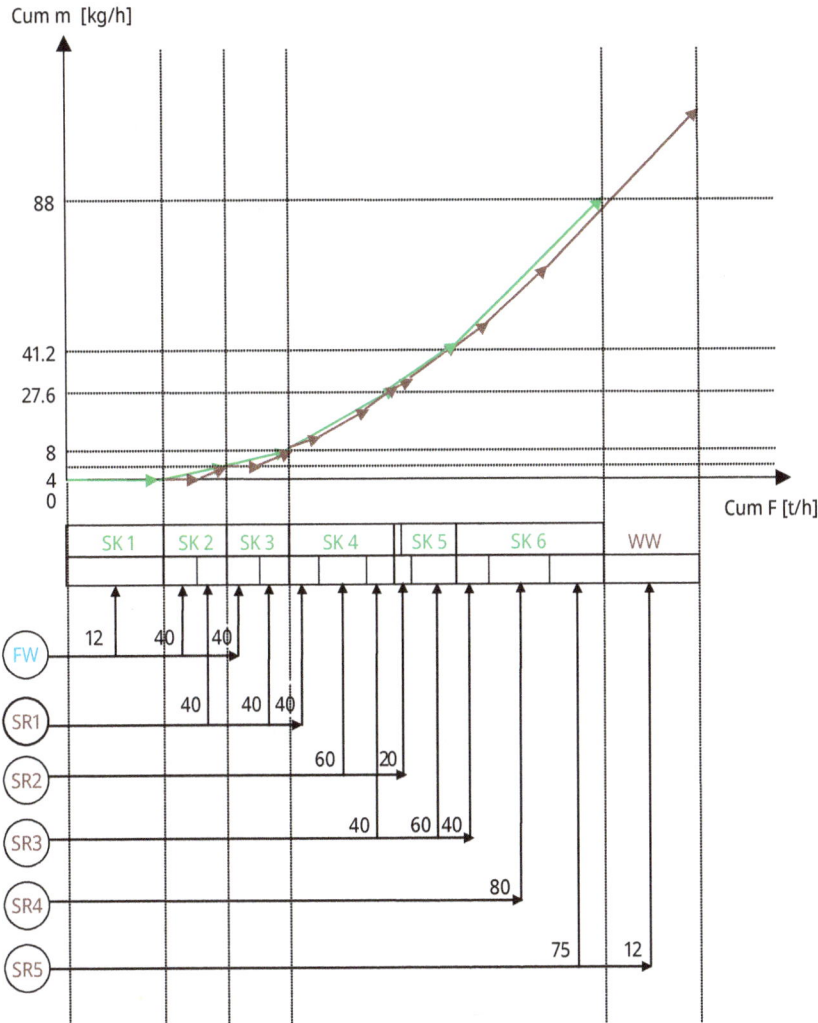

Fig. 8.8: Final Network Allocation Diagram.

SSAC. This represented the exact amount of source allocated for the respective sink. Finally, arrows were drawn linking the vertically aligned sources to the horizontally aligned sources to show how the sources were distributed into the sinks. The final network diagram is as in Fig. 8.8. Note that without the SSAC procedure as a guide, a network diagram may miss the minimum freshwater and wastewater targets for certain cases that were designed using the cleanest to the cleanest rule proposed by Polley and Polley (2000).

8.3.2 Example 8.1: Freshwater purity not superior to all other streams

Table 8.1 shows the limiting data for Example 8.1 to illustrate the case where freshwater purity is not superior to all other streams. In this case, freshwater has a concentration of 30 ppm. A real-life example of cases in which some sinks or sources exist at a concentration below freshwater concentration is a semiconductor plant where ultrapure water is used widely, and the wastewater produced is cleaner than freshwater.

Tab. 8.1: Limiting data for Example 8.1.

Sink	F, t/h	C, ppm	m, kg/h
SK1	70	20	1.4
SK2	70	30	2.1
SK3	100	120	12
Source			
SR1	50	10	0.5
SR2	75	50	3.75
SR3	100	150	15

Step 1: Plot the sources and sinks from the cleanest to the dirtiest cumulatively to form an SSCC.
Drawing the SSCC, cumulative water sink lines are plotted first in ascending order of concentration. The cumulative water source(s) that have concentrations lower than the freshwater concentration are plotted next, corresponding to SR1. A locus of the utility line is then drawn starting from the end of the SR1 line, with a slope of 30 ppm. The remaining water source line (also arranged in ascending order), that is, SR2 and SR3, is then shifted to the right along the utility line until all the water source lines are on the right-hand side of the sink line. The point where the water source line touches the water sink line is the Pinch Point. The new utility flowrate is given by the length of the horizontal freshwater line after shifting the remaining water source lines to the right. The minimum wastewater target is the overshoot of the source line. The Source and Sink Composite Curves for Example 8.1 are shown in Fig. 8.9. The

Pinch Point is noted at 50 ppm. This defines the Region Below and Above the Pinch. The freshwater and wastewater targets are 75 t/h and 60 t/h.

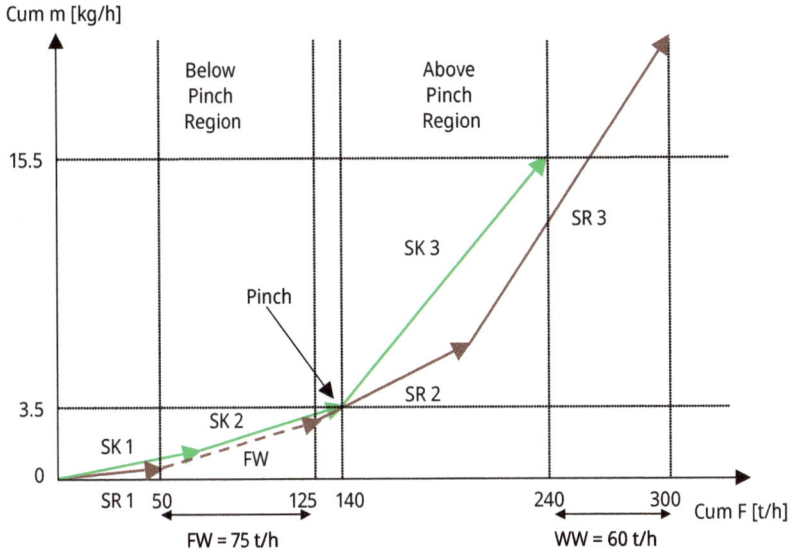

Fig. 8.9: Source and Sink Composite Curves for Example 8.1.

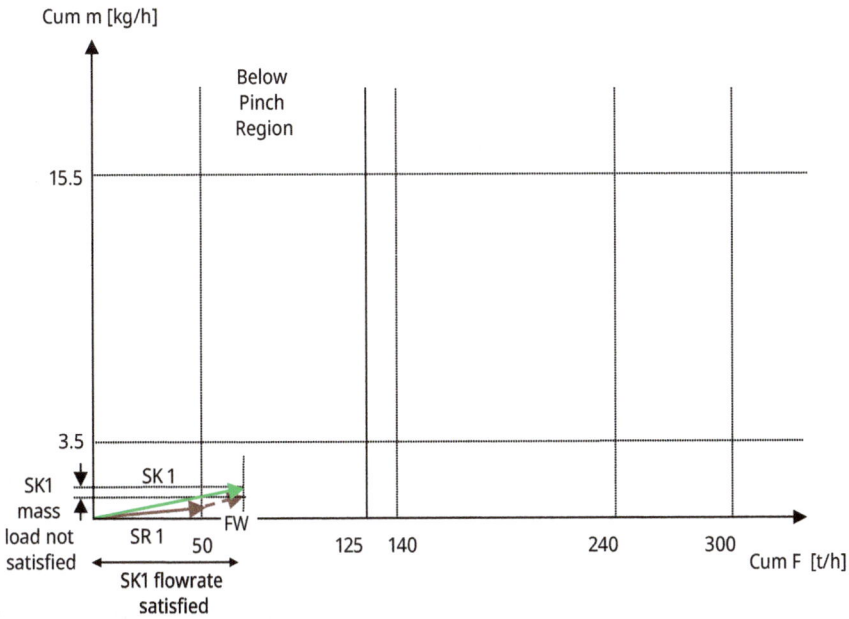

Fig. 8.10: SSAC for Example 8.1 using Rule 1 satisfying SK1 for the Region Below Pinches.

Step 2: Draw the SSAC based on the proposed rules.

1. Region Below the Pinch

    There were two sinks located Below the Pinch Region, i.e. SK1 and SK2 (from Fig. 8.9). SK1 required 70 t/h of water with 1.4 kg/h mass loads. Using *Rule 3(i)*, the utility was treated as a source as well since the utility concentration was not superior to all other streams. Using *Rule 1*, the cleanest cumulative source was used first to satisfy the flowrate and mass load of SK1. It can be seen from Fig. 8.10 that the use of all SR1 and part of the utility or freshwater line satisfy only the flowrate but not the mass load of SK1. Based on *Rule 3(ii)*, the dirtiest source in the Region Below the Pinch was used. Below the Pinch Region, the SR2 line was shifted downward along freshwater and SR2. 50 t/h, 5 t/h, and 15 t/h of SR1, freshwater and SR2 were needed to satisfy both SK1 flowrate and mass load. The remaining freshwater line at 70 t/h flowrate was then used to satisfy SK2 as *Rule 3(iii)* stated. The final SSAC for the Region Below the Pinch for Example 8.1 is shown in Fig. 8.11.

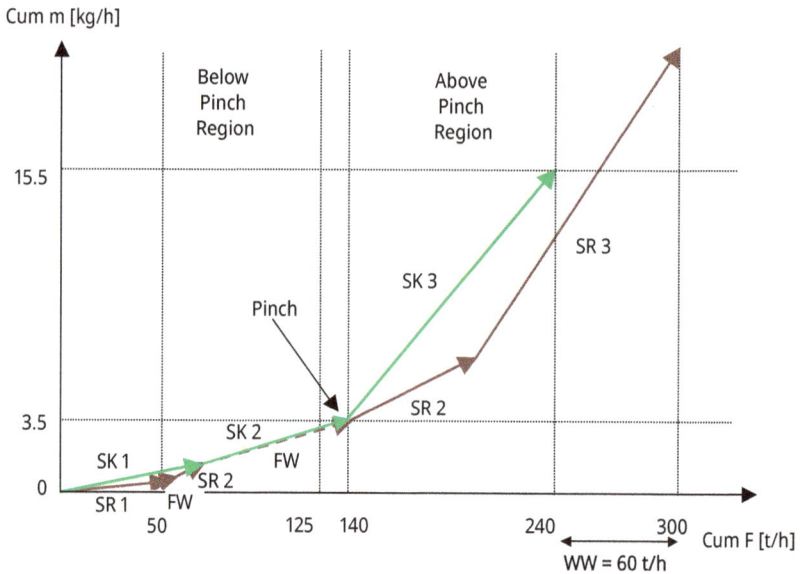

Fig. 8.11: Final SSAC for Example 8.1.

2. The Region Above the Pinch

    Finally, to satisfy the Above Pinch Regions flowrate, the Pinch Sources in the Region were used in ascending order to satisfy the sink flowrate requirement (*Rule 4*). 60 t/h of SR2 and 40 t/h of SR3 were used to satisfy the SK3 flowrate. The amount of mass load cumulated from these two sources only totalled 9 kg/h, but SK3 could actually accept 12 kg/h. The remaining 60 t/h of SR3 were rejected as wastewater since there was no more sink to be satisfied (*Rule 5*). The Above Pinch Source and Sink Allocation are shown in Fig. 8.11.

Step 3: Draw the NAD based on the SSAC.
The NAD was drawn upon completion of Steps 1 and 2. The final network diagram for Example 8.1 is as in Fig. 8.12.

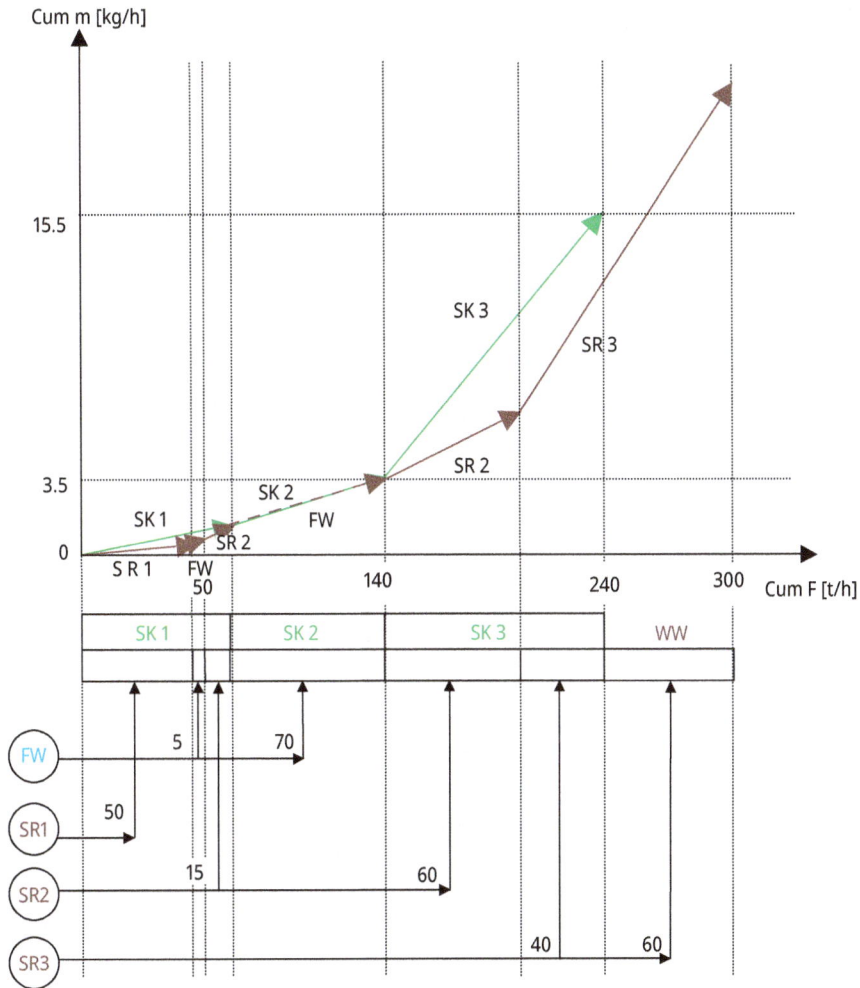

Fig. 8.12: Final Network Allocation Diagram for Example 8.1.

8.3.3 Simplification of a water network or constructing other network possibilities

The network diagram obtained using the step-wise procedure described previously was only one of the many possible network designs that could achieve the minimum freshwater and wastewater targets. The network allocation proposed, however, yielded the minimum freshwater and wastewater targets that resulted in a complex network due to the many mixings of streams and might pose operability problems associated with geographical constraints, safety and cost. The SSAC from Step 2 can actually be further used to explore various other network possibilities in order to reduce the network complexity. The rules for other various source and sink allocation possibilities are as follows:

> 6. *For allocating source and sink without water penalty, shift, cut and allocate the sources line to satisfy any sink in the same Pinch Region provided that the sink mass load and flowrate requirements are fully satisfied.*
>
> 7. *For allocating source and sink with water penalty, satisfy a sink using the minimum number of source streams available either in the same Pinch Region or from other Pinch Regions. A penalty for increased freshwater consumption would be incurred.*

Rule 6 applies to other source and sink allocations that still achieve the minimum freshwater and wastewater targets. *Rule 7* applies for source and sink allocations that yield a simpler water network but result in a penalty of increased freshwater consumption.

For example, Fig. 8.13 is the SSAC constructed using Step 2 from the previous section for Example 7.1 limiting water data. Using *Rule 6*, Fig. 8.14 is another possible SSAC that achieved the minimum freshwater and wastewater targets in addition to Fig. 8.13. In this case, only sources Below the Pinch Region were cut and shifted around. For instance, part of SR2 ($F_{SR2,SK1}$ = 10 t/h) was cut and shifted for use with SK1 instead of SK2 initially. This caused the FW needed to satisfy SK1 to increase to 40 t/h. Next, the network was simplified by shifting all SR1 ($F_{SR1,SK2}$ = 50 t/h) for use only with SK2, with part of SR2 ($F_{SR2,SK2}$ = 10 t/h) and SR3 also cut and shifted ($F_{SR3,SK2}$ = 10 t/h). All the remaining FW were used to satisfy SK2 ($F_{FW,SK2}$ = 30 t/h), and all SK3 was satisfied using the remaining SR2 ($F_{SR2,SK3}$ = 80 t/h).

Fig. 8.15 is a possible SSAC based on *Rule 7* that yielded a simpler structure with fewer splits. However, penalties for freshwater and wastewater were incurred, with the new freshwater target of 100 t/h and 80 t/h compared to the initial targets of 70 t/h and 50 t/h. The sources from Above and Below the Pinch were used in any region to satisfy the sink flowrate but not necessarily the sink mass load. The sink was either fed with sources that had the same total mass load as the sink needed or lesser. From Fig. 8.15 it can be seen that:

- All SK1 are fulfilled using freshwater;
- All SR1 are used to satisfy SK2, and the remaining flowrate of SK2 was satisfied using freshwater;
- All SR2 are used to satisfy SK3, and the remaining SR2 is sent to wastewater;
- All SR3 are used for SK4, and the remaining SR4 becomes wastewater.

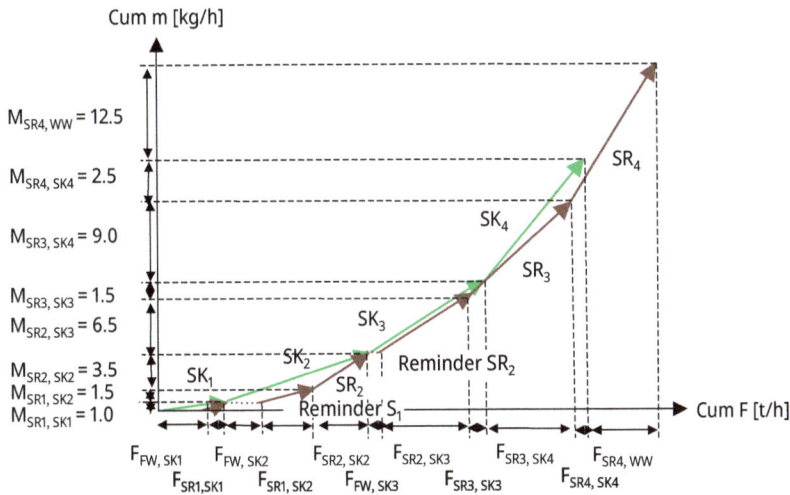

Fig. 8.13: The SSAC using Step 2 for Example 7.1.

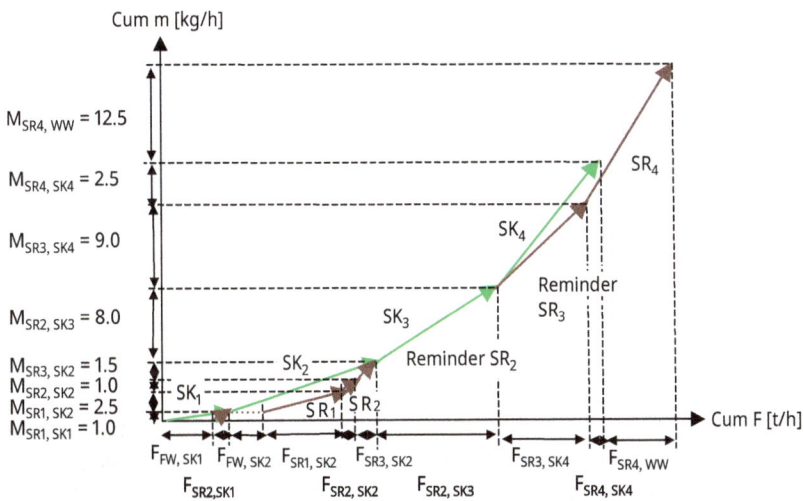

Fig. 8.14: Another possible SSAC achieving the same minimum freshwater and wastewater flowrate targets.

Based on the two examples, it can be clearly seen that SSAC could provide a powerful guideline for network modifications based on the two proposed rules. A designer could opt to design a network with or without water penalty using the SSAC. Once satisfied with the SSAC, the designer could proceed to network design using Step 3.

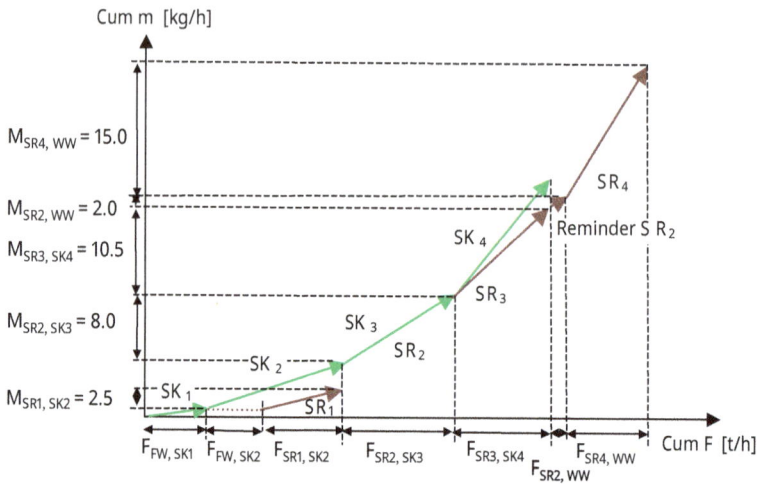

Fig. 8.15: A possible SSAC with freshwater and wastewater penalty.

8.4 Working session

As an assignment, construct the water network design for Example 6.1 by using SSMD and SSAC methods.

8.5 Solution

1. Source/Sink Mapping Diagram
   Fig. 8.16 shows the completed Source/Sink Mapping Diagram for Example 6.1. The sources are arranged in ascending order vertically on the left, while the sinks are arranged in ascending order horizontally on the top.

   SK3 and SK4 are both satisfied by using freshwater since it requires a mass load of 0 kg/h, and there are no available sources with $\Delta m = 0$ kg/h. SK5, which requires Δm of 0.15 kg/h, is satisfied by using 10 t/h of SR4 ($\Delta m = 0.1$ kg/h) and 0.5 t/h of SR5 ($\Delta m = 0.05$ kg/h). The remaining flowrate of 4.5 t/h required by SK5 is satisfied by using freshwater. SK1, which requires Δm of 8 kg/h, is satisfied by using 4.5 t/h of SR5 ($\Delta m = 0.45$ kg/h), 40 t/h of SR3 ($\Delta m = 4$ kg/h) and 5.07 t/h of SR2 ($\Delta m = 3.55$ kg/h). The remaining required SK1 flowrate of 30.43 t/h is satisfied by using freshwater.

   SK2, which requires Δm of 10 kg/h, is satisfied with 14.29 t/h of SR2 ($\Delta m = 10$ kg/h). The remaining flowrate of 35.71 t/h required by SK2 is satisfied by using freshwater. Since all sinks have been satisfied, the remaining 30.64 t/h of SR2 and 20 t/h of SR1 are sent to the wastewater treatment plant. The summation for all

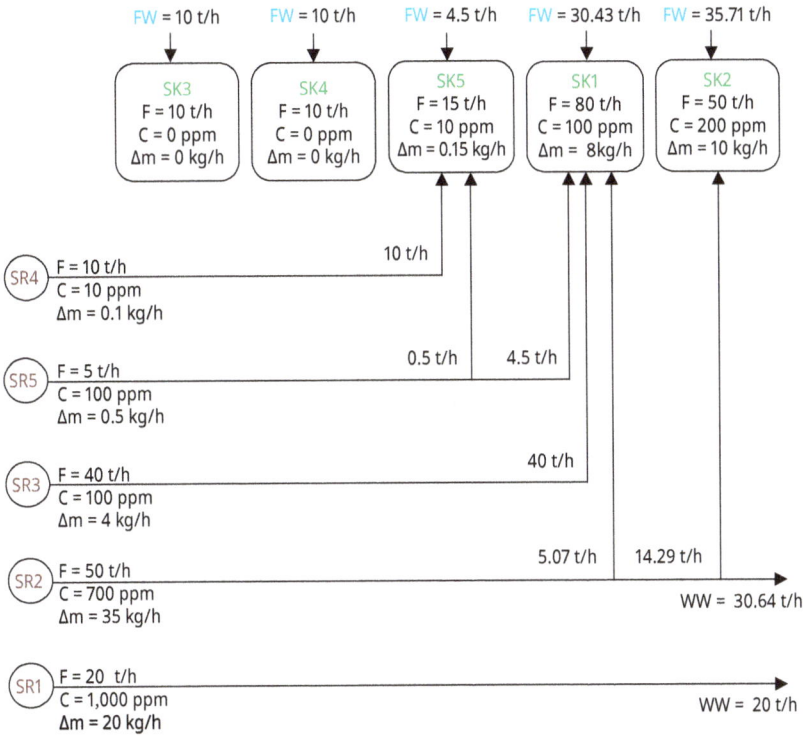

Fig. 8.16: Source/Sink Mapping Diagram for Example 6.1.

freshwater is 90.64 t/h and for all wastewater is 50.64 t/h. This is similar to the targeted value in Chapter 7, and the network design is correct.

2. Source and Sink Allocation Curve

The minimum freshwater and wastewater targets are first targeted by using the SSCC, as described in the previous chapter. The Pinch location is noted to be at SR2 with a concentration of 700 ppm. Next, SSAC is constructed for Example 6.1 by using the methodology described in Section 8.3. Below the Pinch Region, SSAC is performed first. For SK3 and SK4, freshwater is used since there are no sources with zero mass load. SK5 is matched with SR4 and part of SR5. Since this is a *flowrate deficit case*, therefore, freshwater is added until the sink flowrate is satisfied. The remaining sinks are satisfied with the remaining sources. Since all source and sink matches are *flowrate deficit cases*, freshwater is added in order to satisfy the flowrate requirements. For the Above Pinch Region, since there are no more sinks, all the remaining water sources are sent to the wastewater treatment plant.

The SSAC results are then directly translated into NAD, which looks almost similar to SSMD, but with the length of the sources and sinks representing the actual amount of flowrate transferred. Fig. 8.17 shows the SSAC and NAD for Example 6.1. The freshwater and wastewater targets are similar to the previous method.

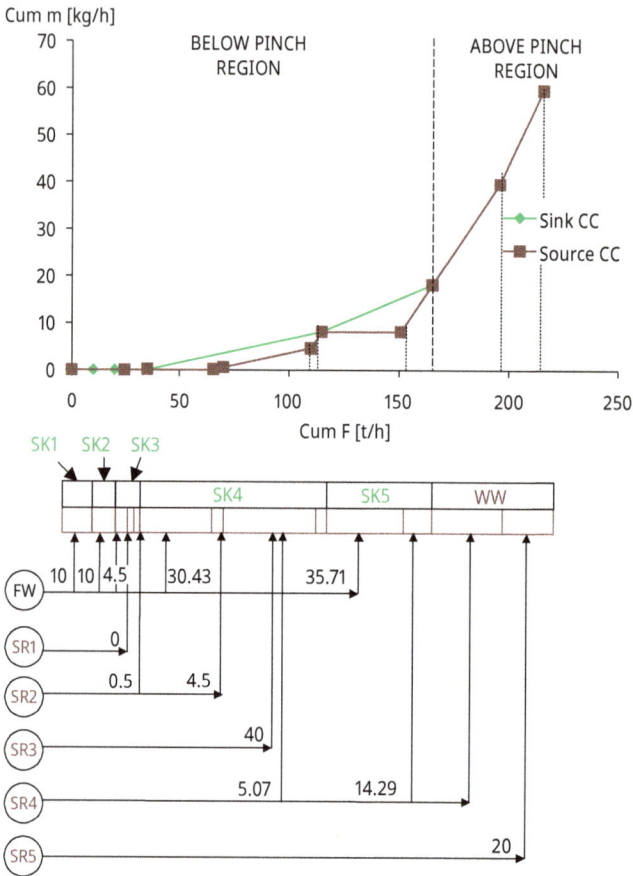

Fig. 8.17: Source/Sink Allocation Diagram and Network Allocation Diagram for Example 6.1.

8.6 Optimal Water© software

Optimal Water© (2006) (formerly known as Water MATRIX©) is software developed by Process Systems Engineering Centre (PROSPECT), Universiti Teknologi Malaysia. The software can target and design the Maximum Water Recovery network using the Water Pinch Analysis method. It incorporates the Balanced Composite Curves, Water Surplus Diagram, Water Cascade Analysis, and Source/Sink Mapping Diagram methods. A user only needs to key in the source and sink flowrates and contaminant concentrations.

The software tool automatically generates the BCC, WSD, WCA, and SSMD. Figs. 8.18 and 8.19 show some of the software's features.

Conc, C (ppm)	ΔP	Sum F Source (kg/s)	Total F (kg/s)	Cum. water flowrate (kg/s)	Cum. water surplus (kg/s)
				373.296	
0			0		
	0.00002			373.296	
20			−466.62		0.00746592
	0.00008			−93.324	
100		201.84	201.84		0
	0.00005			108.516	
150		1,131.54	1,131.54		0.0054258
	0.00002			1,240.056	
170		390.96	−860.88		0.03022692
	0.00005			379.176	
220		265.2	265.2		0.04918572
	0.00006			644.376	
280			−68.7		0.08784828
	0.00002			575.676	
300		68.7	68.7		0.0993618
	0.9997			644.376	
					644.282049

Fig. 8.18: Optimal Water: Water Cascade Table.

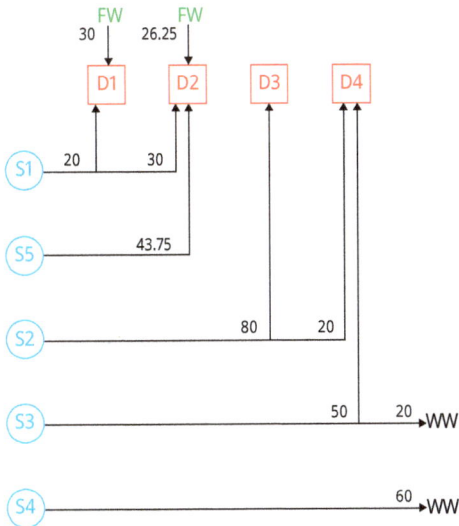

Fig. 8.19: Optimal Water: Source/Sink Mapping Diagram.

References

El-Halwagi M M, Gabriel F, Harell D. (2003). Rigorous Graphical Targeting for Resource Conservation via Material Recycle/Reuse Networks, Industrial & Engineering Chemistry Research, 42, 4319–4328. doi: 10.1021/ie030318a.

Hallale N. (2002). A New Graphical Targeting Method for Water Minimization, Advances in Environmental Research, 6(3), 377–390. doi: 10.1016/S1093–0191(01)00116–00112.

Huang T, Zhou W, Zhang Y, Jia X, Zhang D, Li Z. (2023). An Insight-based Approach for Batch Water Network with Flexible Production Scheduling Framework, Journal of Cleaner Production, 385, 135664.

Kazantzi V, M E-H M. (2005). Targeting Material Reuse via Property Integration, Chemical Engineering Progress, 101(8), 28–37.

Misrol M A, Wan Alwi S R, Lim J S, Manan Z A. (2021). An Optimal Resource Recovery of Biogas, Water Regeneration, and Reuse Network Integrating Domestic and Industrial Sources, Journal of Cleaner Production, 286, 125372.

Moghaddam A M F, Sahlodin A M, Sarrafzadeh M-H. (2022). A Backoff Approach to Design of Optimally Flexible Water Networks under Uncertainty, Journal of Cleaner Production, 371, 133396.

Optimal Water©. (2006). Software for Water Integration in Industries. Copyright of Universiti Teknologi, Malaysia, Johor, Malaysia. http://optimalsystems.my, accessed on 16/10/2022.

Polley G T, Polley H L. (2000). Design Better Water Networks, Chemical Engineering Progress, 96(2), 47–52.

Wan Alwi S R, Ismail A, Manan Z A, Handani Z B. (2011). A New Graphical Approach for Simultaneous Mass and Energy Minimisation, Applied Thermal Engineering, 31(6–7), 1021–1030.

Wan Alwi S R, Manan Z A. (2008). Generic Graphical Technique for Simultaneous Targeting and Design of Water Networks, Ind, Engineering Chemistry Research, 47(8), 2762–2777. doi: 10.1021/ie071487o.

9 Design of Cost-Effective Minimum Water Network (CEMWN)

9.1 Introduction

It is important to note that the concept of Maximum Water Recovery (MWR) only relates to the maximum reuse, recycling, and regeneration (partial treatment before reuse) of spent water. To achieve the minimum water target, all conceivable methods to reduce water usage through elimination, reduction, reuse/recycling, outsourcing, and regeneration need to be considered (Manan and Wan Alwi, 2006). Regenerating wastewater without considering the possibility of elimination and reduction may lead to unnecessary treatment units. The use of water minimisation strategies beyond recycling was first introduced by El-Halwagi (1997), who proposed a targeting technique involving water elimination, segregation, recycling, interception, and sink/source manipulation. Hallale (2002) introduced guidelines for reduction and regeneration based on WPA. However, the piecemeal water minimisation strategies proposed do not consider interactions among the process change options as well as the "knock-on effects" of process modifications on the overall process balances, stream data and the economics. The work of Wan Alwi and Manan (2008) has overcome this limitation. The authors have proposed a new framework by using the Water Management Hierarchy (WMH) as a guide to prioritise process changes and the *Systematic Hierarchical Approach for Resilient Process Screening* (*SHARPS*) strategies as a new cost-screening technique to select the most cost-optimum solution. In Section 9.1, we began by explaining WMH as a foundation for the holistic framework. This is followed by descriptions of a five-step methodology for designing a Cost-Effective Minimum Water Utilisation Network (CEMWN) in Section 9.2. The stepwise application of CEMWN methodology on a semiconductor plant is demonstrated in Section 9.3.

9.2 Water Management Hierarchy

Fig. 9.1 shows the Water Management Hierarchy (WMH) introduced by Wan Alwi and Manan (2008) consisting of five levels: 1) source elimination, 2) source reduction, 3) direct reuse/outsourcing of external water, 4) regeneration, and 5) use of freshwater. Each level represents various water management options. The levels are arranged in order of preference, from the most preferred option at the top of the hierarchy (level 1) to the least preferred at the bottom (level 5). Water minimisation is concerned with the first to the fourth levels of the hierarchy.

    Source elimination at the top of the hierarchy is concerned with the complete avoidance of freshwater usage. Sometimes it is possible to eliminate water rather than to reduce, reuse or recycle water. Examples include using alternative cooling media such as

https://doi.org/10.1515/9783110782981-009

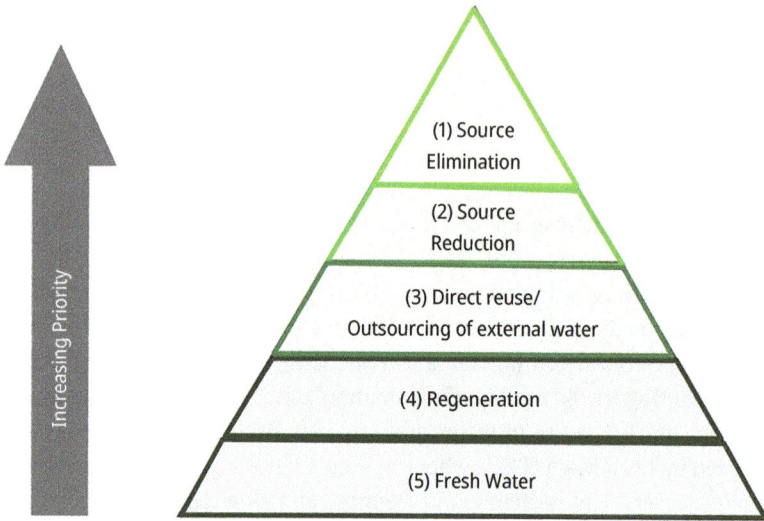

Fig. 9.1: The Water Management Hierarchy (Wan Alwi and Manan, 2008).

air instead of water. Even though source elimination is the ultimate goal, it is often not possible to completely eliminate water. It has to be tried to reduce the amount of water being used at the source of water usage, i.e. certain equipment or processes. Such measures are referred to as *source reduction*, which is the next best option in the WM hierarchy (level 2). Examples of source reduction equipment include water-saving toilet flushing systems and automatic taps. *Source elimination* or *reduction* can be achieved in many ways, e.g. technology and equipment changes, input material changes, good operation practices, product composition changes and reaction changes (Perry and Green, 1998). In the case of good operation practices, it may involve no cost at all, and with all the other changes, the cost varies from low to high.

When it is not possible to eliminate or reduce freshwater at the source, wastewater recycling should be considered. Levels 3 and 4 in the WM hierarchy represent two different modes of water recycling: *direct reuse/outsourcing* (level 3) and *regeneration reuse* (level 4). Direct reuse or outsourcing may involve using spent water from within a building or using an available external water source (e.g. rainwater or river water). Through direct reuse (level 3), spent water or an external water source is utilised to perform tasks which can accept lower quality water. For example, wastewater from a bathroom wash basin may be directly channelled to a toilet bowl for toilet flushing. Rainwater, on the other hand, may be used for tasks which need higher quality water.

Regeneration involves partial or total removal of water impurities through the treatment of wastewater to allow the regenerated water to be reused. Examples of regeneration units are gravity settling, filtration, membranes, activated carbon, biological treatment, etc. These regeneration units can be used in isolation or in combination. There are two possible cases of regeneration. Regeneration-recycling involves the reuse of treated water in the same equipment or process after treatment. Regeneration-reuse involves the reuse of treated water in other equipment after treatment. To increase water availability, the Water Composite Curves and the Pinch concentration can be used to guide the regeneration of water sources as follows (Hallale 2002):

1. *Regeneration Above the Pinch:*
    Water source(s) in the region Above the Pinch are partially treated to upgrade their purity.
2. *Regeneration Across the Pinch:*
    Water source(s) in the region Below the Pinch are partially treated to achieve purity higher than the Pinch purity.
3. *Regeneration Below the Pinch:*
    Water source(s) in the region Below the Pinch are partially treated to upgrade their purity. However, the resulting water source is still maintained Below the Pinch.

Note that regeneration Below and Across the Pinch would reduce freshwater consumption and wastewater generation, while regeneration Above the Pinch would only reduce wastewater generation. A water regeneration unit can be categorised as a fixed outlet concentration (C_{Rout}) and removal ratio (RR) type (Wang and Smith, 1994). For the fixed C_{Rout} type, the wastewater concentration outlet is assumed to be the same regardless of the initial inlet concentration (C_{Rin}) of the water source. For the RR type, the RR value is fixed, C_{Rout} is affected by C_{Rin} and can be calculated by using eq. (9.1), where f_{in} and f_{out} are the inlet and outlet flowrate of the regeneration unit.

$$RR = \frac{f_{in}C_{Rout} - f_{out}C_{Rin}}{f_{in}C_{Rin}} \tag{9.1}$$

Freshwater usage (Level 5) should only be considered when wastewater cannot be recycled or when wastewater needs to be diluted to obtain the desired purity. Note that wastewater has to undergo *end-of-pipe* treatment before discharge to meet the environmental guidelines. The use of freshwater is the least desirable option from the water minimisation point of view and is to be avoided whenever possible. Through the WM hierarchy, the use of freshwater may not be eliminated, but it becomes economically legitimate.

9.3 Cost-Effective Minimum Water Network (CEMWN)

The Cost-Effective Minimum Water Network (CEMWN) design procedure is a holistic framework for water management applicable to industry and urban sectors introduced by Wan Alwi and Manan (2008). Fig. 9.2 illustrates five key steps involved in generating the CEMWN, i.e. 1) specify limiting water data, 2) determine the Maximum Water Recovery (MWR) targets, 3) screen process changes using WMH, 4) apply SHARPS strategies, and 5) design CEMWN. The first step is to identify the appropriate water sources and water sinks having the potential for integration. The next step is to establish the MWR targets using the Water Cascade Analysis (WCA) technique of Manan et al. (2004). The WM hierarchy, along with a set of new process screening heuristics, is then used to guide process changes to achieve the minimum water targets. The fourth step is to use SHARPS strategies to economically screen inferior process changes. The CEMWN is finally designed using established techniques for the design of water networks. The stepwise approach is described in detail next.

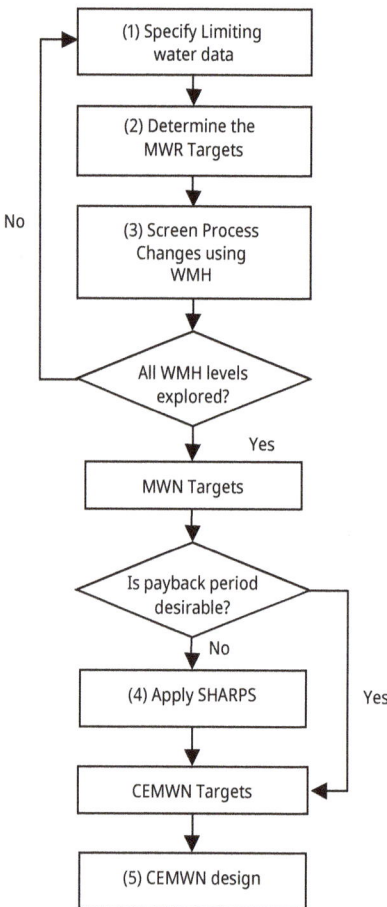

Fig. 9.2: A holistic framework to achieve CEMWN (Wan Alwi and Manan, 2008).

Step 1: Specify the limiting water data

The first step is to specify the limiting water data as described in Chapter 6.

Step 2: Determine the MWR targets

The second step is to establish the *base-case* MWR targets, i.e. the overall freshwater requirement and wastewater generation as described in Chapter 7. Note that the *base-*

case MWR targets exclude other levels of WMH except reuse and recycling of available water sources and mixing of water sources with freshwater to satisfy water sinks.

Step 3: Screen process changes using WMH

Changes can be made to the flowrates and concentrations of water sources and sinks to reduce the MWR targets and ultimately achieve the MWN benchmark. This was done by observing the basic Pinch rules for process changes and by prioritising as well as assessing all possible process change options according to the WM hierarchy. The fundamental rules to change a process depend on the location of water sources and sinks relative to the Pinch Point of a system:

1. Below the Pinch: Beneficial changes can be achieved by either increasing the flowrate or purity of a source or by decreasing the flowrate or purity requirements of a sink. These changes will increase the water surplus Below the Pinch and reduce the amount of freshwater required.
2. Above the Pinch: There is already a surplus of water Above the Pinch, and any flowrate change made there will not affect the target. An exception to this rule of thumb is for the case where source purity is increased so that it moves to the region Below the Pinch, as in the case of regeneration.
3. At the Pinch Point: Increasing the flowrate of a source at the Pinch concentration will not reduce the targets.

It is vital to note that implementation of each process change option yields new Pinch Points and MWR targets. In addition, interactions and "knock-on effects" between the process change options should also be carefully considered. It is therefore, important that each process change be systematically prioritised and assessed with reference to the revised Pinch Points instead of the original Pinch Point so as to obey the fundamental rules for the previously listed process changes and to guarantee that the MWN benchmark is attained. Bearing in mind these constraints, the core of step 3 is the level-wise hierarchical screening and prioritisation of process change options using the WMH and the following four option-screening heuristics, which were sequentially applied to prioritise process changes at each level of WMH. As described below, not all four heuristics are applicable at each level of WMH.

Heuristic 9.1: Begin process changes at the core of a process.

Heuristic 9.1 was formulated from the *Onion Model* for process creation shown in Fig. 9.3 (Smith, 1995). Due to interactions among reaction, separation and recycle, heat and mass exchange network and utility layers, any changes, such as sink elimination should be implemented beginning from the core of a process (reaction system) to the most outer layer (utilities). Excessive water usage at the core of the system causes wastage at the outer layers. Hence improving the core of the system first will eliminate or reduce wastage downstream.

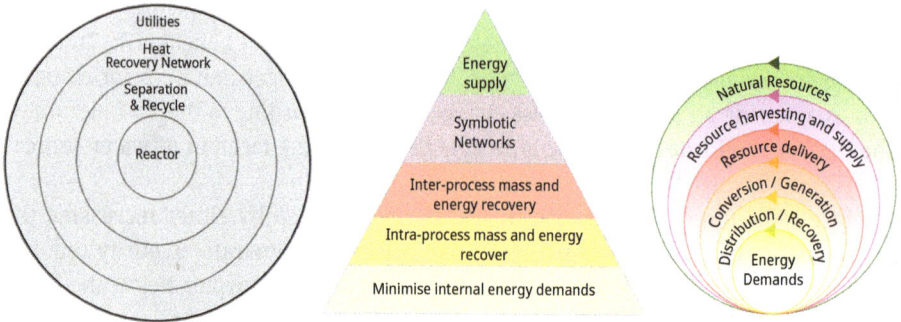

(a) Industrial Heat Integration process (Linnhoff et al. 1982) (b) A generalised energy management hierarchy (Seferlis et al., 2021)

Fig. 9.3: The Onion Diagram as a conceptual model of the hierarchy of Process Design and Optimisation.

Heuristic 9.1 is strictly applied to the process change options at levels 1 and 2 of WMH. Applying Heuristic 9.1 to various source elimination options at level 1 of the WMH will lead to new targets and Pinch Points. For mutually exclusive options, the one giving the lowest revised MWR targets was selected. Heuristic 9.1 was repeated to reduce water at WMH level 2 once all elimination options were explored.

Note also that it is quite common for processes to have independent and non-interacting sources and sinks at various concentrations. For example, reducing the water sink for a scrubber in a waste treatment system does not affect the cooling tower sink. In such a case, the sink flowrate Above the Pinch (see Rule (i) for process changes mentioned previously) can be reduced using Heuristic 9.2.

> **Heuristic 9.2:** Successively reduce all available sinks with a concentration lower than the Pinch Point, beginning from the cleanest sink.

Note that if a dirtier sink were reduced first, followed by a cleaner sink, it might be found later that subsequent reduction of a cleaner sink might cause the dirtier sink to lie Below the New Pinch Point. Such a situation makes the earlier changes to the dirtier sink meaningless.

If a few sinks exist at the same concentration, it is best to begin by reducing the sink that yields the most flowrate reduction to achieve the biggest savings. The next step proceeds to reduce the remaining sinks that exist at concentrations lower than the revised

Pinch concentration, as stated in Heuristic 9.3. Heuristics 9.2 and 9.3 were applicable to levels 1 and 2 of the WMH.

Heuristic 9.3: Successively reduce the sinks starting from the one giving the biggest flowrate reduction if several sinks exist at the same concentration.

There exists a maximum limit for adding new water sources (utilities) either obtained externally, such as rainwater, river water, snow and borehole water or by regenerating wastewater in order to minimise freshwater in a water distribution system. It is, therefore, necessary to:

Heuristic 9.4: Harvest outsourced water or regenerate wastewater only as needed.

Note that the limit for adding utilities through outsourcing and regeneration corresponds to the minimum utility flowrate (F_{MU}), which leads to minimum freshwater flowrate. To calculate F_{MU}, refer to Section 7.4. Heuristic 9.4 only applies to levels 3 and 4 of the WMH.

The revised MWR targets, as well as the four new option-screening heuristics, were used as process selection criteria. The screening and selection procedure was hierarchically repeated down the WMH levels to establish the minimum water network (MWN) targets which yielded the maximum scope for water savings. SHARPS strategy was used next to ensure that the savings achieved were cost-effective and affordable.

Step 4: Apply the SHARPS strategy
The SHARPS screening technique involves cost estimations associated with water management (WM) options prior to detailed design. It includes a profitability measure in terms of the payback period, i.e. the duration for capital investment to be fully recovered. Note that the payback period calculations for SHARPS, as given by eq. (9.2), only concern the economics associated with the design of a minimum water network as opposed to the design of an entire plant:

$$\text{Payback period (y)} = \frac{\text{Net Capital Investment (\$)}}{\text{Net Annual Savings (\$/y)}}. \tag{9.2}$$

Since *SHARPS* is a cost-screening tool, standard plant design *preliminary cost estimation* techniques are used to assess the capital and operating costs of a proposed water system. The equipment, piping and pumping costs built in eq. (9.3) are the three main cost components considered for a building or a plant water recovery system:

$$\sum CC = C_{PE} + C_{PEI} + C_{piping} + C_{IC} \tag{9.3}$$

Where

C_{PE} = Total capital cost for the equipment in $,
C_{PEI} = Equipment installation cost in $,
C_{piping} = Water reuse piping cost investment in $, and
C_{IC} = Instrumentation and controls cost investment in $.

The economics of employing the WM options for grassroots design as well as retrofit cases were evaluated by calculating the Net Capital Investment (NCI) for the MWN using eqs. (9.2) and (9.3) as well as the net annual savings (NAS) using eqs. (9.4) and (9.5). In the context of SHARPS, the NCI for grassroots refers to the cost difference between the new (substitute) equipment and the base-case equipment. The base-case equipment is the initial equipment used before CEMWN analysis. For the retrofit case, the NCI covers only the newly installed (substitute) system:

$$\text{New Capital Investment, \$ (grassroots)} = \sum CC_{\text{new system}} - \sum CC_{\text{base case}} \qquad (9.4)$$

$$\text{New Capital Investment, \$ (retrofit)} = \sum CC_{\text{new system}} \qquad (9.5)$$

Where $CC_{\text{new system}}$ = Capital cost associated with new equipment in $ and $CC_{\text{base-case}}$ = Capital cost for base-case equipment in $. For example, a new $300, 6 L toilet flush gives water savings of 6 L per flush as compared to a $200, 12 L toilet flush (base-case system). For grassroots design, the payback period is therefore based on the NCI given by eq. (9.4), i.e. $100. For retrofit, the payback period is based on the NCI given by eq. (9.5), i.e. $300.

The net annual savings (NAS) is the difference between the base-case water operating cost and the water operating costs after employing WM options as in eq. (9.6):

$$NAS = OC_{\text{base-case}} - OC_{\text{new}} \qquad (9.6)$$

Where NAS = Net annual savings ($/y), $OC_{\text{base-case}}$ = base-case expenses on water ($/y), and OC_{new} = new expenses on water after modifications ($/y).

The total operating cost of a water system includes freshwater cost, effluent disposal charges, the energy cost for water processing and the chemical cost as given by eq. (9.7):

$$OC = C_{FW} + C_{WW} + C_{EOC} + C_c \qquad (9.7)$$

Where

OC = Total water operating cost,
C_{FW} = cost per unit time for freshwater,
C_{FD} = cost per unit time for wastewater disposal,
C_{EOC} = cost per unit time for energy for water processing, and
C_c = cost per unit time for chemicals used by the water system.

In order to obtain a cost-effective and affordable water network that achieves the minimum water targets (hereby termed the *Cost-Effective Minimum Water Network*

(CEMWN)) within the desired payback period, the new *SHARPS* technique was implemented as follows:

Step 1:
Set the desired Total Payback Period (TPP_{set}). The desired payback period can be an investment payback limit set by a plant owner, e.g. two years.

Step 2:
Generate an Investment vs Annual Savings (IAS) composite plot covering all levels of the WM hierarchy. Figure 9.4 shows a sample of the IAS plot. The gradient of the plot gives the payback period for each process change. The steepest positive gradient (m_4), giving the highest investment per unit of savings, represents the most costly scheme. On the other hand, a negative slope (m_3) indicates that the new process modification scheme requires lower investment as compared to the grassroots equipment.

Note that since most equipment cost is related to equipment capacity through a power law, one is more likely to generate a curved line such as m_5. Hence, in such cases, several data points should be taken to plot a curve for each process change. As in the case of a linear line, a curve moving upward shows that more investment is needed and a curve moving downward shows less investment is needed for an increase in annual savings.

Step 3:
Draw a straight line connecting the starting point and the endpoint of the IAS plot (Fig. 9.4). The gradient of this line is a preliminary cost estimate of the Total Payback Period (TPP) for implementing all options in line with the WM hierarchy. The TPP_{BS} is the Total Payback Period before implementing SHARPS.

Step 4:
Compare the TPP_{BS} with the TPP_{set} (the desired Total Payback Period set by a designer). The Total Payback Period (TPP_{BS}) should match the maximum desired payback period set (TPP_{set}) by a designer. Thus, it is possible to tailor the minimum water network as per the requirement of a plant/building owner:
–   if $TPP_{BS} \leq TPP_{set}$, proceed with network design;
–   if $TPP_{BS} > TPP_{set}$, two strategies may be implemented.

Strategy 1 – Substitution: This strategy involved replacing the equipment/process that resulted in the steepest positive gradient with an equipment/process that gave a less steep gradient. Note that this strategy does not apply to the reuse line since there is no equipment to replace. To initialise the composite plot, the option with the highest total annual water savings should be used regardless of the total investment needed. Hence, to reduce the steepest gradient according to Strategy 1, the process change option giving the next highest total annual saving but with lesser total investment was selected to

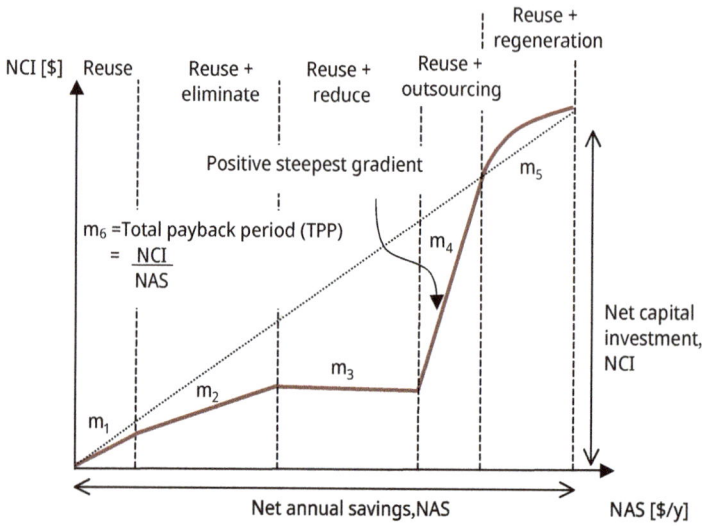

Fig. 9.4: IAS plot covering all levels of WM hierarchy; m_4 is the positive steepest gradient, and TPP is the Total Payback Period for a water network (Wan Alwi and Manan, 2008).

substitute the initial process option and trim the steepest gradient. Figure 9.5 shows that substituting the option causing the steepest positive gradient (m_4) with an option that gives a less steep gradient (m_4') yields a smaller TPP value. For example, a separation toilet may be changed to a much cheaper dual-flush toilet that uses a bit more water. TPP_{AS} is the TPP after implementing SHARPS strategies.

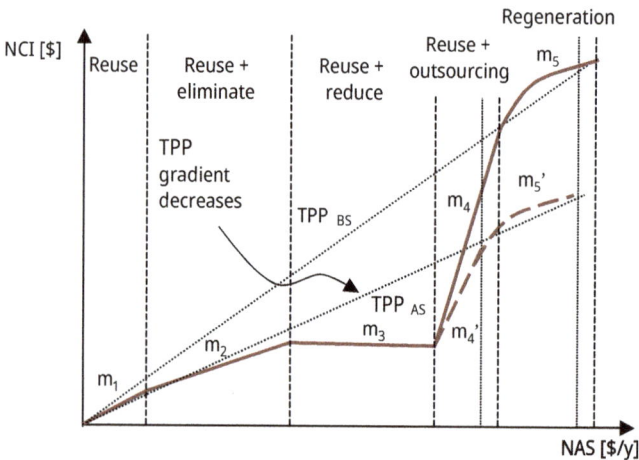

Fig. 9.5: IAS plot showing the revised Total Payback Period when the magnitude of the steepest gradient is reduced using the *SHARPS* substitution strategy (Wan Alwi and Manan, 2008).

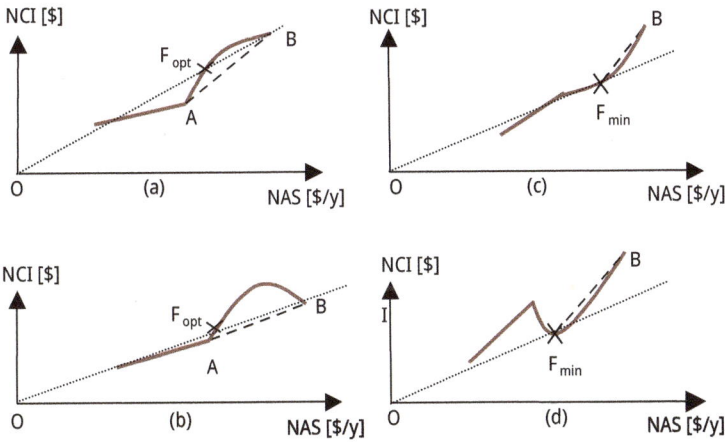

Fig. 9.6: Linearisation of concave curves moving upwards (a) without peak (b) with peak. Convex curves were moving upwards linearisation (c) without valley (d) with valley (Wan Alwi and Manan, 2008).

In the case of a curvature, linearisation is necessary to determine the line of the steepest gradient. For a projecting concave or convex curve, the linearisation of a curve moving upwards is as follows:

1. Concave curves
   Connect a straight line to the start (point A) and end (point B) points of the concave curve to obtain a positive gradient (line AB in Fig. 9.6 a,b). Connect a line from the graph origin (point O) to the end point (point B) of the concave curve. The point where line OB intersects the concave curve is the F_{opt} point. To have beneficial TPP reduction, the concave curve has to be reduced below point F_{opt} (for strategy 2).

2. Convex curve
   Connect a line from the graph origin (point O) to the minimum point of the convex curve (F_{min}). Connect a line from F_{min} to the end point (point B) of the convex curve to obtain a positive gradient (line F_{min}-B in Fig. 9.6c,d). Do not reduce further the line on the left-hand side of F_{min} since this would increase TPP (for strategy 2).

When the linearised line is the steepest positive gradient, *Strategy 1* is implemented to yield a linearised line with a smaller gradient. Note that the proposed linearisation is only a preliminary guide to screen the most cost-effective option that satisfies a preset payback period.

Strategy 2 – Intensification: The second strategy involves reducing the length of the steepest positive gradient until TPP_{AS} is equal to TPP_{set}. This second strategy is also not applicable to the reuse line since there is no equipment to replace. Figure 9.7 shows that when the length of the steepest positive gradient (m_4) is reduced, the new gradient line (m_4') gives a less steep gradient and hence, a smaller TPP. This means that instead of completely applying each process change, one can consider eliminating

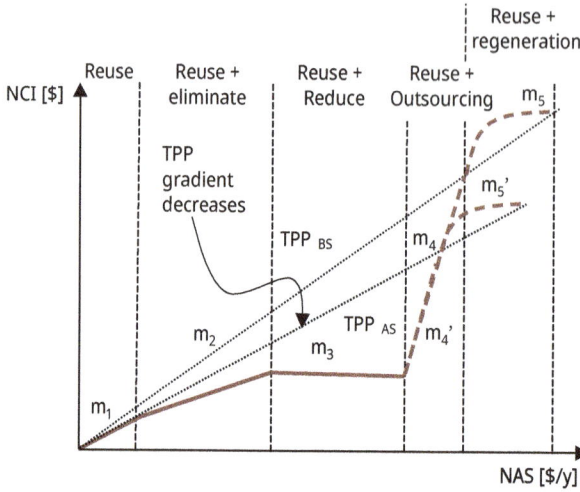

Fig. 9.7: IAS plot showing the revised Total Payback Period with a shorter steepest gradient curve (Wan Alwi and Manan, 2008).

or partially applying the process change that gives the steepest positive gradient and a small annual saving compared to the amount of investment. For example, instead of changing all normal water taps to infrared-type, only 50% of the water taps are changed. If TPP_{AS} is still more than the TPP_{set} even after adjusting the steepest gradient, the length for the next steepest gradient is reduced until TPP is equal to TPP_{set}.

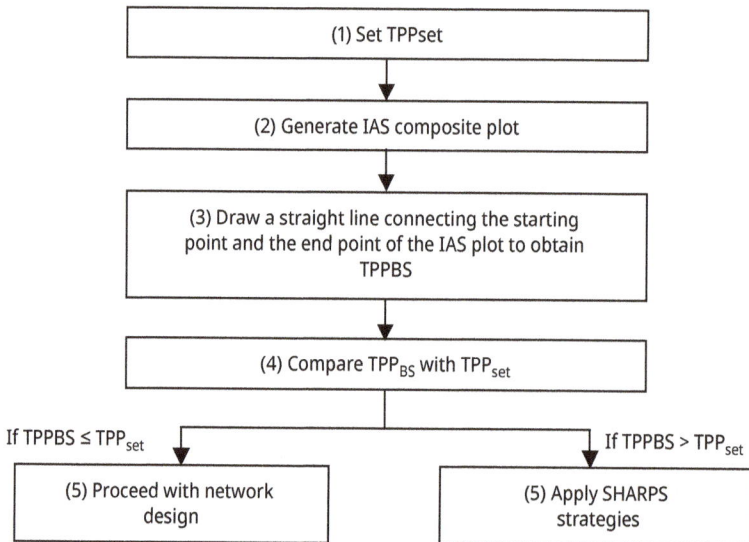

Fig. 9.8: The overall *SHARPS* procedure (Wan Alwi and Manan, 2008).

Similarly, for the case of projecting concave and convex curves moving upward, it is desirable to reduce the length of the curve until TPP_{set} is achieved if linearisation of the curve gives the steepest gradient. Both strategies 1 and 2 should be tested or applied together to yield the best savings. The overall procedure for *SHARPS* is summarised in Fig. 9.8.

Step 5: Network design

Once the CEMWN targets have been established, the next step is to design the Cost-Effective Minimum Water Network (CEMWN) to achieve the CEMWN targets. The water network can be designed using one of the established techniques as described in Chapter 8.

9.4 Industrial case study: a semiconductor plant

The CEMWN method has been implemented in a semiconductor plant in Malaysia by Wan Alwi and Manan (2008). The plant is known as MySem. MySem uses extensive amounts of water to produce wafers. A large amount of water is used to produce DI or ultrapure water. Creating DI water is among the most expensive and energy-intensive steps in the fabrication process, so any decrease in water sinks can result in significant savings. For MySem, creating DI water cost approximately €2.40/m^3. Currently, MySem is paying €15,952/month (34,618 m^3/month) to produce 118 units of the wafer. This amounts to 293 m^3 of water needed for each wafer. It further demonstrated the step-by-step procedure to obtain the CEMWN solution by using this case study.

Step 1: Specify the limiting water data

This step involved a detailed process survey and line tracing, establishing process stream material balances and conducting water quality tests. Stream flowrates were either extracted from plant-distributed control system (DCS) data or from online data logging using an ultrasonic flow meter. Depending on the stream audited, tests for total suspended solids (TSS), biological oxygen demand (BOD), chemical oxygen demand (COD), and total dissolved solids (TDS) were made on-site. MySem processes comprised entirely of ultra-pure water, and TSS were found to be very negligible. BOD was eliminated since there were no biological contaminations. COD was a component of TDS. TDS was ultimately chosen as the dominant water quality parameter for MySem. TDS was monitored using a conductivity meter. Some of the key constraints considered included the following:

– Water streams with hydrogen fluoride (HF), isopropyl butanol (IPA), and dangerous solvents were not considered water sources;
– Multimedia filter (MMF) backwash was not considered a water source since it contained high TSS;

- Wet bench WB202 and 203 cooling were not reused since they involved acid spillage;
- Black water, i.e. toilet pipes, toilet flushing, and office cleaning wastewater, were not reused;
- Grey water could only be reused for processes which did not involve body contact.

Portable ultrasonic flow meters and MySem online monitoring systems were used to establish water balances. Figures 9.9 and 9.10 show the water-using processes in terms of total flowrate in m^3/h for MySem. The Process Flow Diagram of MySem was divided into three sections, including i) DI water balance, ii) domestic water balance, and iii) fabrications (Fab) water balances.

The relevant water streams having potential for recycling were extracted into a table of *limiting water data* comprising process sources and sinks. Table 9.1 shows the water sources and sinks extracted for MySem listed in terms of flowrate and contaminant concentration. The contaminant (C, in ppm) in Tab. 9.1 represents TDS.

Step 2: Determine the Maximum Water Recovery (MWR) targets

This step involved establishing the base-case MWR targets using the WCA technique that was incorporated in the Optimal Water$^{©}$ (2006) software developed at the Universiti Teknologi Malaysia (UTM). Table 9.2 is the Water Cascade Table (WCT) generated by Optimal Water$^{©}$ (2006) for MySem, showing the freshwater and wastewater flowrate targets at $F_{FW} = 11.04$ t/h and $F_{IWT} = 0.02$ t/h. Note from Tab. 9.2 that the cleanest water targeted at 0 ppm concentration actually referred to DI water (F_{DI}) to be supplied to the blend water tank instead of freshwater. This was because freshwater for MySem had a concentration of 30 ppm. The source water flowrate at 30 ppm is shown in Tab. 9.2 was actually the amount of freshwater supply needed. Optimal Water$^{©}$ (2006) computed $F_{DI} = 0$ t/h and $F_{FW} = 11.04$ t/h.

The F_{IWT} of 0.02 t/h is shown in Tab. 9.2 only comprised the IWT wastewater that was considered for reuse. The total IWT should also include 5.69 t/h IWT wastewater not available for reuse due to chemical contamination, giving a total IWT of 5.71 t/h to be discharged for MySem, as shown in Tab. 9.3. Considering the present integration schemes implemented by MySem, the initial FW and total IWT flowrates were at 39.94 t/h and 34.45 t/h. Equations (9.8) and (9.9) gave FW and IWT reductions of 72.4% and 83.4%.

$$\text{FW savings, \%} = \frac{\text{FW flowrate before WPA} - \text{FW flowrate targets after WPA}}{\text{FW flowrate before WPA}} \times 100 \quad (9.8)$$

$$\text{IWT savings, \%} = \frac{\text{IWT flowrate before WPA} - \text{Total IWT to be discharged}}{\text{IWT flowrate before WPA}} \times 100 \quad (9.9)$$

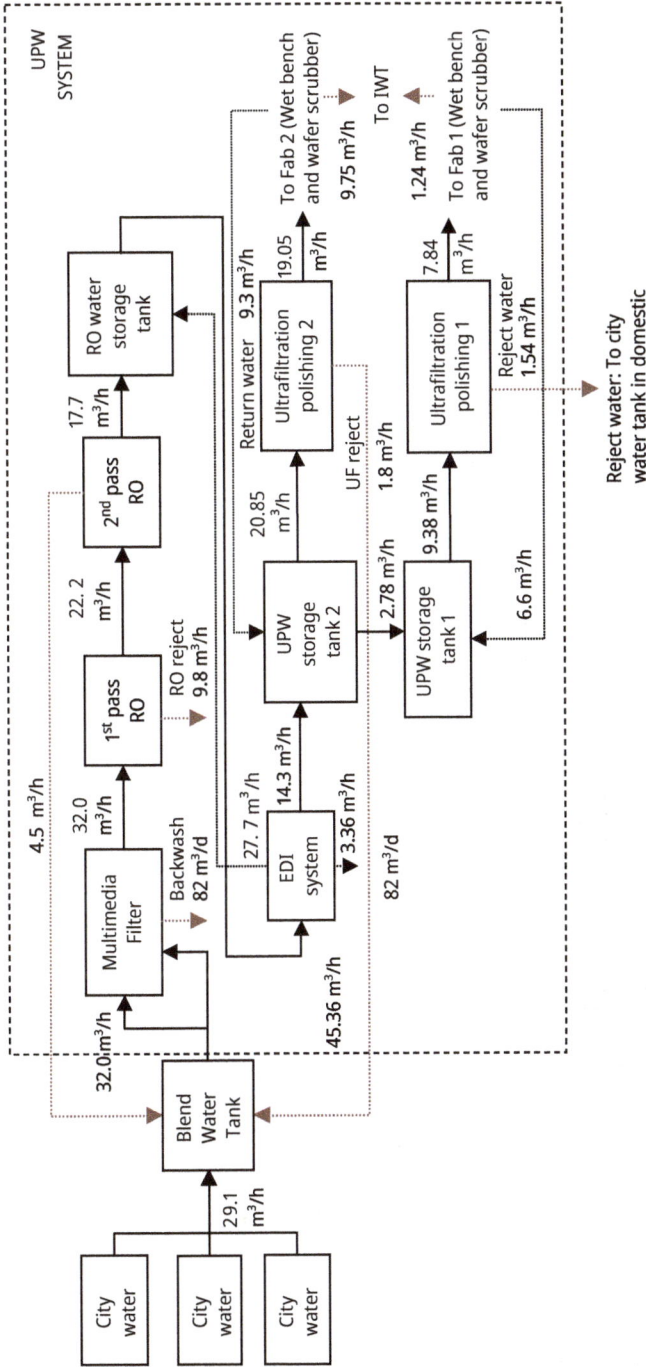

Fig. 9.9: MySem DI water balance.

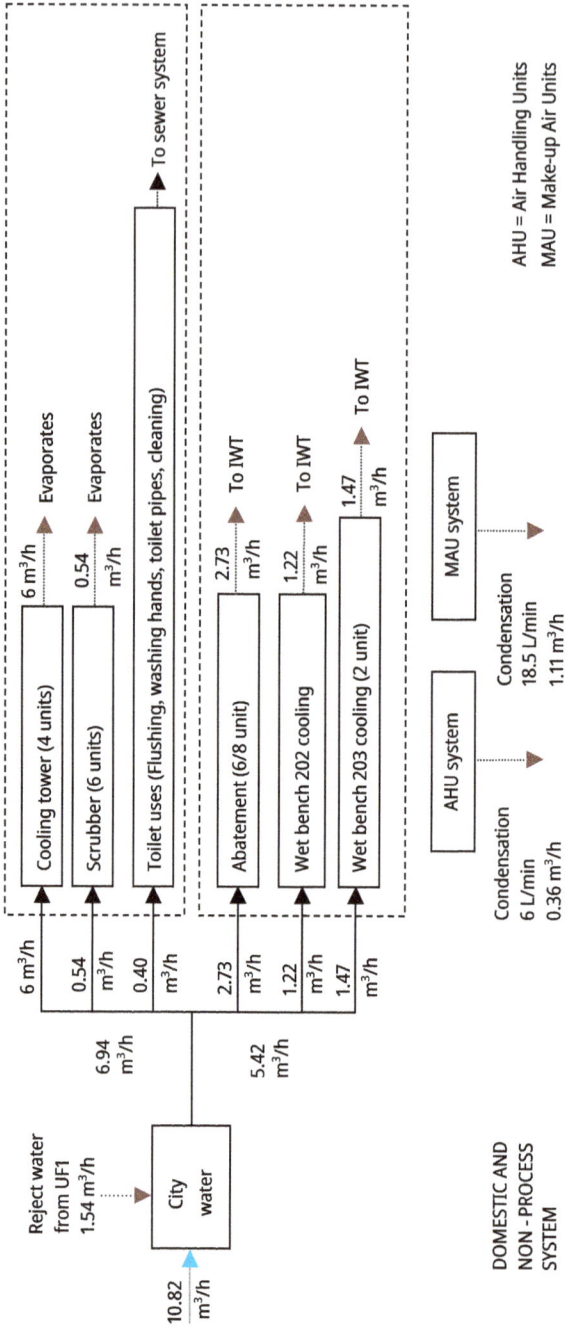

Fig. 9.10: MySem non-process water balance (October–November).

Tab. 9.1: Limiting water data for MySem.

	Sink	F, t/h	C, ppm		Source	F, t/h	C, ppm
SK1	MMF inlet	32.0	52	SR1	MMF rinse	1.33	48.0
SK2	Cooling tower	6.00	100	SR2	RO reject 1st pass	9.80	70.4
SK3	Abatement	2.73	100	SR3	EDI reject	3.36	48.6
SK4	Scrubber	0.54	100	SR4	WB101 rinse water, idle	0.38	0
SK5	Toilet flushing	0.08	100	SR5	WB101 rinse water, operation	0.07	4,608
SK6	Wash basin	0.01	52	SR6	WB102 rinse water, idle	0.22	0
SK7	Ablution	0.15	52	SR7	WB102 rinse water, operation	0.07	4,480
SK8	Toilet pipes	0.12	52	SR8	WB201 rinse water, idle	0.76	0
SK9	Office cleaning	0.05	52	SR9	WB201 rinse water, operation	0.03	23,360
SK10	MMF backwash	2.08	52	SR10	WB202 rinse water, idle	3.48	0
SK11	MMF rinse	1.33	52	SR11	WB202 rinse water, operation	0.07	163.2
SK12	WB203 cooling	1.47	52	SR12	WB203 rinse water, idle	3.63	0
SK13	WB202 cooling	1.22	52	SR13	WB203 rinse water, operation	0.28	928
				SR14	MAU	1.11	6.4
				SR15	AHU	0.36	11.5
				SR16	Cassette cleaner	0.08	0
				SR17	Abatement	2.73	105.6
				SR18	Wafer scrubber	0.54	12.8
				SR19	RO reject 2nd pass	4.50	19.2
				SR20	UF1 reject	1.54	19.2
				SR21	UF2 reject	1.80	0
				SR22	Heater WB101	0.46	0
				SR23	Wash basin	0.01	60
				SR24	Ablution	0.15	40

Step 3: WMH-guided screening and selection of process options

After calculating the base-case MWR targets, all potential process changes to improve the MySem water system were listed according to the various WMH levels as shown in Tab. 9.4. Central to the MWN approach is the level-wise hierarchical screening and prioritisation of process change options using the Water Management Hierarchy (WMH) and three new option-screening heuristics which were sequentially applied to prioritise process changes. MySem had initially selected process changes based on the net annual savings (NAS) associated with each process change, as shown with check marks in column 3 (MySem, 2005) of Tab. 9.4. The ultimate minimum water targets obtained after MWN analysis were given check marks in column 4 of Tab. 9.4. The steps for screening the options according to the WMH are described next.

1. Source elimination

The Pinch Point obtained from the base-case MWR targeting stage was at 23,360 ppm (see the Water Cascade Table, i.e., Tab. 9.2). In order to maximise freshwater savings, the top priority was to consider eliminating water sinks Above the Pinch Point in the cascade table, i.e. any sink with concentration less than 23,360 ppm. Note that all the

Tab. 9.2: Base-case Maximum Water Recovery targets for MySem (without process changes).

C_k, ppm	ΔC_k, ppm	$\sum F_{SKi}$, t/h	$\sum F_{SRj}$, t/h	$\sum F_{SKi} + \sum F_{SRj}$, t/h	F_C, t/h	Δm, kg/h	Cum. Δm, kg/h	$F_{FW, cum}$, t/h	F_C, t/h
					0.00				$F_{DI} = 0.00$
0			10.81	10.81			0.00		
	6.40				10.81	69.17			10.81
6.4			1.11	1.11			69.17	10.81	
	5.12				11.92	61.02			11.92
11.52			0.36	0.36			130.19	11.30	
	1.28				12.28	15.72			12.28
12.8			0.54	0.54			145.91	11.40	
	6.40				12.82	82.04			12.82
19.2			6.04	6.04			227.94	11.87	
	10.80				18.86	203.67			18.86
30			11.04	$F_{FW} = 11.04$			431.61	14.39	
	10.00				29.90	298.98			29.90
40			0.15	0.15			730.59	18.26	
	8.00				30.05	240.38			30.05
48			1.33	1.33			970.97	20.23	
	0.64				31.38	20.08			31.38
48.64			3.36	3.36			991.05	20.38	
	3.36				34.74	116.72			34.74
52		−38.43	0.00	−38.43			1,107.77	21.30	
	8.00				−3.69	−29.54			−3.69
60			0.01	0.01			1,078.24	17.97	
	10.40				−3.68	−38.29			−3.68
70.4			9.80	9.80			1,039.95	14.77	
	29.60				6.12	181.09			6.12
100		−9.35		−9.35			1,221.04	12.21	
	5.60				−3.23	−18.10			−3.23
105.6			2.73	2.73			1,202.94	11.39	
	58.40				−0.50	−29.32			−0.50
164			0.07	0.07			1,173.62	7.16	
	764.00				−0.43	−330.81			−0.43
928			0.28	0.28			842.81	0.91	
	3552.00				−0.16	−550.56			−0.15
4,480			0.07	0.07			292.25	0.07	
	128.00				−0.09	−11.01			−0.09
4,608			0.07	0.07			281.24	0.06	
	18752.00				−0.02	−281.28			−0.01
23,360			0.03	0.03			−0.04	**0.00**	(Pinch)
					0.02	0.00			$F_{IWT} = 0.02$

water sinks in Tab. 9.1 met this criterion. All possible means to change processes or the existing equipment to new equipment to eliminate water sinks were considered. From Tab. 9.4 it was possible to eliminate SK12 and SK13 by changing wet bench 202 and 203 quartz tanks, which initially needed continuous water for cooling, to Teflon

Tab. 9.3: Amount of IWT and domestic wastewater before and after integration.

Utility	Before MWR (t/h)	After MWR (t/h)	% reduction
Total freshwater	39.94	11.04	72.4
Total IWT wastewater	34.45	5.71	83.4
Total domestic WW	0.41	0.25	39.0

tanks. Not only did this option eliminate water requirements, but it also avoided tank cracking as a result of a sudden temperature drop. Elimination of SK12 and SK13 resulted in new water targets at 8.3525 t/h freshwater and 0.0215 t/h IWT (see the third row of Fig. 9.11). The Pinch Point was maintained at 23,360 ppm.

2. Source reduction

After eliminating SK12 and SK13, the next process change considered according to the WM hierarchy was to reduce sinks Above the Pinch Point in the cascade table, i.e. any sink with a concentration lower than 23,360 ppm. Tab. 9.4 lists a few process change options related to source reduction. Following Heuristic 9.2, the sink at the lowest contaminant concentration (52 ppm) was reduced first. There are possible source reduction process changes (listed in Tab. 9.4) affecting sinks SK1, SK10 and SK11 (all located at 52 ppm) and sources SR1 to SR13 and SR19 to SR22 simultaneously due to the interactions between equipment in the water system of the DI plant:

- Wet bench flowrate reduction to the minimum during idle mode;
- Recirculating hot water and switching the heater on the sink for heater WB201;
- Reduction of Fab 1 return flowrate by changing to variable speed pump;
- Decommissioning three EDI units instead of running four units. Note that a sharp decrease in flowrate due to upstream process changes made it possible to reduce by three EDI units (option 2 from Tab. 9.4 was rejected due to the increase in FW and IWT targets. Option 1, i.e. decommissioning three EDI units, reduced the FW and IWT targets to 6.3038 t/h and 0.0378 t/h (Tab. 9.5) and was implemented);
- Increase rate of recovery for reverse osmosis system;
- Decrease multimedia filter backwash and rinsing time.

The FW and IWT after the application of each process change are summarised in Fig. 9.11. Using Heuristic 9.1, the source reduction process changes for the DI water system shown in Tab. 9.4 were implemented from the core of the process (wet bench systems) to the most outer layer (multimedia filter). Excessive water usage at the core of the system was the main reason for water wastage at the outer layers. Improving the core of the system first reduces wastage downstream. Implementation of the entire range of process changes, from wet bench to MMF (multimedia filter), listed in Tab. 9.4 gave revised freshwater and wastewater targets at 6.0857 t/h and 0.0387 t/h and a new

Tab. 9.4: Various process change options applicable for MySem.

WMH	Strategy	Option selected based on NAS	Option selected based on MWN procedure
Elimination	Abatement		
	Option 2 (decommissioning)	X	X
	WB 202 and 203 cooling	√	√
Reduction	WB reduction in Fab 1 and 2	√	√
	Heater reduction	√	√
	Fab 1 return reduction	√	√
	Abatement		
	Option 1 (0.5 GPM during idle)	X	X
	Option 3 (recirculation)	√	X
	Option 4 (on demand)	X	√
	Option 5 (pH analysis)	X	X
	Increase RO system recovery/ install 3rd stage	√	√
	EDI return reduction		
	Option 1 (decommissioning)	X	√
	Option 2 (run intermittently)	√	X
	Domestic reduction	√	X
	Cooling tower reduction using N$_2$	√	√
	MMF reduction by NTU analysis	√	√
Reuse	Total reuse	√	√
Outsourcing	RW harvesting	√	√
Regeneration	Treat all WB water	X	√

(√) for selected option, (X) for eliminated option by MySem.

Pinch concentration at 4,608 ppm (see the ninth row of Fig. 9.9). Since there were no other sinks at 52 ppm, following Heuristic 9.2, the sinks with the next lowest contaminant concentration (100 ppm) were considered next. For MySem, SK2, SK3, and SK6 existed at the same concentration of 100 ppm. SK3, which yielded the biggest flowrate reduction, was chosen first, followed by SK2 and SK6 according to Heuristic 9.3.

The pollution abatement system sink (SK3) existed at 100 ppm. Initially, the pollution abatement system sank 2.73 t/h of water (SK3) and produced 2.73 t/h of IWT (SR17). Tab. 9.6 shows five possible options to reduce the abatement system sink. Option 3, which was predicted to yield the highest savings, was initially chosen by MySem prior to the MWN approach (Tab. 9.6 column 3). However, as shown in Tab. 9.6, option 3 actually increased the freshwater target by 2.2% to 6.2183 t/h. This was because the introduction of a recirculation system that produces no wastewater but relied on the makeup water sink (option 3) reduced the amount of wastewater that could potentially be reused for MySem as a whole, thereby leading to an increased freshwater target. Option 4 in

WMH levels	Specific process changes considered	New FW target, t/h	New IWT+WW (based on limiting data) target, t/h	New Pinch Point concentration, ppm	New total IWT (all IWT considered) target, t/h
Initial	None	39.94	34.85	22,360	34.450
Reuse	Base case	11.040	0.0190	22,360	5.7090
Elimination	Eliminate WB cooling 202 (SK 13) and 203 (SK 12)	8.3525	0.0215	22,360	3.0215
Reduction	WB reduction in Fab 1 and Fab 2	6.7518	0.0258	4,608	1.4216
	Heater WB201	6.7314	0.0264	4,608	1.4454
	Fab 1 return	6.6094	0.0354	4,608	1.3109
	Option 1: EDI decommissioning	6.3038	0.0378	4,608	1.0112
	Increase RO rate of recovery Reduce multimedia filter	6.2110	0.0380	4,608	0.9132
	backwash and rinsing time	6.0857	0.0387	4,608	0.7879
	Option 4: Reduce abatement pollution system (SK 3 = 0.57 t/h and SR 17 = 0.57 t/h).	6.0831	0.0361	4,608	0.7853
	Cooling tower reduction (SK 2 = 5.86 t/h)	5.9452	0.0382	4,608	0.7874
Outsourcing	Add a source, SR 25 = 0.11 t/h of C = 16 ppm by harvesting rainwater	5.8349	0.0379	4,608	0.7871
Regeneration	Regenerate remaining IWT to the maximum flowrate for a source from to C = 52 ppm.	5.7970	0	4,608	0.7492
Minimum water network (MWN) targets		5.7970	0	4,608	0.7492

Fig. 9.11: The effects of WMH-guided process changes on the maximum water reuse/recovery targets and Pinch location.

Tab. 9.5: Various effects of EDI options on water targets.

EDI system	FW target, t/h	IWT target, t/h
Initial EDI flow rate	6.6094	0.0354
Option 1 (decommission 3 EDI unit)	6.3038	0.0378
Option 2 (run intermittently)	6.9281	0.0351

Tab. 9.6 gave the highest freshwater and IWT savings. Choosing option 4 led to new freshwater and IWT targets at 6.0831 t/h and 0.0361 t/h with Pinch Point maintained at 4,608 ppm (the tenth row of Fig. 9.9). It was also possible to reduce sink SK2, which also existed at 100 ppm. SK2, which was the cooling tower makeup, had the second-highest flowrate reduction. Heat Exchange between the cooling tower circuit and the liquid

Tab. 9.6: Effects of abatement options on water targets.

Abatement system	F_D, t/h	F_S, t/h	FW target, t/h	IWT target, t/h
Initial abatement flow rate	2.73	2.73	6.0857	0.0387
Option 1 (0.5 gpm during idle)	1.11	1.11	6.0837	0.0367
Option 2 (decommissioning)	1.36	1.36	6.0840	0.0370
Option 3 (recirculation)	0.14	0.00	6.2183	0.0333
Option 4 (on demand)	0.57	0.57	6.0831	0.0361
Option 5 (pH analysis)	0.79	0.79	6.0833	0.0363

nitrogen circuit had the potential to reduce SK2 to 5.86 t/h, and ultimately the water targets to 5.9452 t/h freshwater and 0.0382 t/h wastewater (see the eleventh row of Fig. 9.9). The Pinch Point was maintained at 4,608 ppm.

Sink SK6 (wash basin) and SK7 (ablution) were reduced to 0.002 t/h and 0.035 t/h by changing the normal water taps to laminar taps. This also reduced sources SR23 and SR24. However, when targeted using Optimal Water© (2006), the freshwater and wastewater targets increased slightly to 5.9455 t/h and 0.0385 t/h. Hence, this process change was rejected.

3. External water sources

The next process change, according to the WM hierarchy, was to add external water sources at a concentration lower than the new Pinch Point concentration of 4,608 ppm. Based on MySem's available roof area and the rain distribution, it was possible to harvest 0.11 t/h (maximum design limit, $F_{maxdesign}$) of rainwater at a concentration of 16 ppm as a new water source, SR25. This option had the potential to reduce the freshwater and IWT targets to 5.8349 t/h and 0.0379 t/h (refer to the twelfth row of Fig. 9.11). The Pinch Point was maintained at 4,608 ppm.

4. Regeneration

Regeneration was the final process change considered according to the WM hierarchy. Freshwater savings could only be realised through regeneration Above or Across the Pinch. Regenerating all "WB201 in-operation" (SR9) at 23,360 ppm and 0.0201 t/h (maximum utility flowrate, F_{MU}, obtained using trial and error method in WCA) of "WB101 in-operation" (SR5) at 4,608 ppm to 52 ppm (the regenerated sources were named SR26) by carbon bed, EDI, and ultraviolet (UV) treatment systems reduced the freshwater and IWT targets to 5.797 t/h and 0 t/h, respectively (Tab. 9.7). Considering the IWT excluded from integration, the new IWT flowrate after regeneration was 0.7492 t/h. This corresponded to 85.5% freshwater and 97.8% industrial wastewater reductions. The Pinch Point was maintained at 4,608 ppm.

The *minimum water targets* were ultimately obtained after considering all options for process changes according to the WM hierarchy. Note that targeting the MWR only through reuse and regeneration resulted in savings of up to 72.4% freshwater and 83.4% wastewater for MySem. Instead, following the holistic framework guided by the WM hierarchy enabled potential freshwater and wastewater reductions of up to 85.5% and 97.8% towards achieving the minimum water network (MWN) design.

Step 4: Apply SHARPS

The desired payback period (TPP$_{set}$) was set at 4 months (0.33 y) by MySem management Fig. 9.10 shows the IAS plot after MWN analysis for the MySem retrofit. Since this is a retrofit case, eqs. (9.1), (9.2), and (9.5) mentioned previously were used. The Total Payback Period to attain the MWN targets before *SHARPS* screening was 0.38 y. Since the initial total payback periods were more than the TPP$_{set}$, i.e. 0.33 y, SHARPS strategies were applied to fulfil the TPP$_{set}$ specified.

Fig. 9.12 shows that *regeneration* process change gives the steepest gradient. Focusing on Strategy 1, there was no other option for the *regeneration* process change. Strategy 1, which called for equipment substitution, could not be implemented. Focusing on Strategy 2, when no *regeneration* was applied, the Total Payback Period reduces to 0.36 y as illustrated in Fig. 9.13, which still does not achieve TPP$_{set}$. Consequently, the next steepest gradient was observed.

The next steepest gradient was noted to be the *cooling tower* process change. Again, Strategy 1 cannot be applied since only one option exists for that process change. It can be noted that the cooling tower line is a concave curve without a peak. F$_{opt}$ is noted to be at the end of the curve, as shown in Fig. 9.14. Hence, based on the linearisation rule explained earlier, any reduction of the cooling tower curve will be beneficial. Using Strategy 2, reducing the *cooling tower* process change curve by only partially applying nitrogen cooling yielded the final IAS plot shown in Fig. 9.15, which achieved the specified payback of 0.33 y. SK2 was reduced only to 5.966 t/h instead of 5.86 t/h initially. This scheme reduced 5.94 t/h of freshwater and 0.79 t/h IWT total flowrate.

Hence, the application of SHARPS screening has successively achieved the TPP$_{set}$ of 4 months with 85.1% and 97.7% of freshwater and wastewater reduction prior to design. This is the final cost CEMWN target for MySem. The effect of applying each scheme on freshwater and IWT are illustrated in Fig. 9.16. Using the pre-design cost estimate method, the system needs approximately a net total investment of $64,500 and would give a net annual savings of $193,550.

Step 5: Network design

Network design based on Source and Sink Allocation Curves (SSAC) methodology proposed in Chapter 8 was used. The Source and Sink Composite Curves (SSCC) and Source and Sink Allocation Curves (SSAC) for the MySem retrofit system are shown in Figs. 9.17 and 9.18. It can be seen that the SSAC are very complicated, with many streams. By directing the reuse wastewater to existing tanks, the system was divided into three

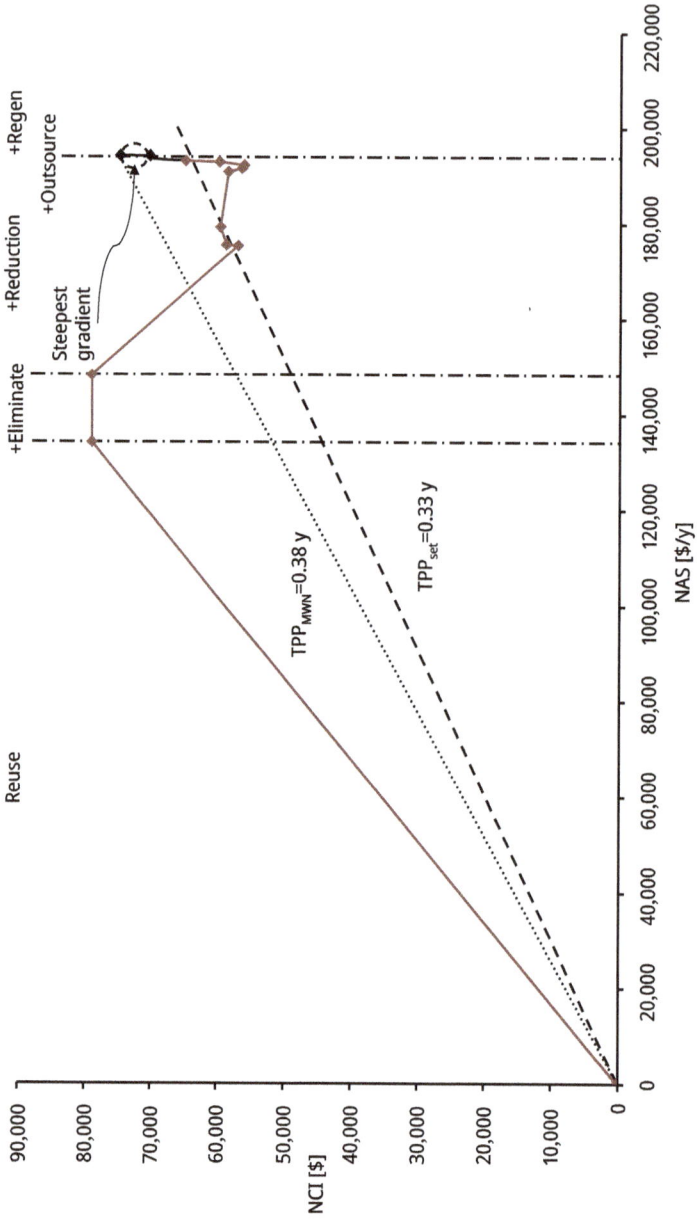

Fig. 9.12: IAS plot for MWN retrofit.

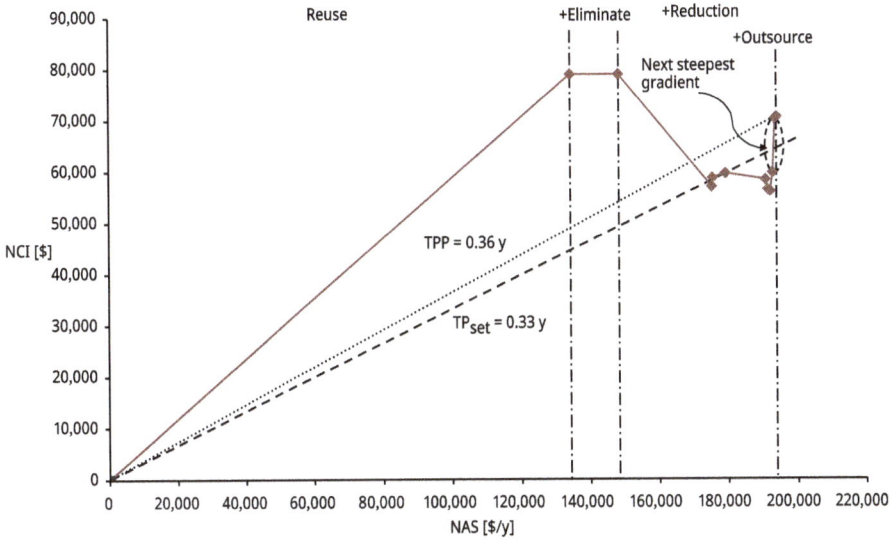

Fig. 9.13: IAS plot after eliminating regeneration curve.

Tab. 9.7: MySem water targets after implementation of MWN technique.

C_k, ppm	ΔC_k, ppm	$\sum F_{SKi}$, t/h	$\sum F_{SRj}$, t/h	$\sum F_{SKi} + \sum FSR$, t/hj	F_C, t/h	Δm, kg/h	Cum. Δm, kg/h	$F_{FW,cum}$, t/h	F_C, t/h
					0.00				$F_{DI} = 0.00$
0			1.727	1.73			0.00		
	6.40				1.73	11.05			1.73
6.4			1.11	1.11			11.05	1.73	
	5.12				2.84	14.53			2.84
11.52			0.36	0.36			25.58	2.22	
	1.28				3.20	4.09			3.20
12.8			0.54	0.54			29.67	2.32	
	3.20				3.74	11.96			3.74
16			0.11	0.11			41.63	2.60	
	3.20				3.85	12.31			3.85
19.2			0.776	0.78			53.94	2.81	
	10.80				4.62	49.93			4.62
30			$F_{FW} = 5.797$	5.80			103.87	3.46	
	10.00				10.42	104.20			10.42
40			0.15	0.15			208.07	5.20	
	8.00				10.57	84.56			10.57
48			0.169	0.17			292.63	6.10	
	0.64				10.74	6.87			10.74
48.64			0.84	0.84			299.50	6.16	
	3.36				11.58	38.91			11.58

Tab. 9.7 (continued)

C_k, ppm	ΔC_k, ppm	$\sum F_{SKi}$, t/h	$\sum F_{SRj}$, t/h	$\sum F_{SKi} + \sum F_{SR}$, t/hj	F_C, t/h	Δm, kg/h	Cum. Δm, kg/h	$F_{FW,cum}$, t/h	F_C, t/h
52		−6.528	0.0381	−6.49			338.41	6.51	
	8.00				5.09	40.71			5.09
60			0.01	0.01			379.12	6.32	
	10.40				5.10	53.03			5.10
70.4			1.153	1.15			432.15	6.14	
	29.60				6.25	185.06			6.25
100		−7.05		−7.05			617.21	6.17	
	5.60				−0.80	−4.47			−0.80
105.6			0.57	0.57			612.74	5.80	
	58.40				−0.23	−13.31			−0.23
164			0.024	0.02			599.43	3.66	
	764.00				−0.20	−155.78			−0.20
928			0.081	0.08			443.65	0.48	
	3552.00				−0.12	−436.54			−0.12
4,480			0.069	0.07			7.11	0.00	
	128.00				−0.05	−6.90			−0.05
4,608			0.0539	0.05			0.21	0.00	(Pinch)
					0.00	0.00			$F_{IWT} = 0.00$

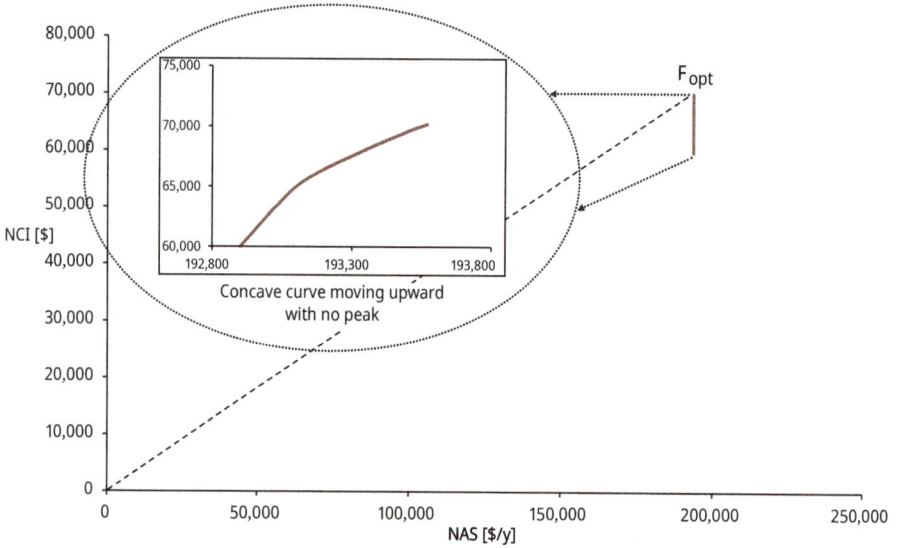

Fig. 9.14: F_{opt} for cooling tower concave curve moving upwards (without peak).

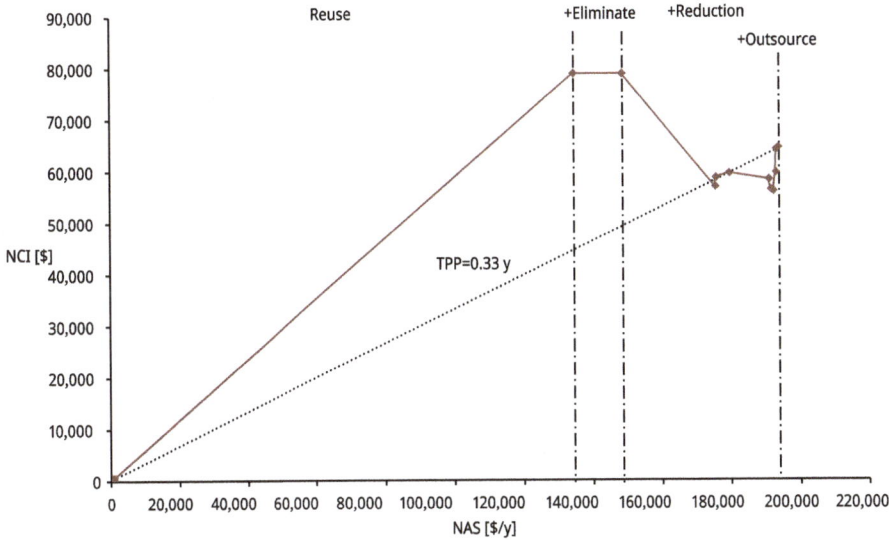

Fig. 9.15: Final IAS plot after *SHARPS* analysis.

categories: domestic city water tank (for sinks SK6 to SK9), process city water tank (for sinks SK2 to SK5) and DI blend water tank (for sinks SK1, SK10, and SK11). The new simplified retrofit SSAC and Network Mapping Diagram are shown in Fig. 9.19. The final network that achieved the CEMWN target for MySem is shown in Fig. 9.20.

9.4.1 Using CEMWN targets as reference benchmarks

CEMWN targets can be used for own performance and international water reduction benchmarking guides. For example, the CEMWN target for MySem, which also corresponds to the best achievable benchmark targets, is a target of freshwater flowrate of 5.94 t/h and a total IWT flowrate of 0.79 t/h. This represented 85.1% freshwater and 97.7% IWT reduction. Hence, these were the best performance benchmark targets (Fig. 9.21) that MySem needed to achieve. The application of total reuse only using Water Pinch Analysis (WPA) method yielded lower water savings potentials of 72.4% freshwater and 83.4% wastewater reduction with a 0.59 y payback period. November 2005 water bills show that all the conventional water reduction strategies applied by MySem had only managed to reduce freshwater usage from 42.6 t/h to 40.24 t/h, representing a saving of $880/month. An estimated total saving of 193,550/y was predicted with the implementation of the CEMWN method. A preliminary cost estimate indicated that this best performance required an investment of approximately $64,507 with a payback period of 0.33 y.

WMH levels	Specific process changes considered	New FW target, t/h	New IWT+WW (based on limiting data) target, t/h	New P inch Point concentra-tion, ppm	New total IWT (all IWT considered) target, t/h
Initial	None	39.94	34.85	22,360	34.4500
Reuse	Base case	11.0400	0.0190	22,360	5.7090
Elimination	Eliminate WB cooling 202 (SK 13) and 203 (SK 12)	8.3525	0.0215	22,360	3.0215
Reduction	WB reduction in Fab 1 and Fab 2	6.7518	0.0258	4,608	1.4216
	Heater WB201 reduction	6.7314	0.0264	4,608	1.4454
	Fab 1 return reduction	6.6094	0.0354	4,608	1.3109
	Option 1: EDI decommissioning	6.3038	0.0378	4,608	1.0112
	Increase RO rate of recovery	6.2110	0.0380	4,608	0.9132
	Reduce multimedia filter backwash and rinsing time	6.0857	0.0387	4,608	0.7879
	Option 4: Reduce abatement pollution system (SK 3 = 0.57 t/h and SR 17 = 0.57 t/h).	6.0831	0.0361	4,608	0.7853
	Cooling tower reduction (SK 2 = 5.966 t/h)	6.0496	0.0366	4,608	0.7858
Outsourcing	Add a source, S 25 = 0.11 t/h of C = 16 ppm by harvesting rainwater	5.9392	0.0362	4,608	0.7854
CEMWN targets		5.9392	0.0362	4608	0.7854

Fig. 9.16: Final CEMWN targets after SHARPS analysis.

Note that the schemes proposed and listed in Fig. 9.16 could also be gradually imple-mented as part of the company's longer-term utility savings program in line with its quality management practices.

Once the best performance benchmark was established through the CEMWN method, the predicted maximum savings of MySem were then compared with the in-ternational benchmark. The International Technology Roadmap for Semiconductors (ITRS 2001) aimed to reduce high-purity water (HPW) consumption from the current rate of 6–8 m^3 in 2005 to 4–6 m^3 per wafer by 2007 (Wu et al., 2004). After CEMWN analysis, MySem had the potential to use 4.06 m^3 of DI water per wafer for Fab 1 and 13.73 m^3 of DI water per wafer for Fab 2, down from its previous consumption of 6.3 and 72.4 m^3 of DI water per wafer. Fab 1 had the potential to meet the ITRS 2001 target. Fab 2, however, was far from this ITRS target due to its wafer production rate of well below the design capacity.

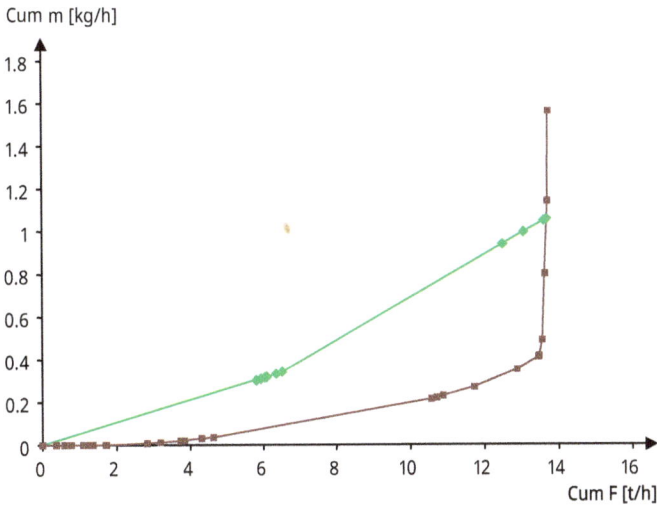

Fig. 9.17: Source and Sink Composite Curves for MySem.

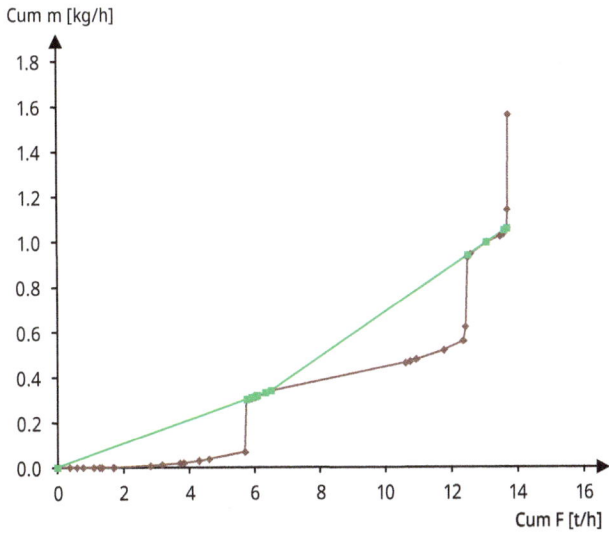

Fig. 9.18: Source and Sink Allocation Curves for MySem.

For more information on the latest WPA developments, please refer to Chapter 10 "Sources of Further Information".

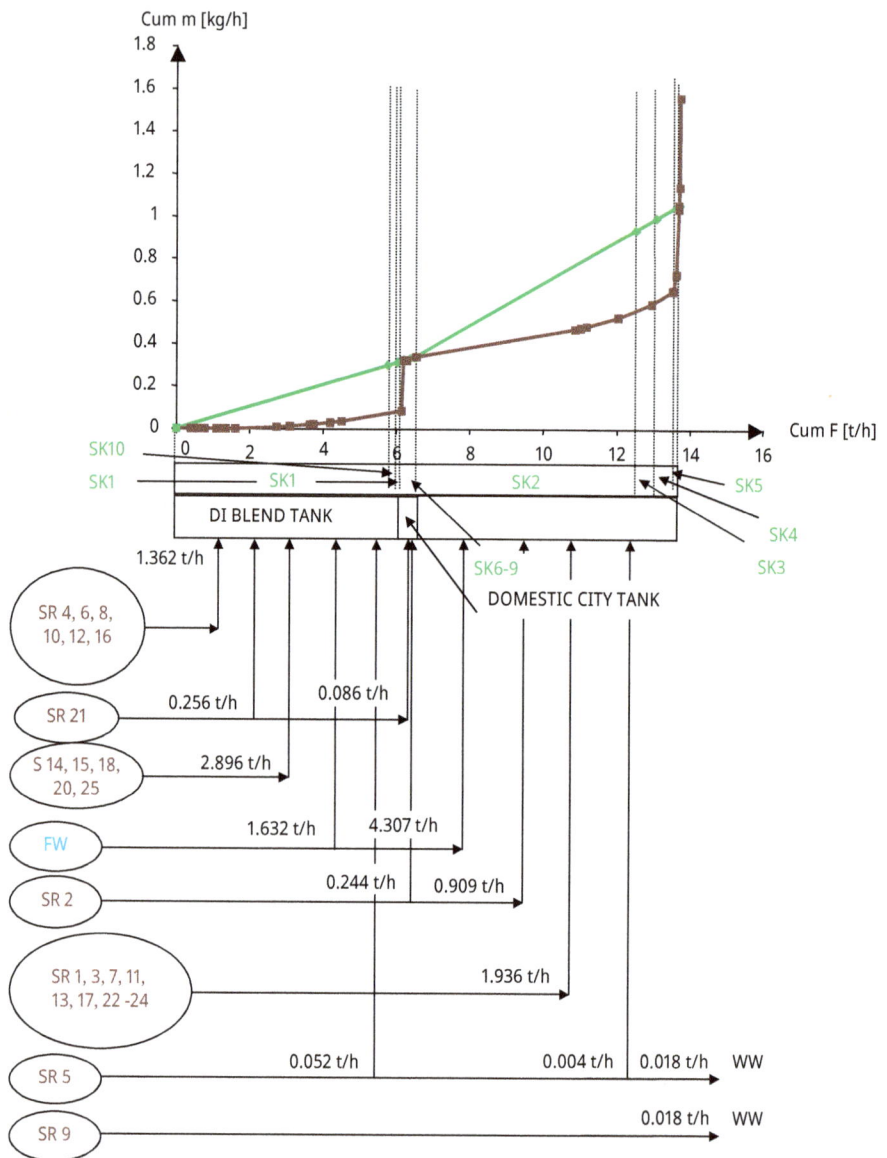

Fig. 9.19: Network Allocation Diagram based on simplified SSAC for MySem retrofit.

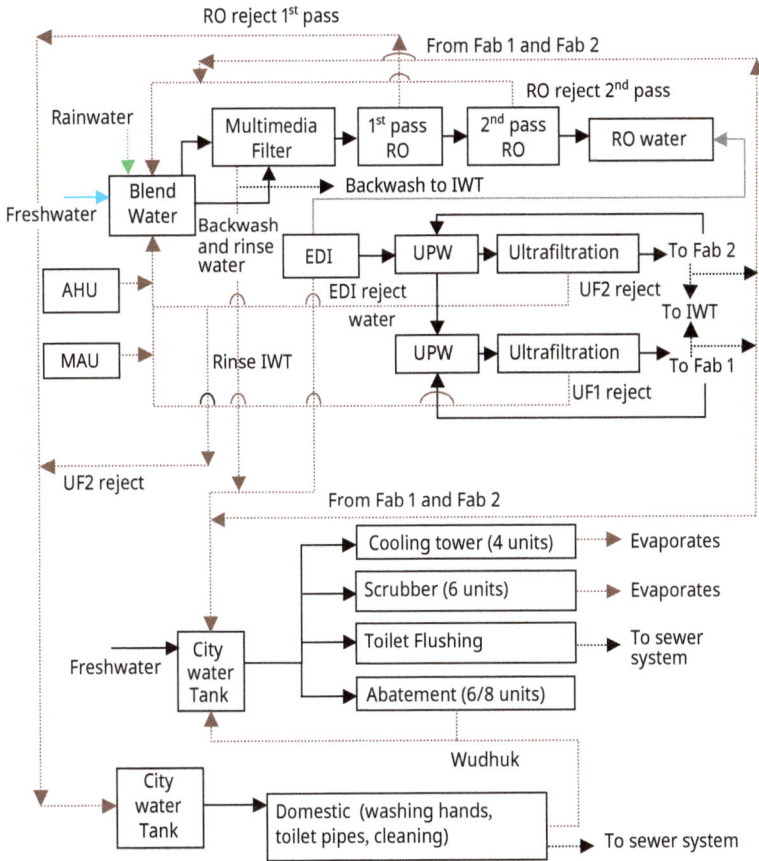

Fig. 9.20: MySem retrofit DI water balance and non-process water balance after CEMWN analysis, achieving 85.5% freshwater and 97.7% IWT reductions within 4 months of payback.

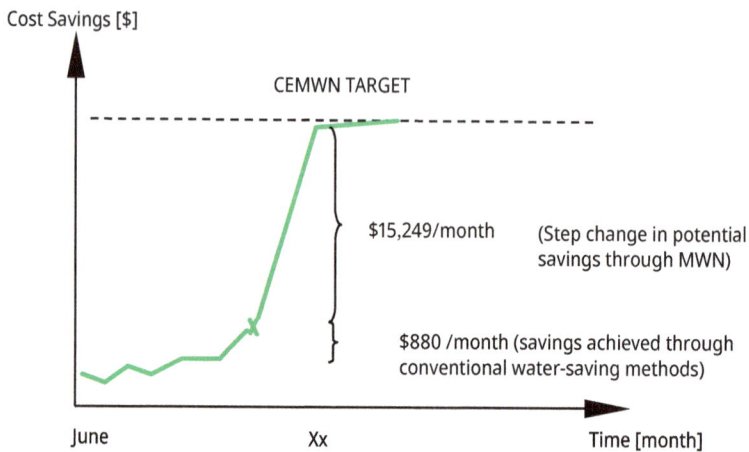

Fig. 9.21: Savings achieved by MySem in comparison to savings predicted through the CEMWN approach (Wan Alwi and Manan, 2008).

References

El-Halwagi M M. (1997). Pollution Prevention through Process Integration: Systematic Design Tools, Academic Press, San Diego, USA.

Hallale N. (2002). A New Graphical Targeting Method for Water Minimization, Advances in Environmental Research, 6(3), 377–390.

International Technology Roadmap for Semiconductors (ITRS). (2001). Process Integration, Devices, and Structures and Emerging Research Devices. www.itrs.net, accessed on 16 June 2016.

Manan Z A, Tan Y L, Foo D C Y. (2004). Targeting the Minimum Water Flowrate Using Water Cascade Analysis Technique, AIChE Journal, 50(12), 3169–3183.

MySem. (2005). MIMOS Semiconductor (M) Sdn. Bhd. UTM-MIMOS Water Minimization Project (UMWMP). Unpublished report.

Optimal Water©. (2006). Software for Water Integration in Industries. Copyright of Universiti Teknologi, Malaysia, Johor, Malaysia. http://optimalsystems.my, accessed on 16/10/2022.

Perry R H, Green D W. (1998). Perry's Chemical Engineers' Handbook, 7th ed. McGraw-Hill, New York, USA.

Seferlis P, Varbanov P S, Papadopoulos A I, Chin H H, Klemeš J J. (2021). Sustainable Design, Integration, and Operation for Energy High-performance Process Systems, Energy, 224, 120158. doi: 10.1016/j.energy.2021.120158.

Smith R. (1995). Chemical Process Design, McGraw-Hill, Inc, Chicester, UK.

Wan Alwi S R, Manan Z A. (2006). SHARPS – A New Cost-screening Technique to Attain Cost-effective Minimum Water Utilisation Network, AIChE Journal, 11(52), 3981–3988.

Wan Alwi S R, Manan Z A. (2008). A New Holistic Framework for Cost Effective Minimum Water Network in Industrial and Urban Sector, Journal of Environmental Management, 88, 219–252.

Wang Y P, Smith R. (1994). Wastewater Minimisation, Chemical Engineering Science, 49, 981–1006.

Wang Y P, Smith R. (1995). Wastewater Minimization with Flowrate Constraints, Transactions of IChemE Part A, 73, 889–904.

Wu M, Sun D, Tay J H. (2004). Process-to-process Recycling of High-purity Water for Semiconductor Wafer Backgrinding Wastes, Resources, Conservation and Recycling, 41, 119–132.

10 Extension of Water Integration to Water Mains, Total Site targeting and multiple quality problems

10.1 Introduction

Using Water Mains helps reduce the complexity of water networks and contributes to better controllability of the flow and properties of the water sources. However, for the Pinch-based approach, the number of headers and the inlet limiting concentrations of these headers are usually specified or fixed based on certain heuristics (Klemeš et al., 2018a). The approaches to Water Integration can be extended to site-level water management where water sources from multiple plants are involved. Total Site Water Integration plays a crucial role in achieving minimum site or regional fresh resource usage, providing a solution benchmark for practitioners. The concept of Total Site Water Integration is similar to Total Site Heat Integration (Klemeš et al., 1997), where the recycling of resources should be emphasised first inside the individual plants. The sources remaining after plant-level integration can be potentially used for other plants' demands. The Pinch-based concept should also extend to multiple qualities problem. This chapter introduces a Pinch Analysis methodology to target the parameters of Water Mains/headers including the number of mains, source flows, and quality levels of the header sources (Klemeš et al., 2018b). This chapter also introduces a method to target the minimum fresh resources in Total Sites as a guide to determine the optimal resource exchange strategies. Section 10.2 provides a guided description of Water Mains targeting. Section 10.3 then focuses on the Total Site framework, which also utilises the Water Mains framework to target Total Sites and Water Mains. Section 10.4 explains the Pinch-based methodology that targets resources involving multiple quality properties.

10.2 Water Mains/headers

Water sources (including freshwater) can be mixed into Water headers/Mains, and the sinks can be satisfied by the header mix. This allows better controllability of the water flow and quality as the water flow network becomes less complex. Fig. 10.1 shows the typical configuration of the headers for the design of water allocation networks. However, excessive mixing of the sources could lead to unnecessary water quality deterioration, which may result in additional fresh resource intake. Proper header flowrate and concentration determination are required to ensure that the minimum freshwater target is still satisfied.

It is assumed in this text that the water-using processes or units can be represented with source-sink representation. A sink is a water stream demand of some

https://doi.org/10.1515/9783110782981-010

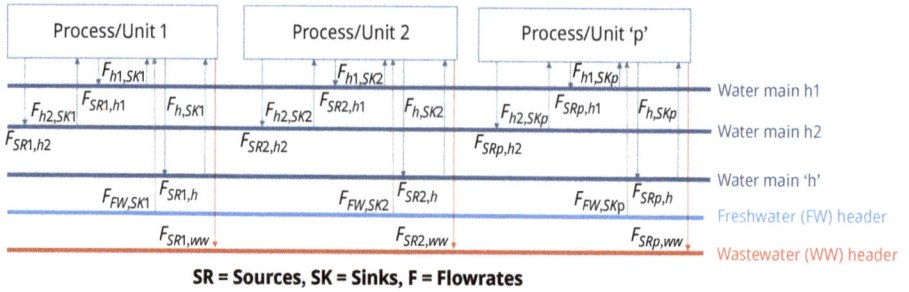

Fig. 10.1: Typical configuration of Water Mains/headers in a water allocation network, which can be within the plant or in a Total Site.

process units, while a source is a water supply outlet from process units. For both fixed-flowrate and fixed-load water-using operations, it is assumed that they can be represented with the source-sink operation mapping. The maximum flowrates and concentrations are used for both sources and sinks.

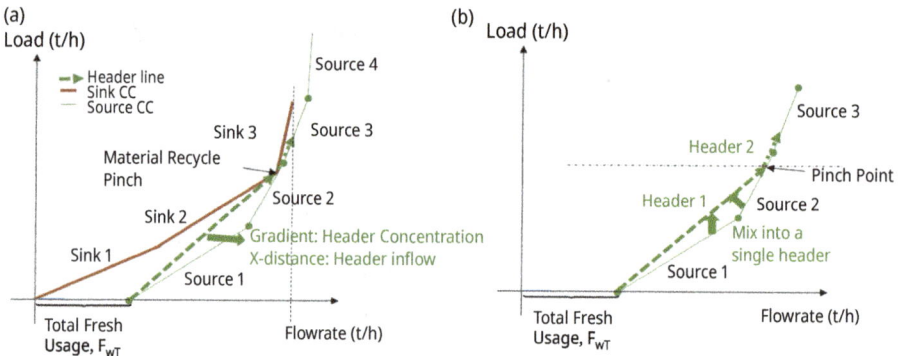

Fig. 10.2: (a) Constructing header line for the illustrative example; (b) allocation of sources to the header.

The Composite Curves provide information on source mixing options for the system (see Chapter 7). Fig. 10.2 shows an example of already-constructed Composite Curves for a water allocation problem. Fig. 10.2a illustrates how to draw the line representing a water header. The header line is drawn starting from the beginning of the Source Composite Curve to the point of touching the Sink Composite Curve. Note that the user can adjust the length or slope of the header lines, depending on the preference on the flowrates or quality.

Based on Fig. 10.2a, it can be seen that Sinks 1 and 2 are fulfilled by a mixture of freshwater, Source 1 and part of Source 2. Source 1 and part of Source 2 can be mixed and form a single source header. This possibility is represented in Fig. 10.2b. It can be seen that a single header line (arrow dashed line) can be drawn below the Pinch Point.

The line of Header 1 is constructed by representing the mixing of Source 1 and part of Source 2, as shown in the drawing in Fig. 10.2b. The header line is feasible, as the total freshwater target identified is still the same without violating the contamination load limits for each sink. Considering individual source-to-sink allocation, Sink 1 may require more freshwater as compared to the variant without a header design, but Sink 2 requires less freshwater compared to the variant without a header design due to the mixing of dirtier and cleaner sources.

The total freshwater requirements for both Sinks 1 and 2 are still identical compared to those without header design. The horizontal length of the Header 1 line represents the flowrate inside the header, while the gradient is the contaminant concentration inside the header. Similarly, another header line can be constructed dedicated to the sinks above the Pinch (i.e. Sink 3). In Fig. 10.2a–b, it can be seen that the header line for Sink 3 (Header 2) represents the remaining part of Source 2 and part of Source 3. For this example, the total number of headers required is 4 (1 header Below the Pinch, 1 header Above the Pinch, freshwater header and wastewater header).

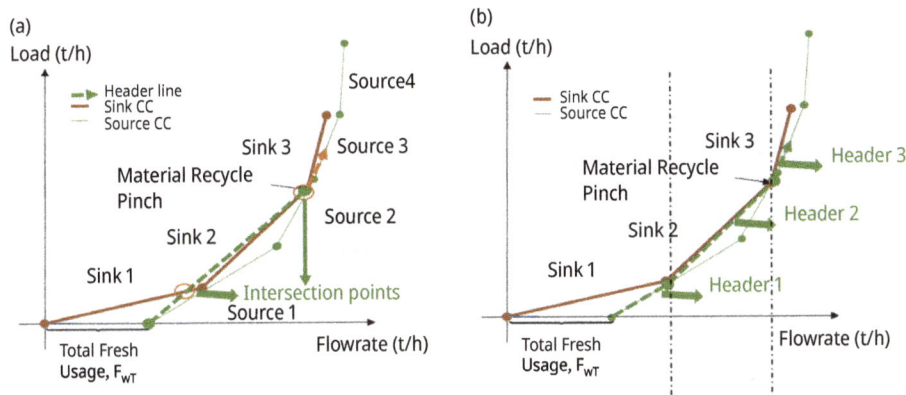

Fig. 10.3: (a) Single header Below the Pinch is not enough; (b) another header is drawn by splitting the header line.

There are certain cases where a single header line below the Pinch Point is not sufficient. This happens when the header line Below the Pinch crosses at the Sink CC – see Fig. 10.3a. This indicates excessive mixing of the sources, which decrease the quality of the water sources and require extra fresh resource intake. To ensure the quality of the sources, another header line can be drawn by splitting the header line below the Pinch into two – see Fig. 10.3b. Notice that the Headers 1 and 2 lines in Fig. 10.3b can be at different combinations of lengths. They can be formed by drawing lines along with the Source CC and ensuring that both header lines are below the Sink CC. At this point, the following procedure for headers targeting using CC can be derived in various steps:

Step 1: Construct Composite Curves for the process to determine the fresh resource target

Step 2: Draw a single header line by mixing the selected sources below the Pinch Point(s), starting from the shifted Source CC initial point to the Pinch Point(s). If there are Multiple Pinch Points, connect the subsequent Pinch Points with the header lines

Step 3: Draw a single header line by mixing the selected sources above the **last** Pinch Point, starting from the Pinch Point along with the locus of the Source CC until the horizontal length matches where the Sink CC ends. (For a single Pinch problem, Pinch Point = last Pinch Point)

Step 4: Check if intersection(s) exist(s) between the header lines and the Sink CC (the header line should be entirely below the Sink CC). If there are any, split the header lines and draw the header lines with any combinations until the entire Header Curve is at the right side of Sink CC.

Step 5: Determine the number of headers (number of segments in the header lines), each header source's concentration (gradient) and flowrates (horizontal length).

The targeting using headers and CCs is straightforward and user-friendly, where users can easily evaluate the various sources mixing options. The advantages of using CC include the header sources flow and quality being determined directly through the graph, where flow is represented by the horizontal length of the header line, and the quality of the header source is characterised by its gradient (load/flowrate). The minimum number of headers can also be identified in this case. The following subsection demonstrates the step-by-step heuristics in performing headers targeting through several examples.

10.2.1 Working session 1: Water Mains/headers targeting with Pinch Analysis

Example 10.1 is used to demonstrate the method. The sources and sinks with their flowrates and qualities data required for this study are presented in Tab. 10.1 based on Fadzil et al. (2018).

Step 1: Construct Composite Curves for the process to determine the fresh resource target. Fig. 10.4a shows the CC representation for Example 10.1 – see the explanation in Chapter 7, and it can be observed that there are two Pinch Points for the process. The freshwater target is 206.67 t/h.

Step 2: Draw a single header line below the Pinch Point(s), starting from the shifted Source CC initial point to the Pinch Point(s). If there are Multiple Pinch Points, connect the subsequent Pinch Points with the header lines. As the problem has two Pinch Points, a header line for the sinks Below the Pinch can be drawn, and another header line is drawn connecting the two Pinch Points.

Tab. 10.1: Example 10.1 case study plant data.

Sinks	F, t/h	C (ppm)
SK1	200	0
SK2	80	50
SK3	80	50
SK4	140	100
SK5	200	120
SK6	200	200
Sources	F, t/h	C (ppm)
SR1	200	50
SR2	80	100
SR3	80	100
SR4	140	150
SR5	200	200
SR6	200	450

Step 3: Draw a single header line above the **last** Pinch Point, starting from the Pinch Point along with the locus of the Source CC until the horizontal length matches where the Sink CC ends. (For a Single-Pinch problem, Pinch Point = last Pinch Point.) Note that since there are no sinks above the last Pinch Point, no header is required. The procedure is similar if there are sinks above the last Pinch Point, where the header line can be drawn starting from the Pinch Point along with the Source CC locus until the flowrate matches the last point of Sink CC.

Step 4: Check if intersection(s) exist(s) between the header lines and the Sink CC. If there are any, split the header lines and draw the header lines with any combinations until the entire Header Curve is at the right side of Sink CC. One header line below the first Pinch Point is not enough, as the Header 1 line is above the Sink CC below the first Pinch, which means the sources are overmixed. In this case, the header line can be split into two below the Pinch. Header 1 can be drawn under the "elbow" point of the Sink CC, and Header 2 is drawn connecting the end point of Header 1 to the Pinch Point – see Fig. 10.4b.

Step 5: Determine the number of headers (number of segments in the header lines), each header source's concentration (gradient) and flowrates (horizontal length). The flowrates and the qualities of all header sources using the configurations in Fig. 10.4b are: Header 1: (200 t/h, 50 ppm), Header 2: (293.33 t/h, 122.73 ppm), and Header 3: (200 t/h, 200 ppm). Note that Header 1 and Header 2 can be in various combinations as well, which indicates that the sources below the first Pinch can be mixed in various ways, as long as the entire header lines are under the Sink CC. These header configurations also ensure that the freshwater target is not violated.

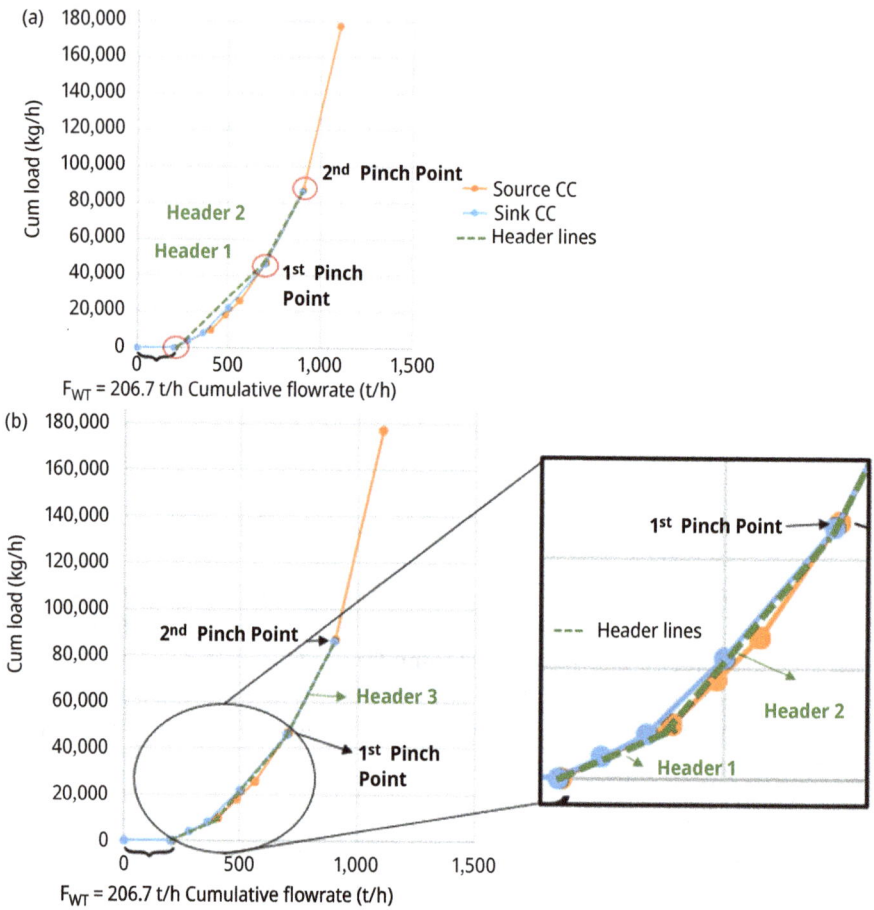

Fig. 10.4: (a) Infeasible header lines for Example 9.1; (b) feasible header lines for Example 9.1.

10.2.2 Working session 2

As an assignment, identify the minimum number of headers sources required with their respective flowrates and quality levels for Example 10.2. The initial data of the example are given in Tab. 10.2 from Chapter 8. The available freshwater is at 30 ppm.

Tab. 10.2: Example 10.2 Case study plant data.

		F (t/h)	C (ppm)			F (t/h)	C (ppm)
SK1	MMF inlet	32.00	52.00	SR1	MMF rinse	1.33	48.00
SK2	Cooling tower	6.00	100.00	SR2	RO reject First pass	9.80	70.40
SK3	Abatement	2.73	100.00	SR3	EDI reject	3.36	48.60
SK4	Scrubber	0.54	100.00	SR4	WB101 rinse water, idle	0.38	0.00
SK5	Toilet flushing	0.08	100.00	SR5	WB101 rinse water, operation	0.07	4,608.00
SK6	Wash basin	0.01	52.00	SR6	WB102 rinse water, idle	0.22	0.00
SK7	Ablution	0.15	52.00	SR7	WB102 rinse water, operation	0.07	4,480.00
SK8	Toilet pipes	0.12	52.00	SR8	WB201 rinse water, idle	0.76	0.00
SK9	Office cleaning	0.05	52.00	SR9	WB201 rinse water, operation	0.03	23,360.00
SK10	MMF backwash	2.08	52.00	SR10	WB202 rinse water, idle	3.48	0.00
SK11	MMF rinse	1.33	52.00	SR11	WB202 rinse water, operation	0.07	163.20
SK12	WB203 cooling	1.47	52.00	SR12	WB203 rinse water, idle	3.63	0.00
SK13	WB202 cooling	1.22	52.00	SR13	WB203 rinse water, operation	0.28	928.00
				SR14	MAU	1.11	6.40
				SR15	AHU	0.36	11.50
				SR16	Cassette cleaner	0.08	0.00
				SR17	Abatement	2.73	105.60
				SR18	Wafer scrubber	0.54	12.80
				SR19	RO reject Second pass	4.50	19.20
				SR20	UF1 reject	1.54	19.20
				SR21	UF2 reject	1.80	0.00
				SR22	Heater WB101	0.46	0.00
				SR23	Wash basin	0.01	60.00
				SR24	Ablution	0.15	40.00

10.2.3 Solution

1. Composite Curves for the problem with a single header Below the Pinch
   Fig. 10.5 shows the Composite Curves for Example 10.2. The freshwater target identified is 29.9 t/h (at 30 ppm). It is clearly shown that a single header line Below the Pinch Point is not enough to cover the sinks without inputting extra fresh resources.
2. Split the header lines
   A better solution is to split the header lines into two. One of the possible header configurations is presented in Fig. 10.6. The header sources parameters are: {Header 1: 44.5 t/h, 31.2 ppm} and {Header 2: 3.23 t/h, 477.4 ppm}. Readers could refer to Chin et al. (2021d) for more examples of headers targeting.

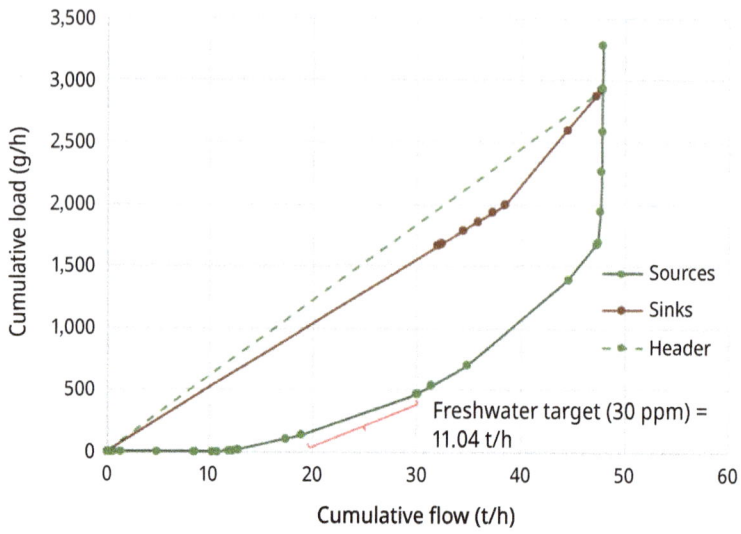

Fig. 10.5: Infeasible header lines for Example 10.2.

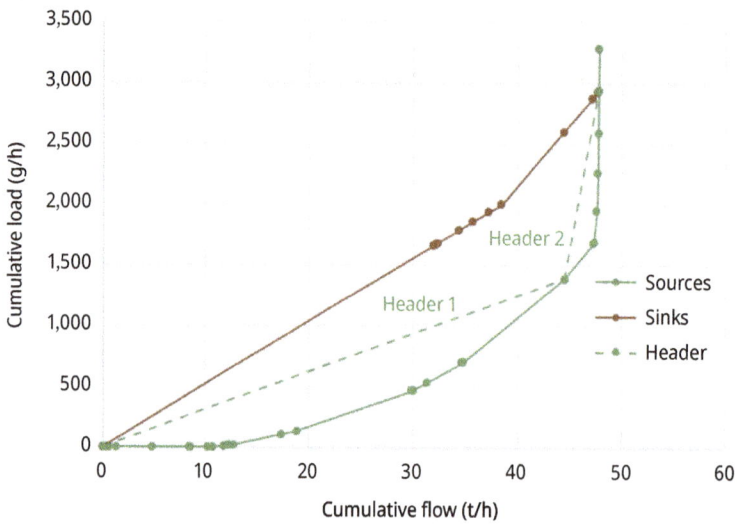

Fig. 10.6: Feasible header targeting solution for Example 10.2.

10.3 Total Site Water Integration

The concept of Total Site Heat Integration focuses on reducing the thermal utility requirements of a site through reusing/recycling in-house resources. A similar concept can be applied to Total Site Mass Integration as well for material resources planning. It relies on the two main principles in resources targeting of a Total Site:
(i)   Each plant should use its own resources first to fulfil its own demands prior to sending or accepting resources to or from other plants.
(ii) The necessary sources transfer from one plant to another should be minimised to save the capital cost

Before diving into the modelling methodology, let's revise the Composite Curves construction for a plant. Recall that the Pinch Method splits the process with a single Pinch Point into two regions/categories, namely the High-Quality Region, i.e. HQR (Below the Pinch Point where fresh resources are needed for all source-sink pairs) and Low-Quality Region, i.e. LQR (Above the Pinch Point where fresh resources are not needed but waste resources are generated). This is also applicable to a site as well (see Fig. 10.7). The objective of Total Site targeting is to ensure the site-level minimum fresh resources are met while minimising the transfer of the cross-plant source. The Pinch Points of each plant should also be lower (better quality) or equal to the Site Pinch Point, as if one plant's Pinch Point has worse quality than the Site Pinch Point, this indicates the allocation of sources is not done properly and violates the Site Pinch, which also means higher overall site-level fresh resources intake is needed.

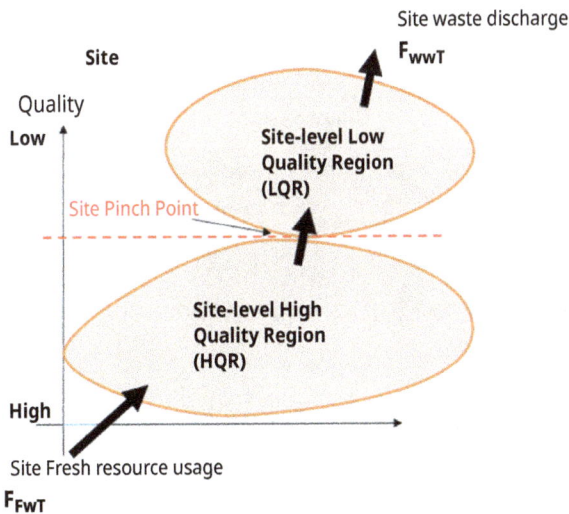

Fig. 10.7: Division of site-level resources profile into two categories.

The CC shows the information on how the cross-plant transfer can be performed to ensure that the fresh resources can be further reduced. Chin et al. (2021b) analysed various options for cross-plant transfer schemes, also adapted from Chew et al. (2010), and recommended three strategies:

(a) Scheme 1: Transfer of sources from a LQR of one plant to another plant's HQR, provided the LQR source supplier has quality below (better than) the Site Pinch Point. Fig. 10.7 shows an illustration of the cross-transfer according to Scheme 1, where the source from the LQR of one plant can be used to supply water to sinks in the HQR of another plant. The Pinch Principles indicate that there should not be resources flowing from LQR to HQR within a plant. However, this is feasible when the recipient plant has its Pinch Point higher than the Site Pinch Point (at lower water quality), and the supplier plant has its Pinch Point lower than the Site Pinch Point (at higher water quality). Note that the Site Pinch Point represents the overall Pinch Point for the site including all plants. This will be explained further in the working session in the following sections. In this example, Plant B has sources in the LQR, which are of better quality than the Site Pinch Point. This makes them suitable to replace sources with quality lower than the Site Pinch Point in Plant A. This could potentially lower the Pinch Point in Plant A so that its Pinch Point could reach the Site Pinch Point.

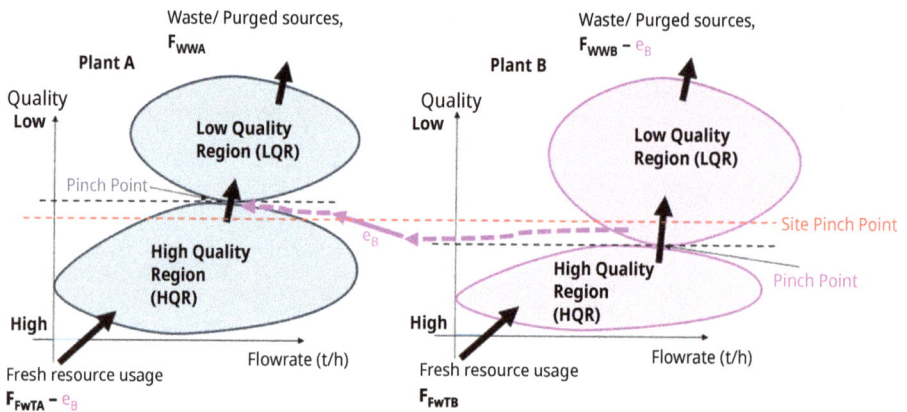

Fig. 10.8: Plant B send its unused sources (LQR) to Plant A (HQR higher than Site Pinch). Pinch Point for plant B is lower than the Site Pinch Point.

(b) Scheme 2: Transfer of sources from the HQR of one plant to another plant's LQR, along with at least one Scheme 1.

(c) Scheme 3: Transfer of sources from the LQR of one plant to another plant's LQR, along with at least one Scheme 1.

Another possibility is to apply Scheme 2, when the transfer of a source in the HQR of one plant to the LQR of another plant is possible, as shown in Fig. 10.8. In this demonstration, the Pinch-causing source (Source at Pinch Point of B) in Plant B is actually wasted in the LQR sinks since the Pinch Point in Plant B is lower than Site Pinch Point. Unlike in Fig. 10.7, this transfer is from precisely the Point Point in Plant B instead of above the Pinch Point as compared to Scheme 1. In this case, a source replacement in Plant B at the LQR (at the Pinch Point) is required. In exchange, the HQR source in Plant A, which has worse quality than the Site Pinch Point, can be used to replace the LQR Pinch-causing source in Plant B. Note that the supplier Plant A can send any source which has a quality lower than the plant's Pinch Point but higher than the Site Pinch Point.

Scheme 3 also follows the same demonstration with Scheme 2, as shown in Fig. 10.9, but the sequence is reversed. In this case, the supplier Plant A should send the Pinch-causing source (source at the Pinch Point of A) in its LQR (instead of HQR) to the LQR of the receiver Plant B. This is to replace the Pinch-causing source in Plant B, so it can be used for Plant A in order to lower the Pinch Point of Plant A. In exchange, the source from LQR Plant B should be replaced and used in HQR Plant A.

In fact, any other cross-plant source exchanges are possible but may not be as effective and could incur more unnecessary transfers that cost more. For example, the supplier plant can give up part of their sources in the HQR/LQR to send to the receiver plant's HQR/LQR, but it is ineffective to the overall fresh resource consumption. By observing the transfer schemes, the plants that have the Pinch Point lower than the Site Pinch Point (higher quality) can send out their sources to other plants that have Pinch Points higher than the Site Pinch (lower quality). This guarantees to reduce the overall fresh resource consumption. Tab. 10.3 shows the full summary of all possible sources transfer options.

Other important facts to know are the sequential source allocation strategy by El-Halwagi et al. (2003) and also the sources mapping diagram (see Chapter 8). This approach is similar to the dynamic programming approach, where the fresh resource is minimised for each sink sequentially, starting with the cleanest sink and followed by the next cleanest. The highest quality sink is satisfied with the highest quality source first, and then the algorithm proceeds to the next better quality sink. If the highest quality source is not sufficient, the next better quality source is then used.

In the CC construction proposed by El-Halwagi et al. (2003), the typical arrangement of sources is based on ascending order of the source's impurity (or descending order of the source's quality), shown in Fig. 10.10a. However, the sources that are in the HQR could be actually arranged in different ways. For example, in Fig. 10.10b, although Source 2 is dirtier (lower quality) than Source 1, swapping the position of Source 2 with Source 1 does not alter the total fresh resource target in Fig. 10.10a. Swapping the order of sources causes Sink 1 to require more fresh resources as compared to the ascending order arrangement of sources, but it also reduces the fresh resource for Sink 2. The increment of fresh resources for Sink 1 and the reduction of fresh resource for Sink 2 are

Tab. 10.3: Summary of possible cross-plant transfer schemes. Note that in this formulation, " >" means "higher than/worst than", " <" means "lower than/better than", "/" means "or" and " =" means "equals to".

Supplier (From)	Receiver (To)	Pinch Point conditions	Remarks
HQR	LQR	Supplier > Site Pinch Receiver ≤ Site Pinch	Desirable, but the supplier requires sources from other plants that are ≤ Site Pinch
		Supplier ≥ Site Pinch Receiver > Site Pinch	Unnecessary
		Supplier < Site Pinch	Not desirable
		Supplier = Site Pinch Receiver < Site Pinch	Possible, the supplier can only give up part of the HQR source so that the supplier's Pinch Point is still ≤ Site Pinch Point
		Supplier = Site Pinch Receiver = Site Pinch	Unnecessary
LQR	HQR	Supplier > Site Pinch Receiver </≥ Site Pinch	Not desirable
		Supplier = Site Pinch Receiver </> Site Pinch	Desirable so that receiver can give up the source for other plants. However, the receiver's Pinch Point eventually should be ≤ Site Pinch Point
		Supplier < Site Pinch Receiver ≥ Site Pinch	Ideal transfer. If receiver = Site Pinch, its LQR source can be used for other plants as well.
		Supplier > Site Pinch Receiver ≥ Site Pinch	Not desirable

Tab. 10.3 (continued)

Supplier (From)	Receiver (To)	Pinch Point conditions	Remarks
HQR	HQR	Supplier = Site Pinch Receiver > Site Pinch	Possible if there are no other options, a supplier can only give up part of the HQR source. This should come along with at least one LQR → HQR transfer
LQR	LQR	Supplier > Site Pinch Receiver ≤ Site Pinch	Possible if there are no other options so that receiver can give up the source for other plants. This should come along with at least one LQR → HQR transfer

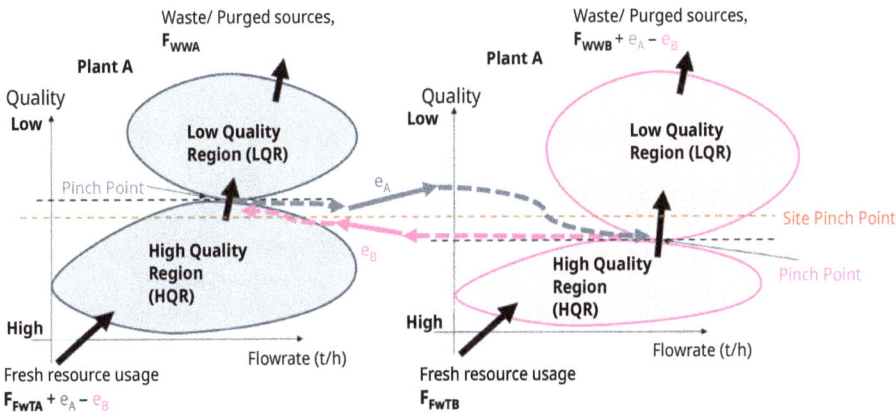

Fig. 10.9: Plant B send its unused sources (LQR) to Plant A (HQR) and Plant A send its sources at the Pinch Point from LQR (higher than the Site Pinch Point) to Plant B (Pinch Point). Plant A would require another source with quality at the 'Site Pinch' Point.

balanced, which do not alter the overall fresh resource consumption. Another feasible arrangement is to split part of Source 2 and put it before Source 1 – see Fig. 10.10c. A similar arrangement also can be made by reversing the typical order of the sources in the HQR, i.e. arranging sources in HQR with descending order of Source CC gradient – see Fig. 10.10d.

The part of Source 3 in the HQR can be arranged as the first source segment in the Source CC. Note that the Source CC crosses the Sink CC in Fig. 10.10d. In the conventional CC construction, the Source CC has to be on the right side of the Sink CC. However, this only applies when the sources are arranged in the ascending order of the impurities. The final load of the Source CC in the HQR still matches with the Sink CC in the HQR, and this indicates the constraints of the quality load of the sinks are not violated. Due to the mixing of sources it is still feasible despite the fact that the order is changed – see the rules

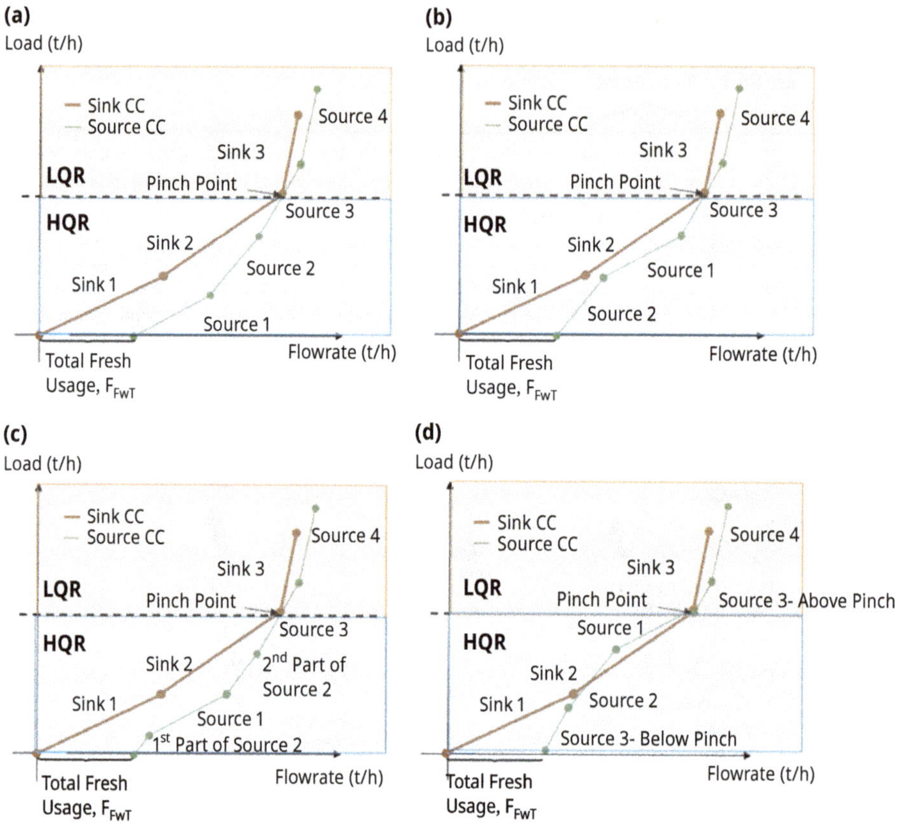

(a)

Load (t/h)

(b)

Load (t/h)

(c)

Load (t/h)

(d)

Load (t/h)

Fig. 10.10: Feasible CC representations with different source arrangements in the HQR region (below Pinch): (a) ascending order of sources; (b) swapping position of Sources 1 and 2; (c) splitting part of the sources and arranging them in different positions; (d) descending order of sources. (Note that the arrangement of sources must satisfy all the quality loads of ALL sinks or else it won't work).

in CC constructions for multiple qualities in Chin et al. (2022). Based on the observations, this implies that the source arrangement in the HQR can be changed in any way without altering the overall fresh resource target or consumption. This only applies where all the contamination limits (quality loads) are fulfilled for each sink in HQR.

This concept of sources order swapping is important when deriving the Total Site source allocation as most of the plants prefer to use their own sources even though the sources from other plants may have better quality. In this case, the site-level Composite Curves may have sources arrangement which is not ordered with their qualities. However, this is not critical to the fresh resources intake as long as the sources all belong to one region (i.e. either Below the Pinch or Above the Pinch). The heuristic to minimise the cross-plant source transfer along with the site-level header is summarised in Fig. 10.11 and is explained in the following working session.

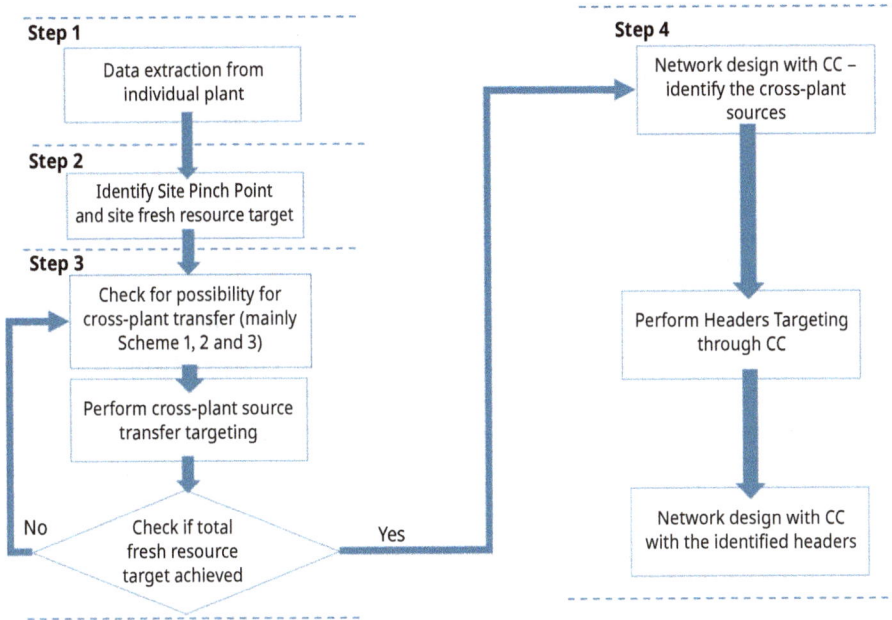

Fig. 10.11: Summarised procedure in Total Site Water Allocation targeting.

10.3.1 Targeting minimum fresh resources in a Total Site

The demonstration of the Total Site resources targeting is presented through Example 10.3, as shown in Tab. 10.4. This study consists of three plants with various types of sources and sinks but with common single quality.

Tab. 10.4: Case study for Example 10.3 – Chew et al. (2010).

Site/plants	Sinks	F (t/h)	C (ppm)	Sources	F (t/h)	C (ppm)
Plant A	SK1	20	0	SR1	66.67	80
	SK2	66.67	50	SR2	20	100
	SK3	100	50	SR3	100	100
	SK4	41.67	80	SR4	41.67	800
	SK5	10	400	SR5	10	800
Plant B	SK1	20	0	SR1	66.67	80
	SK2	66.67	50	SR2	20	100
	SK3	15.63	80	SR3	15.63	400
	SK4	42.86	100	SR4	42.86	800
	SK5	6.67	400	SR5	6.67	1,000

Tab. 10.4 (continued)

Site/plants	Sinks	F (t/h)	C (ppm)	Sources	F (t/h)	C (ppm)
Plant C	SK1	20	0	SR1	80	50
	SK2	80	25	SR2	20	100
	SK3	50	25	SR3	50	125
	SK4	40	50	SR4	300	150
	SK5	300	100	SR5	40	800

Step 1: Data extraction from the individual plants. This step is to identify the "transferrable" sources from each individual plant. Fig. 10.12a–c shows the CC constructed from the stream data of each plant. All sources in their respective HQR and LQR (the waste sources) are potential sources (shaded region). Note that the allocation of sources to sinks in LQR should be conservative by using minimal amount of "cleaner" source so that the good quality source is reserved for other plants – see Step 2.

Step 2: Identify Site Pinch Point and site fresh resource target. This step identifies the Site Pinch Point as a benchmark for determining the possible transfer schemes. The CC construction for the site (considering all sources and sinks for all plants) is shown in Fig. 10.13. The freshwater identified is 316.36 t/h, with the Site Pinch Point at 150 ppm. The total freshwater targets from each plant yield about 339.65 t/h (98.33 + 54.65 + 186.67), which indicates that cross-plant source transfer is needed to further reduce the overall fresh resources.

Step 3: Check for the possibility of cross-plant sources transfer. The proposed schemes in Tab. 10.3 can be referred to, but the main three schemes, as presented in Section 10.2.1, should be prioritised. The first step would be to determine the possibility of Scheme 1. Notice that the Pinch Point for Plant A is 100 ppm, Plant B is 400 ppm, and Plant C is 150 ppm. It is fairly obvious that Plant A or C can transfer the sources in their LQR to HQR of Plant B since both of them have Pinch Points lower than or equal to the Site Pinch Point. Two possibilities of transfer are presented here:
(i)   Plant A can send its Pinch-causing source in the LQR (100 ppm) to Plant B
(ii) Plant C can send their Pinch-causing source in the LQR (150 ppm) to Plant B

In fact, the two transfers are feasible and able to reduce the fresh resources flow. Recall that in Fig. 10.10, it is presented that sources arrangement within either HQR or LQR still yields the same fresh resources target. This is applicable to the site level as well. Since the Site Pinch is at 150 ppm, the sources that have ≤150 ppm can be allocated in various ways to the HQR sinks without violating the fresh resource target. In other words, whether Plant A or Plant C is to transfer their sources first to Plant B, eventually, the fresh resource target can be achieved.

However, the question is, which transfer is more effective? Although minimum fresh resource usage is the ultimate goal, the transfer of the water sources across

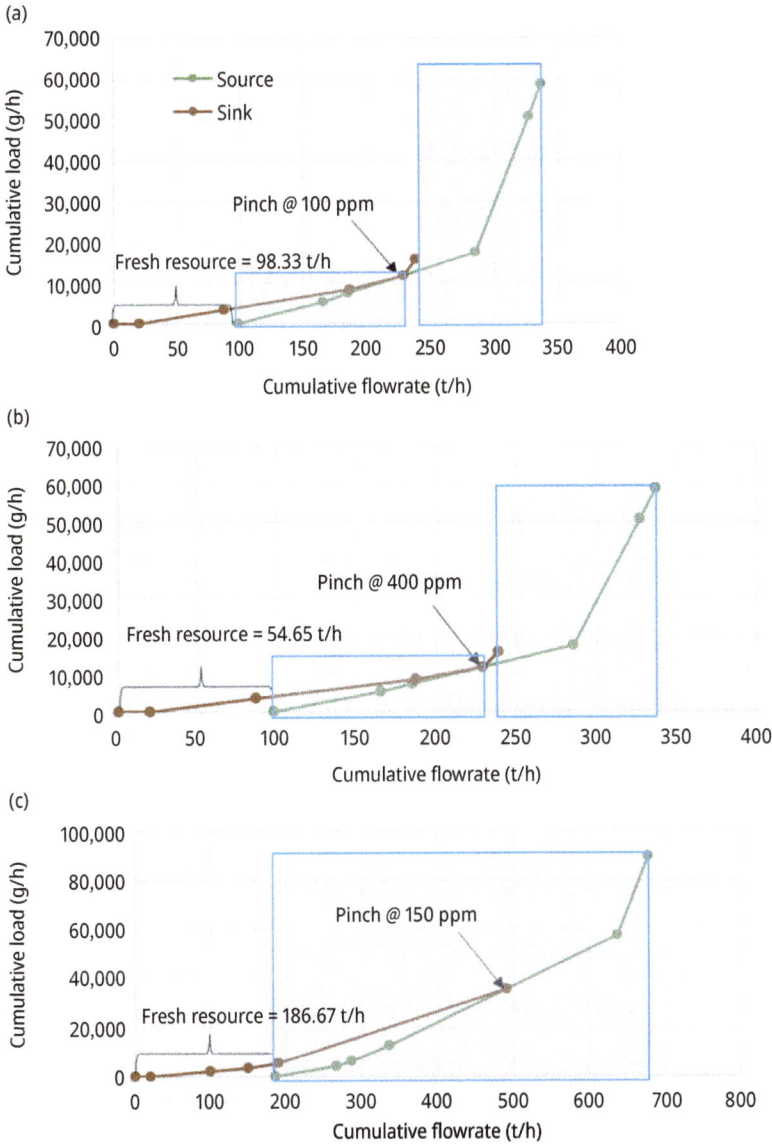

Fig. 10.12: Composite Curves for Example 10.3: (a) Plant A, (b) Plant B, and (c) Plant C.

plants should also be minimised so that significant capital costs can be saved. Since Plant A has a better quality source to offer, it is more effective to use the source from Plant A first before using the source from Plant C, if necessary.

The next step is to determine how much of the 100 ppm source is remained from Plant A. Since they are to give up their 100 ppm source from their LQR, it has to be replaced with another source that has lower quality (higher concentration) within the

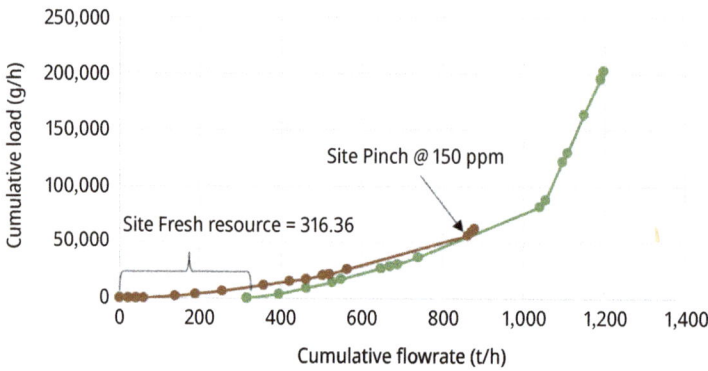

Fig. 10.13: Total Site Composite Curves for Example 10.3.

same Plant. As Plant A has only one LQR sink, which is 400 ppm, it can be fulfilled by the mix of 100 and 800 ppm sources. The objective now here is to use the minimum amount of 100 ppm sources to fulfil the 400 ppm sink. This can be solved by forming a simple optimisation model and can easily be solved using Excel – see Fig. 10.14. The available 100 ppm source originally had 100 t/h. Since about 43.34 t/h is used in the HQR, it is left at about 56.66 t/h in the LQR. By solving the optimisation model, the

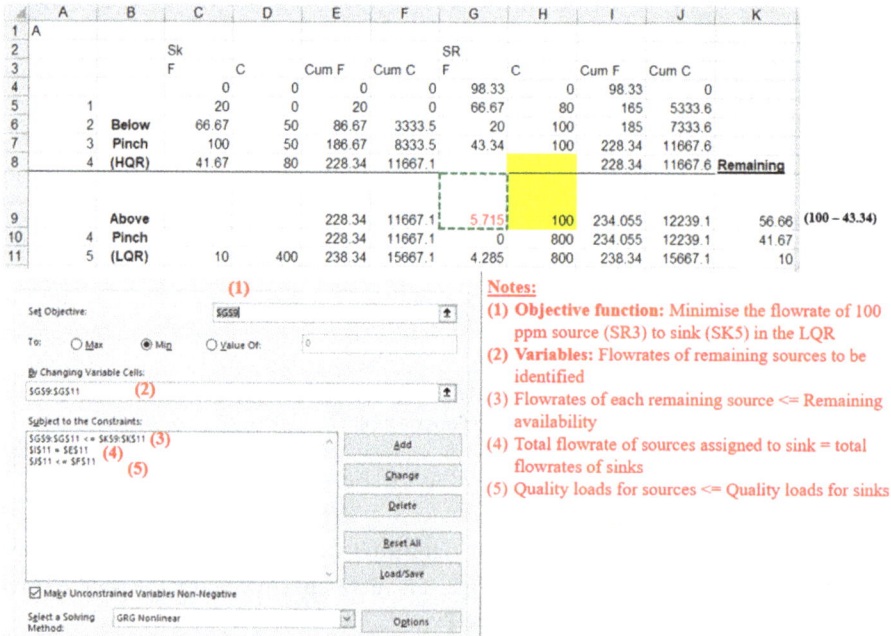

Fig. 10.14: Minimising the usage of good quality sources for LQR sink using Excel solver for Example 10.3.

required use of a 100 ppm source is about 5.72 t/h, which leaves about 50.95 t/h of a 100 ppm source for Plant B.

A similar step can be applied for Plant C as well to identify the remaining available 150 ppm source for Plant B. Since Plant C has no sinks that are in the HQR (≥150 ppm), the remaining 150 ppm source can be easily identified. About 153.33 t/h is used up in the HQR of Plant C, which left about 146.67 t/h of 150 ppm source from Plant C.

By sending all remaining water from the 100 ppm source (50.95 t/h) from Plant A to Plant B, in fact, the Pinch Point of Plant B changed to 100 ppm. Similarly, by sending all remaining 150 ppm source (146.67 t/h) from Plant C along with the source from Plant A, it lowers the Pinch Point of Plant B (100 ppm), and the fresh resources for Plant B are reduced to 43.13 t/h – see Tab. 10.5. The total fresh resources are now reduced to 328.13 t/h, but still not equal to the final target yet. The reason is that the source from Plant A (100 ppm) is used for a sink that is in the LQR (400 ppm > 150 ppm) – see Tab. 10.5, which in this case, is a waste of the resources. In this case, another strategy is to be sought out to achieve further fresh resources reduction.

Step 3: Revise cross-plant transfer schemes. Based on the previous steps, it seems that some of the "good" quality resources (lower than 150 ppm Site Pinch Point) are not used up properly. Recall that for Plants A and B, some of the 100 ppm sources are actually used up for 400 ppm sink (see Tab. 10.5). This means that some of the sources that should be in the site-level HQR are used in the site-level LQR. For Plant A, the 400 ppm sink can actually be satisfied by a combination of 800 ppm source and 150 ppm source as well to ensure that Plant A has a Pinch Point of not higher than 150 ppm. In this case, Plant C can send the remaining 150 ppm source in their LQR to the LQR of Plant A (Scheme 3). As Scheme 3 is to be followed by at least one LQR-HQR transfer, the obvious choices here are that Plant A can send out their resources in LQR (100 ppm) to Plant B (HQR) or even Plant C (HQR) so that more 150 ppm sources can be used if necessary. Similarly, for Plant C, they can send out their LQR resources (150 ppm) to Plant B HQR as well. The overall transfer Scheme 3 (not necessarily in order) is:

(i) Plant C can send their Pinch-causing source in the LQR (150 ppm) to the LQR of Plant A.

(ii) Plant C can send their Pinch-causing source in the LQR (150 ppm) to the HQR of Plant B.

(iii) Plant A can send their Pinch-causing source in the LQR (100 ppm) to the LQR of Plant B/HQR of Plant C

For Plant B, since it is potentially receiving two sources from two plants (100 and 150 ppm), the cross-plant transfer flowrates should be minimised as well. This becomes the problem of targeting multiple impure fresh resources problem. The solving procedure is not shown and follows the same excel procedure as presented in Fig. 10.14. The cleaner source should be minimised first followed by minimising the next source. The solution identified is that Plant B can accept a minimum of about 9.74 t/h of 100 ppm

Tab. 10.5: Results for Example 10.3 – first iteration using only Scheme 1.

Site/plants	Below/above pinch	Sinks	F (t/h)	C (ppm)	Sources	F (t/h)	C (ppm)
Plant A					Fresh	**98.33**	0
	Below Pinch (HQR)	SK1	20	0	SR1	66.67	80
		SK2	66.67	50	SR2	20	100
		SK3	100	50	SR3 (Pinch)	43.34	100
		SK4	41.67	80			
	Above Pinch (LQR)	SK5	10	400	SR3	5.715	100
					SR4	0	800
					SR5	4.285	800
	Waste				SR4 + SR5	47.39	800
Plant B					Fresh	**43.13**	0
	Below Pinch (HQR)	SK1	20	0	SR1	66.67	80
		SK2	66.67	50	SR2	20	100
		SK3	15.63	80	SR3 (Plant A) – Pinch	15.363	100
		SK4	42.86	100			
	Above Pinch (LQR)	SK5	6.67	400	SR3 (Plant A)	6.67	100
	Waste				SR3	15.63	400
					SR4	42.86	800
					SR5	6.67	1,000
Plant C					Fresh	**186.67**	0
	Below Pinch (HQR)	SK1	20	0	SR1	80	50
		SK2	80	25	SR2	20	100
		SK3	50	25	SR3	50	125
		SK4	40	50	SR4 – Pinch	153.33	150
		SK5	300	100			
	Waste				SR4	146.67	150
					SR5	40	800

source from Plant A and 3.75 t/h of 150 ppm source from Plant C. Plant A needs only about 6.15 t/h of 150 ppm source from Plant C to fulfil the 400 ppm sink, along with 3.85 t/h of 800 t/h source within their vicinity. The rest of the 100 ppm source from Plant A (56.66–9.74 = 46.92 t/h) can be transferred to Plant C. The freshwater targets for each plant are now changed to 98.33 t/h at 100 ppm Pinch Point for Plant A, 45 t/h at 150 ppm Pinch Point for Plant B, and 171.03 t/h at 150 ppm Pinch Point for Plant C. The total freshwater target is now about 314.36 t/h, which agrees with the overall fresh resource target. Notice that all plants now have Pinch Points lower than or equal to the Site Pinch Point (150 ppm) as well.

Step 4: Identify the site-level headers using Site CC (Tab. 10.6). Note that the identified cross-plant sources transfer are:

(a) 6.15 t/h of SR4 from Plant C (150 ppm) is sent to Plant A

(b) 9.74 t/h of SR3 from Plant A (100 ppm) is sent to Plant B

(c) 3.75 t/h of SR4 from Plant C (150 ppm) is sent to Plant B
(d) 46.92 t/h of SR3 from Plant A (100 ppm) is sent to Plant C.

Tab. 10.6: Revised Results for Example 10.3: second iteration using Scheme 3.

Site/plants	Below/Above Pinch	Sinks	F (t/h)	C (ppm)	Sources	F (t/h)	C (ppm)
Plant A					Fresh	**98.33**	0
	Below Pinch (HQR)	SK1	20	0	SR1	66.67	80
		SK2	66.67	50	SR2	20	100
		SK3	100	50	SR3 (Pinch)	43.34	100
		SK4	41.67	80			
	Above Pinch (LQR)	SK5	10	400	SR4 (Plant C)	6.15	150
					SR4	0	800
					SR5	3.85	800
	Waste				SR4 + SR5	47.82	800
Plant B					Fresh	**45**	0
	Below Pinch (HQR)	SK1	20	0	SR1 (Pinch)	41.67	80
		SK2	66.67	50			
	Above Pinch (LQR)	SK3	15.63	80	SR1	25.00	80
		SK4	42.86	100	SR2	20	100
		SK5	6.67	400	SR3 (Plant A)	9.74	100
					SR4 (Plant C)	3.75	150
					SR3	6.67	400
	Waste				SR3	8.96	400
					SR4	42.86	800
					SR5	6.67	1,000
Plant C					Fresh	**171.03**	0
	Below Pinch (HQR)	SK1	20	0	SR1	80	50
		SK2	80	25	SR2	20	100
		SK3	50	25	SR3 (Plant A)	46.92	100
		SK4	40	50	SR3	50	125
		SK5	300	100	SR4 – Pinch	122.05	150
	Waste				SR4	168.04	150
					SR5	40	800

All of these sources are plotted in the site-level CC as shown in Fig. 10.15.

Following the headers constructing procedure as presented in Fig. 10.11, the site-level headers can be identified by mixing the cross-plant sources. Only the sources, either below or above the Site Pinch, can be mixed within their own region. The cross-plant sources below the Site Pinch can be mixed together (100 ppm), i.e. 46.92 + 9.74 t/h. Notice that for the cross-plant transfer (c), although it is used for LQR in Plant B, it is used for a sink that is below the Site Pinch Point (150 ppm). In this case, it is still in the site HQR. The sources in site HQR can be mixed to form a header

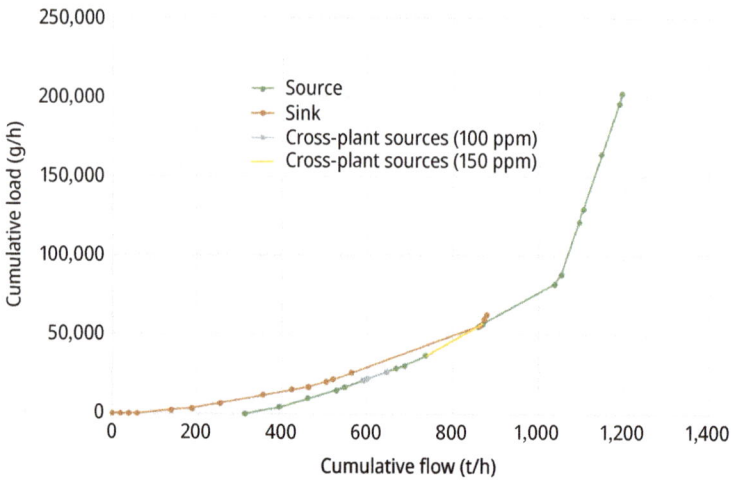

Fig. 10.15: Site Composite Curves for Example 10, with cross-plant sources.

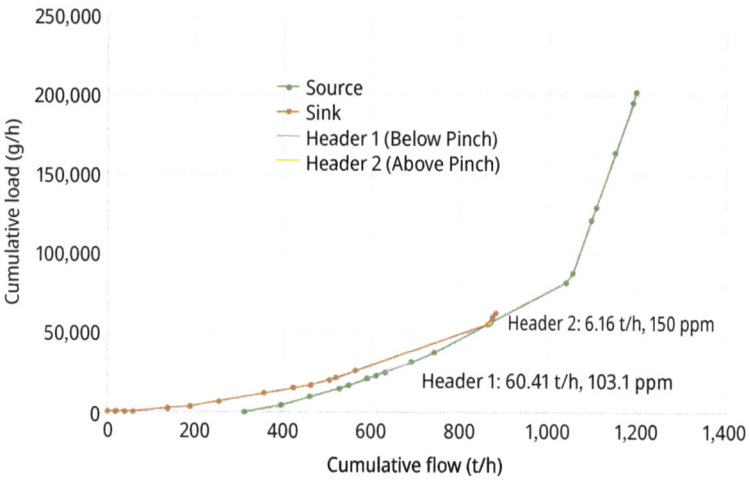

Fig. 10.16: Site Composite Curves for Example 10.3, with cross-plant headers.

$(46.92 + 9.74 + 3.75 = 60.41\ t/h)$ with a quality of 103.1 ppm. The rest of the cross-plant source in the site LQR can be its own header $(6.15\ t/h, 150\ ppm)$ – see Fig. 10.16.

It is to be noticed that since the cross-plant sources are mixed, the freshwater targets for each individual plant may be changed too. However, the overall freshwater resources at the site level should remain the same. For more examples and working sessions, the readers can refer to Chin et al. (2021c).

10.4 Targeting fresh resources for multiple qualities

The key concepts of Pinch Analysis rely on two main steps:
(i) Prioritisation of sinks based on quality, where better quality is prioritised
(ii) Allocation of sources to fulfil the sinks, where the allocation follows the same prioritisation rules.

In order to understand how the prioritisation and allocation rules for multiple quality problems, it is important to check the source-sink allocation model. The typical source-sink allocation model of the water network gives rise to four governing equations – eqs. (10.1)–(10.4). The problem maps to an optimisation formulation. Its objective function is expressed in eq. (10.1), which stipulates the minimisation of the total required freshwater, which is the freshwater target.

Equations (10.2) and (10.3) express the mass balances for sources and sinks, while eq. (10.4) represents the contaminant constraints for individual sinks. Note that as proven by El-Halwagi et al. (2003), the composition of the sink should represent the maximum contaminant concentration to minimise the use of a fresh resource. These equations are crucial for understanding the model characteristics to derive the proper procedures for obtaining the optimal freshwater target:

$$\text{Min } F_{\text{FWT}} = \sum_{j} F_{\text{Fw, SK}j} \tag{10.1}$$

$$F_{\text{SR}i} = \sum_{j} F_{\text{SR}i,\text{SK}j} + F_{\text{SR}i,\text{WW}} \quad \forall i \tag{10.2}$$

$$F_{\text{SK}j} = \sum_{i} F_{\text{SR}i,\text{SK}j} + F_{\text{Fw,SK}j} \quad \forall j \tag{10.3}$$

$$\sum_{i} F_{\text{SR}i,\text{SK}j} C_{k,\text{SR}i} + F_{\text{Fw,SK}j} C_{k,\text{Fw}} \leq F_{\text{SK}j} Z_{k,\text{SK}j} \quad \forall j \ \forall k \tag{10.4}$$

where F_{FWT} represents the total freshwater flowrate, $F_{\text{Fw,SK}j}$ is freshwater to sink "j" flowrate, $F_{\text{SR}i}$ is source "i" flowrate, $F_{\text{SR}i,\text{SK}j}$ is source "i" to sink "j" flowrate, $F_{\text{SR}i,\text{WW}}$ is source "i" to waste flowrate, $F_{\text{SK}j}$ is the sink "j" flowrate, $C_{k,\text{SR}i}$ is the concentration of contaminant 'k' in source 'i', $C_{k,\text{Fw}}$ is the concentration of contaminant 'k' in freshwater, and $Z_{k,\text{SK}j}$ is the concentration of contaminant "k' in sink "j".

As mentioned, the mass-based Pinch Analysis relies on the ranking or prioritisation of sources/sinks to determine the fresh resource target. The sinks with lower concentration are prioritised because they are harder to be fulfilled, and they should be fulfilled first with better quality/lower concentration sources. This follows the traditional Process Integration principle to resolve the most constrained part of the problem first. For multi-contaminant problems, the sources and sinks are not constrained by a single contaminant anymore. Each sink is controlled by the most limiting contaminants. The identification of the limiting contaminant plays an important role in determining the source prioritisation pattern. The following sub-sections are devoted to utilising eqs. (10.1)–(10.4) for understanding the multi-contaminant problem and deriving certain heuristics for fresh resource targeting purposes.

Although there are multiple contaminants, each sink is probably limited by one type of limiting contaminant. Limiting contaminants are those whose constraints sink the most and determine the real freshwater flowrate required based on the available sources. Let's consider two contaminant problems (contaminant "k1" and "k2") with one source (SR1) and one sink (SK1). Using eqs. (10.3) and (10.4), the mass balance for the sink and its contaminant constraints is

$$F_{SK1} = F_{SR1,SK1} + F_{FW,SK1} \tag{10.5}$$

$$F_{FW,SK1Ck1,Fw} + F_{SR1,SK1Ck1,SR1} \le F_{SK1Zk1,SK1} \ \forall k \tag{10.6}$$

Equations (10.5) and (10.6) can be rearranged to find the expression of the minimum freshwater target. To eliminate the unknown variables $F_{SR1,SK1}$, express $F_{SR1,SK1}$ in terms of F_{SK1} and F_{FW1} in eq. (10.5) and substitute into eq. (10.6) and move F_{FW1} to the left-hand side, the minimum freshwater requirement for both contaminants are formulated – see eqs. (10.7a–c):

$$F_{FW,SK1} \ge F_{SK1}\left[1 - \frac{Z_{k1,SK1} - C_{k1,FW}}{C_{k1,SR1} - C_{k1,FW}}\right], \text{for Contaminant } k1 \tag{10.7a}$$

$$F_{FW,SK1} \ge F_{SK1}\left[1 - \frac{Z_{k2,SK1} - C_{k2,FW}}{C_{k2,SR1} - C_{k2,FW}}\right], \text{for Contaminant } k2 \tag{10.7b}$$

The freshwater requirement is equal to the maximum value among the requirements across all of the contaminants – expression (10.7c). The maximum value is used as one contaminant may require more dilution than another and becomes limiting. For example, if contaminant "k1" requires 1,000 t/h of freshwater to dilute the sources to fulfil sink constraints, but contaminant "k2" requires 2,000 t/h of freshwater to dilute, then the freshwater flowrate is 2,000 t/h of freshwater (maximum amount):

$$F_{FW,SK1} \ge F_{SK1}\text{Max}_k\left(\left[1 - \frac{Z_{k,SK1} - C_{k,FW}}{C_{k,SR1} - C_{k,FW}}\right]\right) \tag{10.7c–1}$$

$$F_{FW,SK1} \ge F_{SK1}\left[1 - \text{Min}_k\left(\frac{Z_{k,SK1} - C_{k,FW}}{C_{k,SR1} - C_{k,FW}}\right)\right] \tag{10.7c–2}$$

Based on expression (10.7c), the maximum value of the bracketed term represents the real minimum water target if only a single source is used. As the flowrate of sink SK1 (F_{SK1}) is constant, the minimum value of the concentration ratio of the sink to the source corresponds to the maximum amount of the term, indicating more fresh resource is needed. This also means that if contaminant "k" corresponds to the minimum value of the ratio, then the sink is limited by contaminant "k". The method agrees with the concentration potential concept proposed by Liu et al. (2009). This concentration ratio also means how much SR1 can be used to fulfil one unit of SK1, as shown in expression (10.7d):

$$F_{\text{Fw, SK1}} \geq F_{\text{SK1}} \left[\text{Min}_k \left(\frac{Z_{k,\text{SK1}} - C_{k,\text{FW}}}{C_{k,\text{SR1}} - C_{k,\text{FW}}} \right) \right] \tag{10.7d}$$

For a problem with "k" contaminants with "N" internal sources, the real freshwater target is shown in expression (10.8) – see derivation in Chin et al. (2021b). To identify the preferred source ranking for the specific sink, the ratio of sink concentration to the source concentration plays an important role. In expression (10.8), it is proposed that the maximum source concentration is the denominator because it also shows exactly the prioritisation of sources if more than two sources are used. For example, if the coefficient of $F_{\text{SR1,SK1}}$ is more negative than $F_{\text{SR2,SK1}}$, then SR1 should be prioritised:

$$F_{\text{Fw, SK1}} \geq \text{Max}_k \left(F_{\text{SR1, SK1}} \left[\frac{Z_{k,\text{SK1}} - C_{k,\text{FW}}}{C_{k,\text{SR}_{\max}} - C_{k,\text{FW}}} - 1 \right] + \ldots + F_{\text{SRN, SK1}} \left[\frac{Z_{k,\text{SK1}} - C_{k,\text{FW}}}{C_{k,\text{SR}_{\max}} - C_{k,\text{FW}}} - 1 \right] \right.$$
$$\left. + F_{\text{SK1}} \left(\left[1 - \frac{Z_{k,\text{SK1}} - C_{k,\text{FW}}}{C_{k,\text{SR1}} - C_{k,\text{FW}}} \right] \right) \right) \tag{10.8}$$

The proposed constant ratio $(Z_{k,\text{SK1}} - C_{k,\text{FW}})/(C_{k,\text{SR\_max}} - C_{k,\text{FW}})$ in the last term helps to identify the source prioritisation as this ratio constitutes the constant term in expression (10.8). If the ratio is minimum for contaminant "k", then the constant term: F_{SK1} $[1 - (Z_{k,\text{SK1}} - C_{k,\text{FW}})/(C_{k,\text{SR\_max}} - C_{k,\text{FW}})]$ has a maximum value which means the source allocation is more likely to follow the prioritisation sequence for contaminant "k". The optimal source allocation is done to minimise this term. However, this does not mean that the sink should follow exactly the source prioritisation sequence for contaminant "k" because different sources are traded-off by other contaminants. The proposed concentration ratio is just to show which source prioritisation order is likely to be followed to achieve the minimum fresh resource target for the specific sink.

For a more general problem with more than one source, the following heuristics are presented to identify the most likely limiting contaminant for a sink but also to identify the assignment of the sinks to the proper cascade:

(a) The ratio of the shifted sink concentration to the shifted maximum of source concentration $(Z*_{k,\text{sk}}/C*_{k,\text{sr\_max}})$ for each contaminant "k" are evaluated, where $Z*_{k,\text{SKj}}$ $= Z_{k,\text{SKj}} - C_{k,\text{FW}}$ and $C*_{k,\text{SR\_max}} = C_{k,\text{SR\_max}} - C_{k,\text{FW}}$. The lower value of the ratio corresponds to tighter limitations of the contaminant on the current sink.

(b) If contaminant "k" corresponds to the minimum value of the ratio, then the sink is limited by contaminant "k" for the specific source "i". It is to be noted that the limiting contaminant identified here is the most likely limiting contaminant, but not all. This is because a sink can be limited by several contaminants.

The concentration ratio $(Z*_{k,\text{sk}}/C*_{k,\text{SR\_max}})$ shows which contaminant's source prioritisation order is likely to be followed for the specific sink. However, source prioritisation cannot be fully followed as other sources might trade off other contaminants as well. The following sub-sections explain the source allocation characteristics for a multi-contaminant problem and infer the source allocation steps to be followed to achieve the minimum freshwater target.

10.4.1 Conflicting sources

The sources that are conflicting are defined as sources which have different sequences when considering different contaminants. By referring to the numerical example in Tab. 10.7, if only contaminant "A" is considered, SR1 is cleaner than SR2, while for contaminant "B", SR2 is cleaner than SR1. The prioritisation sequence of the sources is not obvious in this case. Considering the problem as a whole requires computing the ratio $(Z*_{k,sk}/C*_{k,sr\_max})$. For the example in Tab. 10.7, for sink SK1, it is obtained $Z_{A,SK1}/C_{A,SRmax} = 100/500 = 0.2$ and $Z_{B,SK1}/C_{B,SRmax} = 100/200 = 0.5$. Since $Z_{A,SK1}/C_{A,SRmax} < Z_{B,SK1}/C_{B,SRmax}$, the source prioritisation sequence based on contaminant "A" should be used because contaminant A is the limiting contaminant to fulfil SK1. This means that SR2 is preferred before SR1.

Tab. 10.7: Example: two-contaminant problem with both examples of conflicting and non-conflicting sources.

SR	F_{sr} (t/h)	$C_{A,sr}$ (ppm)		$C_{B,sr}$ (ppm)	
		Example: conflicting	Example: non-conflicting	Example: conflicting	Example: non-conflicting
1	23	500	500	30	500
2	123	50	50	200	200
Fw	–	0	0	0	0
SK	F_{sk} (t/h)	$Z_{A,sk}$ (ppm)		$Z_{B,sk}$ (ppm)	
1	50	100	100	100	100

Reusing eqs. (10.8) by expressing $F_{SR1,SK1}$ as the variable, the minimum freshwater targets for both contaminants "A" and "B" are shown in eqs. (10.9a), (109b). Note that the flowrate of SR2 ($F_{SR2, SK1}$) is dependent on SR1 and F_w, so its flowrate can be computed by performing mass balance around SK1 ($F_{SR2,SK1} = F_{SK1} - F_{FW,SK1} - F_{SR1,SK1}$). These inequalities are plotted in Fig. 10.17:

$$F_{Fw} \geq F_{SR1, SK1}\left[\frac{500}{50} - 1\right] + 50\left[1 - \frac{100}{50}\right], \text{ for contaminant "A"} \qquad (10.9a)$$

$$F_{Fw} \geq F_{SR1, SK1}\left[\frac{30}{200} - 1\right] + 50\left[1 - \frac{100}{200}\right], \text{ for contaminant "B"} \qquad (10.9b)$$

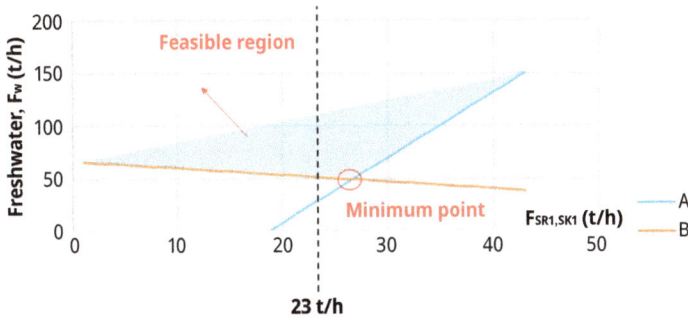

Fig. 10.17: Relationship between freshwater requirement (Fw) vs flowrate of SR1 for numerical example 2, conflicting SR1 and SR2.

Fig. 10.17 shows th e plot of the relationship between the freshwater target for SK1 and the flowrate of SR1 supplied to SK1 for both contaminants. The shaded region represents the feasible region of the freshwater requirement. By observing Fig. 10.17, it is apparent that the minimum freshwater always falls into the boundaries of the feasible region, i.e. one of the impurity constraints would always be active. The interesting point is where both the boundary lines intercept, which represent both impurity constraints, are active. As SR1 and SR2 are conflicting sources in both contaminants "A" and "B", the boundary lines are linear with the respective negative and positive gradients. The minimum freshwater point is always at the point where all impurity constraints are active, assuming no limitation on SR1 and SR2 flowrates.

If one of the sources has a limited flowrate, the minimum point cannot be achieved, then any points along the lowest boundary line are resorted to. In this example, since SR1 has only 23 t/h, all SR1 is used up first and then SR2. This also coincides with the observation earlier, where the sequence of sources based on contaminant "A" should be prioritised (i.e. SR1 → SR2). Note that in this case, the contaminant "B" becomes the limiting now with minimum freshwater requirements, but the sequence of sources ordering should correspond to contaminant "A".

10.4.2 Non-conflicting sources

The previous part explains how the sources with conflicting sequences should be allocated. Using the same example in Tab. 10.7 with non-conflicting source example, it is apparent that SR1 and SR2 are not conflicting anymore as $C_{A,SR1} > C_{A,SR2}$, and $C_{B,SR1} > C_{B,SR2}$. Similarly, the minimum boundary lines for the freshwater target showing the relationship between freshwater and SR1 flowrate are plotted and illustrated in Fig. 10.18. The freshwater target of SK1 varies monotonically with the variation of the flowrate of SR1 supplied to SK1. As a result, the freshwater target is minimal, with the minimum use of SR1.

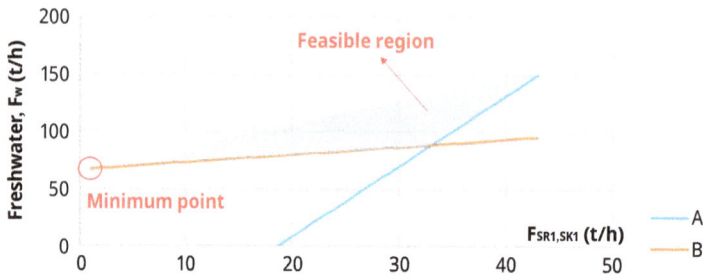

Fig. 10.18: Relationship between freshwater requirement (Fw) vs flowrate of SR1 for non-conflicting SR1 and SR2.

In this case, the intersection point between the boundary lines is no longer the minimum point because both lines are with a positive gradient. The usage of SR1 should be minimised, which means that SR2 should be maximised to reduce the freshwater target. SR2 should be fully used up first before SR1 is used. The source prioritisation becomes similar to the single contaminant case. As $C_{A,SR1} > C_{A,SR2}$, and $C_{B,SR1} > C_{B,SR2}$, the source prioritisation sequence for both contaminants are SR2 → SR1. The following inference can be made: for the "k" contaminants problem, if the source prioritisation sequence is consistent among all the "k" contaminants, then the sources with lower concentrations should be used up first.

10.4.3 Sources with n-contaminants ($n > 2$)

The previous two parts explore different scenarios of source prioritisation for a two-contaminant problem. The source allocation pattern becomes slightly more complicated if more than two contaminants are involved. As established in the two-contaminant problem, the common point where all impurity constraints are active represents the minimum freshwater point. This is similar to any problem with "k" contaminants. Again, a problem with two sources and one sink is illustrated in Fig. 10.19.

Based on the geometric property of straight lines, if there exists a common intersection point among all "k" straight lines, then no other intersection points exist. According to Fig. 10.19, other than the common intersection point, there is no other intersection point for these three straight lines. The converse of the property is also true, i.e. If the intersection point for lines A–B and lines B–C/A–C are the same, then there exists a common intersection point. If such a point exists, then it represents the minimum freshwater point (if all the sources have enough flowrate). The common point is the minimum freshwater point only if there is at least one conflicting sequence of the two sources.

However, not every source allocation problem has a common intersection point. Figure 10.20 below shows the possible scenarios for various lines of contaminant constraints, again for two sources and one sink. The lines for contaminants A and B are

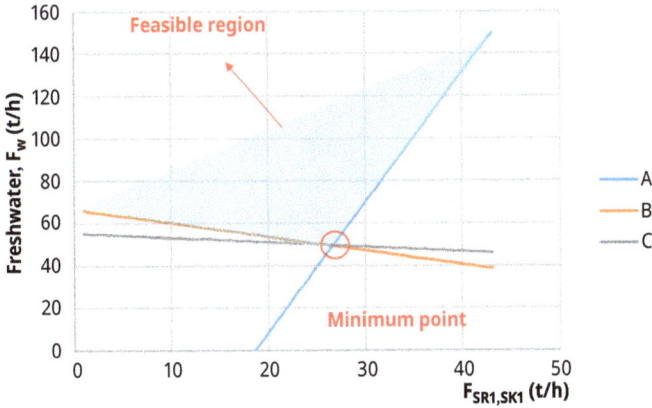

Fig. 10.19: Relationship between freshwater requirement (Fw) vs flowrate of SR1, with a common intersection for all three contaminants.

positive and negative lines, i.e. the two sources have a conflicting sequence for contaminants "*A*" and "*B*", but the gradient of line contaminant "*C*" is varied.

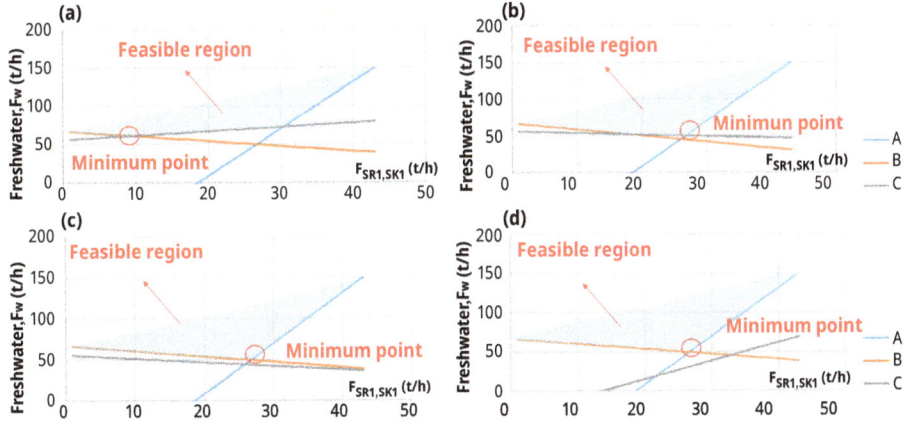

Fig. 10.20: Possible scenarios of source prioritisation sequence in a three contaminants problem (with two sources and one sink): (a) sequence in *A* and *C* conflicts with *B*; (b) sequence in *B* and *C* conflicts with *A*; (c) sequence in *B* and *C* conflicts with *A* but *B*–*C* intersection violates contaminant "*A*" constraint; (d) the sequence in *A* and *C* conflicts with *B*, but the *B*–*C* intersection violates the contaminant "*A*" constraint.

The above demonstration could reveal certain prioritisation rules for a multiple-quality problem:

(i) Prioritisation of sinks based on the limiting contaminant, where the lower the sink-to-source quality ratio is, the contaminant is more limiting

(ii) Allocation of sources should prioritise satisfying the sinks' contamination limits (the more limiting ones) if the sources' sequence is conflicting in various contamination. Otherwise, if the sequence is non-conflicting, the usual way of prioritising a better-quality source is preferred.

Note that the presented rules also apply to single quality as well. The following subsection shows the proposed Pinch-based heuristics in identifying the water allocation network design and target for each sink.

10.4.4 Working session: design and target for multi-contaminants water allocation networks with Pinch-based heuristics

This case study is an illustrative example with two contaminants from Teles et al. (2008). The data is manipulated to explore a scenario that involves changing the concentration or flowrate of sources. This example explores the scenario with a completely conflicting sources sequence, i.e. the source prioritisation for contaminant "A" is completely opposite with contaminant "B". The data is presented in Tab. 10.8:

Tab. 10.8: Data for Example 10.4.

SR	F_{sr} (t/h)	$C_{A,SR}$ (ppm)	$C_{B,SR}$ (ppm)
1	23	50	300
2	47	100	120
3	123	150	100
4	60	250	80
Fw	–	0	0

SK	F_{sk} (t/h)	$Z_{A,Sk}$ (ppm)	$Z_{B,SK}$ (ppm)
1	23	20	60
2	47	50	20
3	123	100	150
4	70	200	80

The first step is to identify which contaminant cascades for each sink should be assigned. The purpose of this step is also to determine the most likely source prioritisation sequence for each sink. This is performed by determining the ratio of shifted sink concentration to the shifted maximum of source concentration ($Z^*_{k,sk}/C^*_{k,sr\_max}$) for each contaminant "$k$". Table 10.9 presents the results, and it shows that SK1 and SK3 should be in contaminant cascade "A", while SK2 and SK4 should be in contaminant "B" cascade.

Tab. 10.9: Identification of source prioritisation sequence/contaminant cascade of each sinks for Example 10.4.

	$Z_{A,SK}/C_{A,SRmax}$ $C_{A,SRmax} = 250$	$Z_{B,SK}/C_{B,SRmax}$ $C_{B,SRmax} = 300$	Minimum ratio	Cascade
SK1	**0.08**	0.20	0.08	A
SK2	0.20	**0.07**	0.07	B
SK3	**0.40**	0.50	0.40	A
SK4	0.8	**0.27**	0.27	B

The second step is to classify the sinks into Above and Below Pinch regions for both contaminants "A" and "B". The source order for contaminant "A" is: SR1→ SR2 → SR3–SR4, while the source order for "B" is: SR4 → SR3 → SR2–SR1.

As the total sink flowrates (SK1 + SK3) in contaminant cascade "A" has the highest flowrate, the design methodology is performed on contaminant cascade "A" first. The design methodology starts with SK1 first. The source ranking order based on contaminant "B" is SR1→ SR2 → SR3 → SR4. Fig. 10.21a first presents the source and sink CCs specifically for SK1. Notice that the Pinch occurs at contaminant "B", with required freshwater as 18.4 t/h.

According to the source allocation strategy presented earlier, since the Pinch does not occur at the sink's main limiting contaminant "A", as well as SR1 and SR2 are conflicting sources, there is still room for freshwater reduction. SR1 allocated to SK1 can be reduced. SR1 can be reduced until the distance of sink CC to source CC for both contaminants is identical. The source CC is then shifted to the left. Fig. 10.21b shows the source and sink CCs with reduced SR1. Since this involves two contaminants, pair with two sources from the arranged source, i.e. SR1 and SR2 determine the flowrates of SR1 and SR2 to be allocated to SK1. This step can also be formulated as an optimisation model to identify the minimum fresh resource for SK1 (similar step as in Fig. 10.14). The determined flowrates are found: SR1 = 3.45 t/h and SR2 = 2.875 t/h. The procedure is then repeated for SK3. It can be noticed that the freshwater requirement is reduced to 16.675 t/h, with only 3.45 t/h of SR1 to be allocated to SK1 – see Fig. 10.21b.

The design methodology is repeated for SK3, and the Composite Curves are stacked above SK1. Fig. 10.22a first presents the source and sink CCs before identifying the freshwater target for SK3. It is worth to be noted that contaminant "B" is not constraining for SK3 if the sources are arranged in ascending order of contaminant "A". After shifting the source CC to the right, the Pinch occurs at the sink's main limiting contaminant "A", see Fig. 10.22b. As the Pinch occurs at the main limiting contaminant, there is no room to further reduce the freshwater target. In this case, the total freshwater target for both SK1 and SK3 is 16.675 + 13.258 = 29.933 t/h.

The step is repeated again for the sinks that are in contaminant cascade "B". The source ranking order based on contaminant "B" is SR4 → SR3 → SR2 → SR1. The design first starts with SK2. If the sources are allocated based solely on the prioritisation of

(a)

(b)

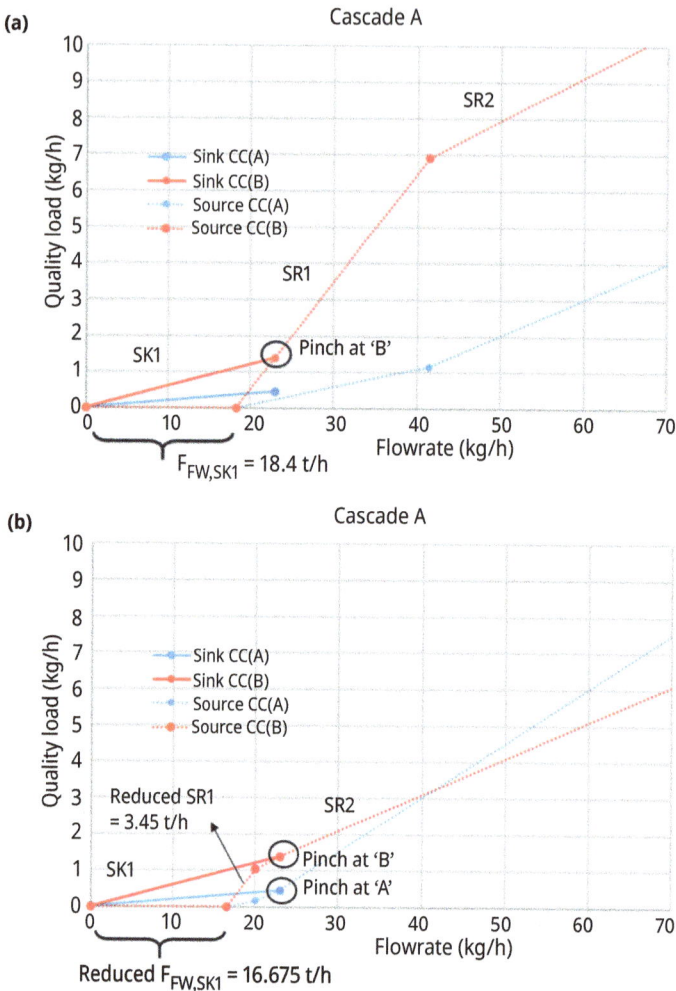

Fig. 10.21: Source and sink CC for SK1: (a) follow the source ranking order; (b) further freshwater reduction.

SR4, the Pinch occurs at contaminant "*A*" (see Fig. 10.23a). This is not optimal as SR3 can still be allocated due to the limit for contaminant "*B*" hasn't been reached, which is the main limiting contaminant for SK4. The flowrate of SR4 is then reduced, and SR3 is used until both contaminants have Pinches (see Fig. 10.23b). An identical scenario is observed for SK4 (Fig. 10.23c and d). Notice that in Fig. 10.23d, a triangle is formed around the source and sink CC for contaminant "*A*". This is feasible as the final load of contaminant "*A*" for SK4 is still satisfied.

The Composite Curves show exactly the sources to be allocated for each sink. This is because the freshwater target and the Pinch are determined sequentially for each

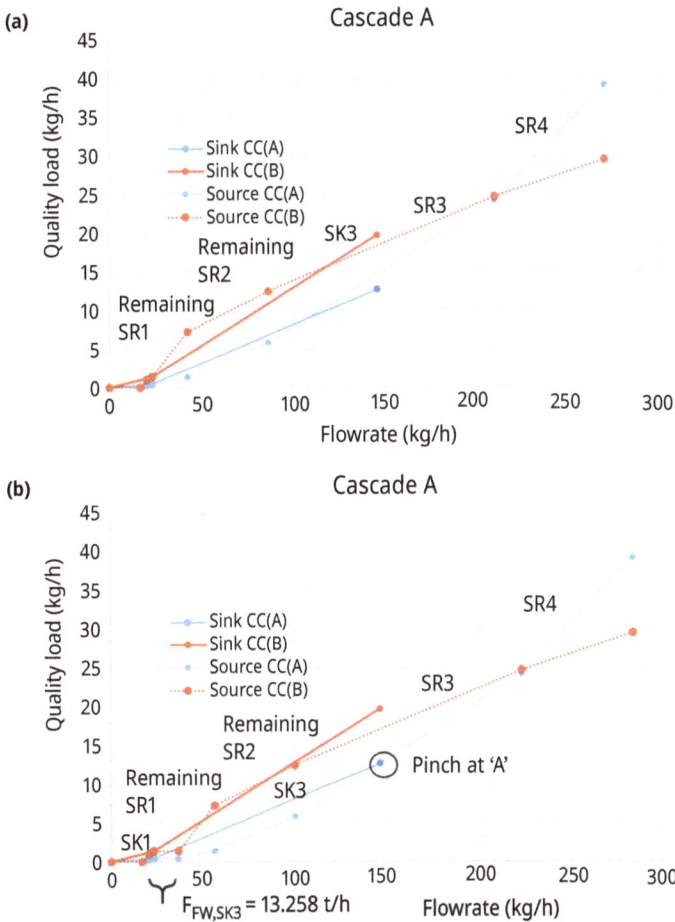

Fig. 10.22: Source and sink CC for SK1 + SK3 in contaminant cascade "A": (a) before shifting source CC for SK3; (b) after shifting source CC for SK3.

sink. By doing this, the simultaneous target and source allocation for each sink can be obtained. The allocation of sources is directly connected to the network design as well. A detailed illustration of the source allocation to the sinks is shown in Fig. 10.24.

The readers could refer to Chin et al. (2021b) or Chin et al. (2022) for more examples or working sessions for multiple quality problems. Note that there are also other well-established methods as well such as concentration potential methods (Liu et al., 2009) or water sources diagram (Calixto et al., 2020). These heuristics provide a general understanding of the strategy for solving multiple quality problems and can always be complemented with Mathematical Optimisation methods to solve the network design effectively.

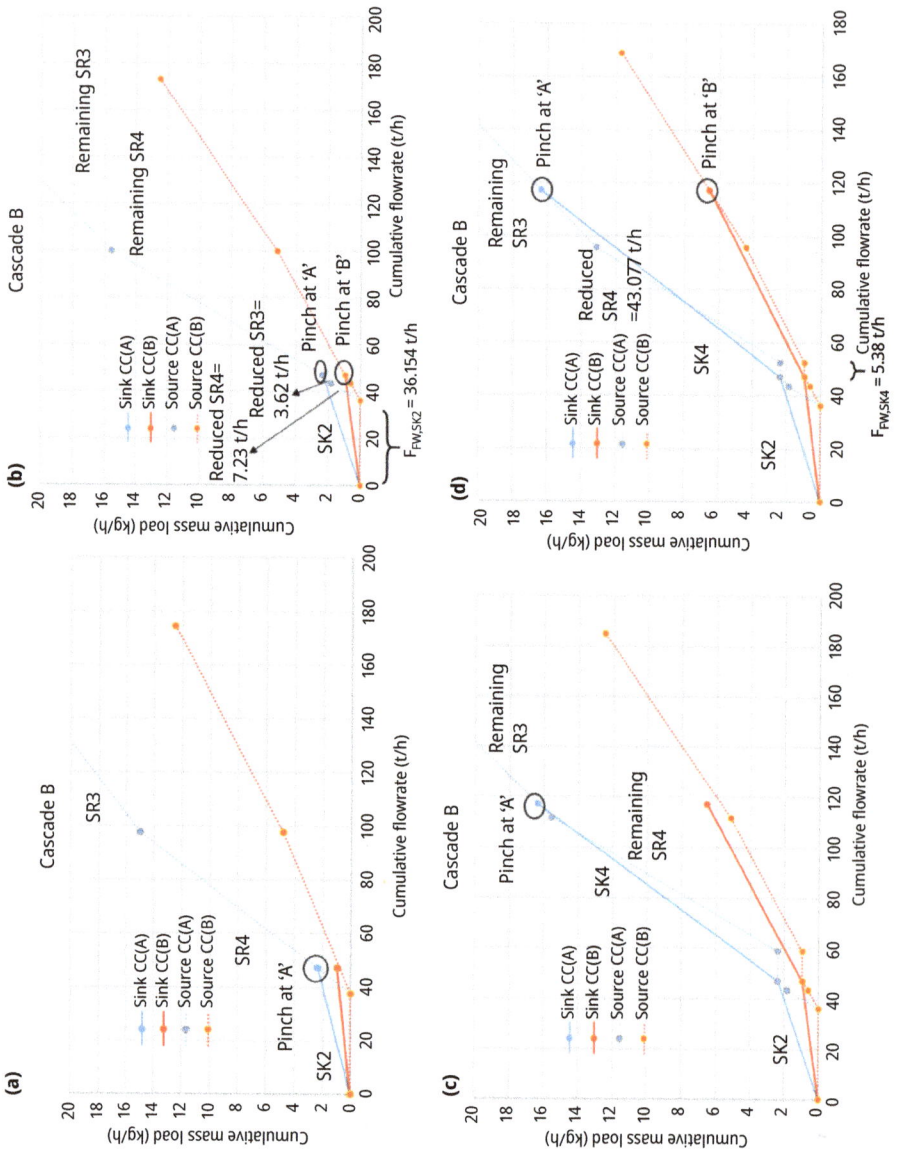

Fig. 10.23: Source and sink CC for scenario 1 case study 1: (a) SK2 but not optimal freshwater target; (b) SK2 with the reduced freshwater target; (c) SK4 but not optimal freshwater target; (d) SK4 with reduced freshwater target.

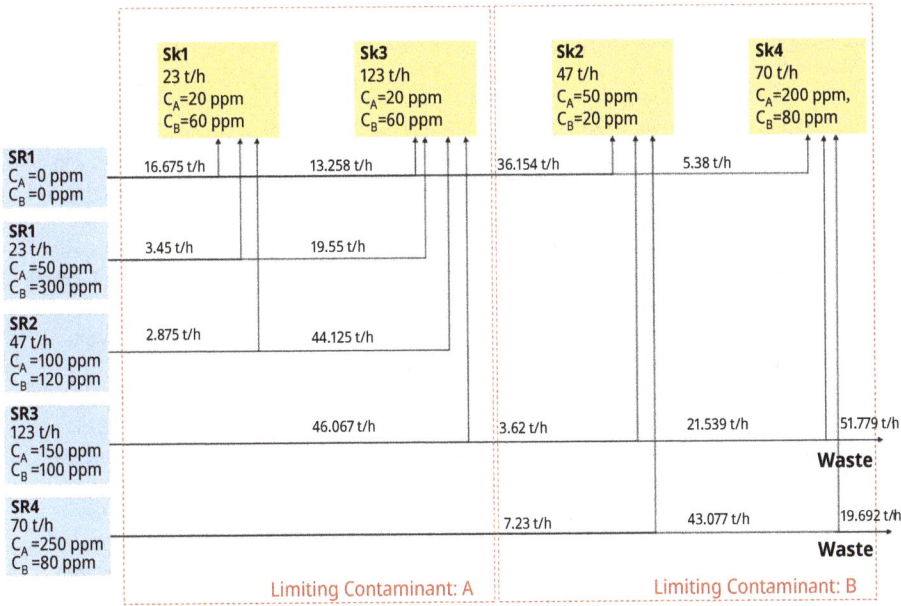

Fig. 10.24: Detailed network design for case study 1.

10.5 Conclusion

For other book resources, the readers could refer to Klemeš et al. (2018b) for the Process Integration textbook for more examples demonstration, Klemeš et al. (2008) for the application of water and energy management in the field of food processing, and Klemeš et al. (2010) for the Process Integration approaches in addressing the sustainability issues in the process industry. The readers could refer to Chin et al. (2021b) for the detailed studies of the multiple qualities Pinch Analysis and Chin et al. (2021a) for utilising the Cascade Table for multiple qualities problem. Chin et al. (2021c) also provide the Total Site Mass Integration considering centralised headers synthesising using Pinch Analysis and Chin et al. (2021d) for different scenarios of headers targeting.

References

Calixto E, Pessoa F L P, Mirre R C, Francisco F, Queiroz E M. (2020). Water Sources Diagram and Its Application, Processes, 8(3), 313.

Chew I M L, Foo D C Y, Ng B K S. (2010). Flowrate Targeting Algorithm for Interplant Resource Conservation Network. Part 1: Unassisted Integration Scheme, Industrial Engineering & Chemistry Research, 49, 6439–6455.

Chin H H, Jia X, Varbanov P S, Klemeš J J, Wan-Alwi S R. (2021d). Targeting Flowrates and Concentrations in Internal or Total Site Water Mains for Single Contaminant, Chemical Engineering Transaction, 86, 895–900.

Chin H H, Liew P Y, Varbanov P S, Klemeš J J. (2022). Extension of Pinch Analysis to Targeting and Synthesis of Water Recycling Networks with Multiple Contaminants, Chemical Engineering Science, 248, 117223.

Chin H H, Varbanov P S, Klemeš J J, Liew P Y. (2021a). Enhanced Cascade Table Analysis to Target and Design Multi-constraint Resource Conservation Networks, Computers & Chemical Engineering, 148, 107262.

Chin H H, Varbanov P S, Klemeš J J, Wan-Alwi S R. (2021c). Total Site Material Recycling Network Design and Headers Targeting Framework with Minimal Cross-Plant Source Transfer, Computers & Chemical Engineering, 151, 107364.

Chin H H, Varbanov P S, Liew P Y, Klemeš J J. (2021b). Pinch-Based Targeting Methodology for Multi-Contaminant Material Recycle/Reuse, Chemical Engineering Science, 230, 116129.

El-Halwagi M M, Gabriel F, Harell D. (2003). Rigorous Graphical Targeting for Resource Conservation via Material Recycle/Reuse Networks, Industrial Engineering Chemistry Research, 42, 4319–4328.

Fadzil A F A, Wan Alwi S R, Manan Z, Klemeš J J. (2018). Industrial Site Water Minimisation via One-way Centralised Water Reuse Header, Journal of Cleaner Production, 200, 174–187.

Klemeš J, Dhole V R, Raissi K, Perry S J, Puigjaner L. (1997). Targeting and Design Methodology for Reduction of Fuels, Power and CO_2 on Total Sites, Applied Thermal Engineering, 7, 993–1003.

Klemeš J J, Varbanov P S, Walmsley T G, Jia X. (2018a). New Directions in the Implementation of Pinch Methodology (PM), Renewable and Sustainable Energy Reviews, 98, 439–468.

Klemeš J J, Smith R, Kim J-K. (eds.). (2008). Handbook of Water and Energy Management in Food Processing. Woodhead, Cambridge, UK.

Klemeš J J, Friedler F, Butalov I, Varbanov P S. (2010). Sustainability in the Process Industry: Integration and Optimisation. McGraw-Hill Professional, New York, US.

Klemeš J J, Varbanov P S, Wan-Alwi S R, Manan Z A. (2018b). Sustainable Process Integration and Intensification: Saving Energy, Water and Resources, 2nd ed. DeGruyter, Berlin.Germany.

Liu Z Y, Yang Y, Wan L Z, Wang X, Hou K H. (2009). A Heuristic Design Procedure for Water-using Networks with Multiple Contaminants, AIChE Journal, 55, 374–382.

Teles J, Castro P M, Novais A Q. (2008). LP-based Solution Strategies for the Optimal Design of Industrial Water Networks with Multiple Contaminants, Chemical Engineering Science, 63(2), 376–394.

11 Resources or Solid Waste Pinch Analysis

11.1 Introduction and the concept

Pinch Analysis (Linnhoff et al., 1982) has been applied commonly for Heat and Mass Integration (Klemeš et al., 2018). Water Integration (Wang and Smith, 1994) and Hydrogen Integration (Marques et al., 2017) are typical applications grouped under Mass Integration. They shared a similar algorithm. For example, the Water Integration method has been extended for Hydrogen Integration conducted by Gai et al. (2021) for the refineries case study. Solid waste or resources has different properties and characteristics where direct adaption of the Mass Integration method for water or hydrogen is unattainable. Especially as solid waste has different "quality indicator – y-axis" and "exchanging" methods. The "quality indicator" of Heat Integration is temperature, and the "exchanging" are via heat exchanger; the "quality indicator" of Mass Integration is usually concentration (e.g. contaminant or purity) and the "exchanging" are via pipes. The quality matching and integration of solid waste could hardly be integrated via heat exchangers or pipes due to the state of the matter unless the waste is converted to energy. In this case, it is a Heat or Power Integration (Klemeš et al., 2022) problem.

Several Pinch-based methods have been introduced to facilitate resource planning. They have been the important basis for developing Solid Waste Integration, although they have not been applied directly to waste management problems. One of the pioneer works is by Singhvi et al. (2004) for production planning in supply chains, using time as the "quality indicator". It is capable of determining the minimum production rate, starting inventory, and stock out. A similar concept, time as "quality indicator", has been advanced by Ooi et al. (2013) for carbon capture and storage planning with the additional and excessive storage requires being accessed. The other method inspired by the Pinch concept and could be adapted to waste management problems is by Tan and Foo (2007), known as Carbon Emission Pinch Analysis (CEPA). It is developed to facilitate the planning in the energy sector with CO_2 emission as the "quality indicator", minimising and ensuring the targeted CO_2 limit is met. Ho et al. (2018) first explored the application for solid waste planning following the existing CEPA with the main adaption of changing the x-axis from energy demand to cumulative waste. The developed method, Waste Management Pinch Analysis (WAMPA), is applied by Jia et al. (2018) to waste management planning in China. Fan et al. (2019) extend the original application of WAMPA to the country and regional level, using net greenhouse gas (GHG) emissions of the overall environmental burden from waste treatment as the "quality indicator".

Based on the classification as stated in Fan et al. (2022), the Pinch Analysis method could be divided into two categories, with one following the Heat Pinch Analysis closely where the targets are identified and used for problem decomposition. This method has not been widely used for solid waste planning compared to Heat or Water Integration planning. CEPA and WAMPA belong to the second category, with graphical

https://doi.org/10.1515/9783110782981-011

representation to inform decision-makers as the main feature, and the targets are set beforehand. This chapter introduces the Pinch Analysis with a targeting function for solid waste planning. Time (Fan et al., 2021a) and moisture content (Fan et al., 2021b) are applied as quality indicators in conducting the Pinch-Based Analysis. The other quality indicator, such as C:N ratio, as performed recently by Chee et al. (2022), could be incorporated into the method based on the representability of the "quality indicator" under the assessed scenarios (e.g. type of waste) and conditions.

11.2 Solid waste management targeting: time as quality indicator

This section introduces Solid Waste Pinch Analysis using time (Fan et al., 2021a) as a quality indicator, as described in Section 11.2.1. It is developed to facilitate integrated waste treatment planning, matching the waste amount and the available waste treatment capacity at different places. This could maximise the use of existing waste treatment plants, minimising the need for new waste treatment facilities, which is not in line with the priority of the circular economy. The surplus waste and surplus capacity after the regional or clustered integration could be identified for further planning. However, considering capacity in different places, such integrated waste management design might have incurred an additional environmental footprint contributed by the transporting activities. The environmental footprint related to the travelled distances and the incentives to drive the integrated system where the parties are encouraged to participate in the integrated design have to be optimised after the supply and demand matching/targeting. The applicability of the method is demonstrated through Sections 11.2.2 and 11.2.3.

11.2.1 Step-by-step guidance

Figure 11.1 shows the overall graphical representation and interpretation of the proposed Solid Waste Pinch Analysis method using time as a quality indicator. Figure 11.1a shows the original plot before shifting, as listed in Fig. 11.2 as Steps 1–2. The shape of the Cumulative Curve depends on the incoming waste flow whether continuously or by time batch. In this example, the Cumulative Curve follows a staircase-like line (Fig. 11.1) where the waste need to be treated is assumed to come in batch, especially for waste trading across countries with a time delay for transporting.

Figure 11.1b shows the plot after the Pinch Analysis where the Cumulative Demand Curve, representing the waste that needs to be treated, is shifted to the right until the Pinch Point is formed without compromising the defined constraints listed in Fig. 11.2 as Step 3. It is shifted to the right as the Cumulative Demand Curve is behind the Cumulative Supply Curve. The waste is stored before being treated in the following time interval. Without shifting (Fig. 11.1a), the flow is forbidden as waste treatment capacity cannot be "stored", and waste generated in the future cannot be treated by the past capacity. The constraints could be different based on the concern of the stakeholders, which will be further explained and specified in the Working Session (Section 11.2.2). The constraints further affect the magnitude of the shifting or Cumulative Demand Curve towards the right side. The target identification (Step 4, in Fig. 11.2) includes the amount of waste that could be recovered and deficit/surplus capacity or surplus waste (see Fig. 11.1b). Steps 5 and 6, which show below the dotted line in Fig. 11.2, are the optimising processes in identifying and implementing the integrated waste management design based on the identified targets in Steps 1–4. It was demonstrated by Fan et al. (2022a) through a case study that considers an integration between Germany, Austria, the Czech Republic, Poland, and Slovakia. Section 11.2.3 demonstrates the application of the approach through a numerical example for better understanding.

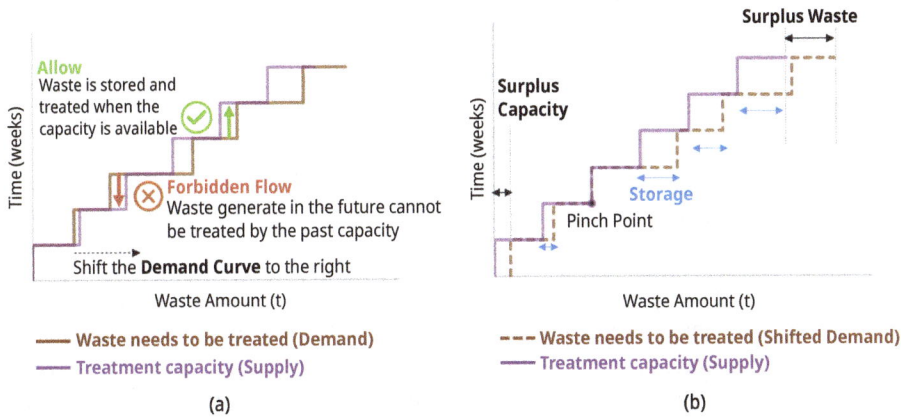

Fig. 11.1: The graphical representation and algorithm of the Solid Waste Pinch Analysis using time as the quality indicator: (a) the original plot – with the infeasible flow where the waste needs to be treated is not able to treat and (b) the shifted plot (adapted from Fan et al. (2021a)).

11.2.2 Working session

A more straightforward case study involving three countries or clusters is demonstrated in this working session for a clearer illustration and understanding. The assignment for this exercise is to:

1. Perform the solid waste management targeting using time as the quality indicator (Steps 1–4, Fig. 11.2), based on the case study and information listed in Tab. 11.1, and identify the targets. The applied constraint is the waste can only be stored for up to one month.

```
                          ┌──────────┐
                          │  Begin   │
                          └──────────┘
                               │
Step 1   ┌──────────────────────────────────────────────┐
         │ Define the axes (x-axis = waste amount; y-axis = time), unit │
         │              and constraint(s)                │
         └──────────────────────────────────────────────┘
                               │
Step 2   ┌──────────────────────────────────────────────┐
         │ Construct the Cumulative Curves: Treatment capacity │
         │   (Supply) and Waste needs to be treated (Demand) │
         └──────────────────────────────────────────────┘
                               │
Step 3   ┌──────────────────────────────────────────────┐
         │ Shift the "Demand" curve until the defined constraint is │
         │                  fulfilled                     │
         └──────────────────────────────────────────────┘
                               │
Step 4   ┌──────────────────────────────────────────────┐
         │ Identify the "targets" – waste could be recovered, surplus │
         │           capacity and surplus waste           │
         └──────────────────────────────────────────────┘
- - - - - - - - - - - - - - - - - - - - - - - - - - - - - - - - -
                               │
Step 5   ┌──────────────────────────────────────────────┐
         │ Identify the possible design that fulfilling the identified │
         │         targets and performed optimisation     │
         └──────────────────────────────────────────────┘
                               │
Step 6   ┌──────────────────────────────────────────────┐
         │ Identify the changes required (e.g. policy – incentives) to │
         │            realise the proposed solution       │
         └──────────────────────────────────────────────┘
```

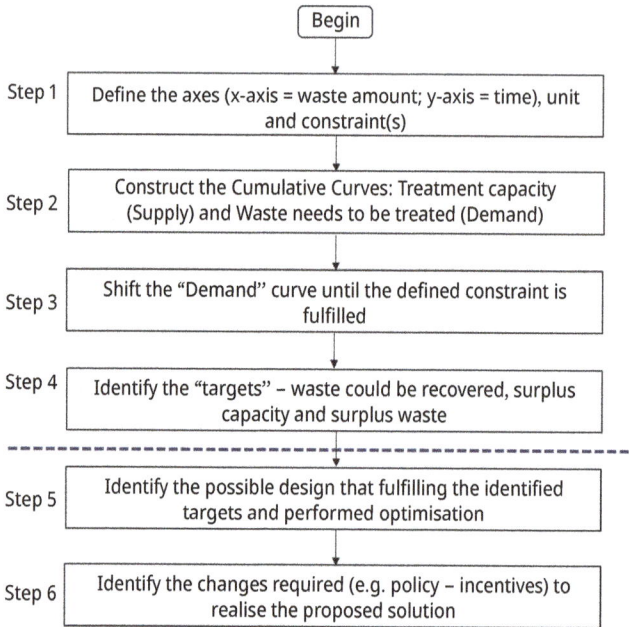

Fig. 11.2: The steps of Solid Waste Pinch Analysis using time as the quality indicator (Steps 1–4) and the steps for subsequent design and implementation.

2. Identify the integrated design (Step 5) before Pinch Point using minimum environmental footprint as the objective function and information stated in Tab. 11.2 to calculate the potential GHG saving from the integrated design. The GHG of waste transporting is assumed to be 304 g/tkm. The emission factor of landfill is assumed as 568 kg GHG/t and 386 kg GHG/t for incineration. The output of incineration is assumed to be 315 kWh/t, and the carbon intensity of electricity is 329 g/kWh. The design for "Before the Pinch Point" is focused, aiming to provide an easier-to-follow representation. "After Pinch Point" design could be developed based on the same approach.

Tab. 11.1: The variation of waste amount (t) in Clusters A, B, and C. Bold value represents the waste amount is more than the treatment capacity in that particular cluster as stated in Tab. 11.2.

Month	Cluster A (t)	Cluster B (t)	Cluster C (t)	Total (t)
1	1,600	**600**	200	2,400
2	1,600	450	200	2,250
3	**2,200**	**600**	300	3,100
4	**2,200**	400	**400**	3,000

Tab. 11.1 (continued)

Month	Cluster A (t)	Cluster B (t)	Cluster C (t)	Total (t)
5	**2,500**	400	200	3,100
6	**2,200**	400	200	2,800
7	**2,100**	300	150	2,550
8	**2,100**	400	300	2,800
9	1,800	**500**	200	2,500
10	1,800	**550**	150	2,500
11	2,000	300	300	2,600
12	1,800	400	250	2,450

Tab. 11.2: The distance between each cluster and the waste treatment capacity per month.

Distance (km)	Cluster A	Cluster B	Cluster C	Total treatment capacity per month (t)
Cluster A	–	200	120	2,000
Cluster B	200	–	60	450
Cluster C	120	60	–	300
				2,750

The transporting distance to the landfill is assumed to be zero in this case study.

11.2.3 Solution

1. The solid waste management targeting using time as a quality indicator has been performed for Clusters A, B and C. The resulting plots are shown in Fig. 11.3. The brown-orange line represents the cumulative curve of the total waste amount in Clusters A, B, and C at a particular time. The purple line represents the cumulative curve of the total waste treatment capacity in Clusters A, B, and C at one specific time. The waste flow to the waste treatment capacity shown by the original plot (Fig. 11.3a) is infeasible/ not matching, as shown in Fig. 11.1. Fig. 11.3b shows the shifted brown-orange line to the right until a Pinch Point is formed. The Pinch point is in the third month (March), where no waste from February is required to be treated in March, with the difference between supply and demand zero (see Fig. 11.4). Fig. 11.4 summarises the differences between the demand (brown-orange) and supply (purple) curves in different months, as in Fig. 11.3b. The overall surplus capacity is 0.85 kt, and the waste deficit is 0.1 kt. The maximum required storage is 1 kt in June (most significant differences), where the waste will be brought forward to July for treatment or recovery.

2. By using minimum environmental footprint (see eq. (11.1)), specifically GHG as the objective function, and focus on "Before the Pinch Point". One of the designs is given as shown in Fig. 11.5, matching the information that could be extracted from Figs. 11.4

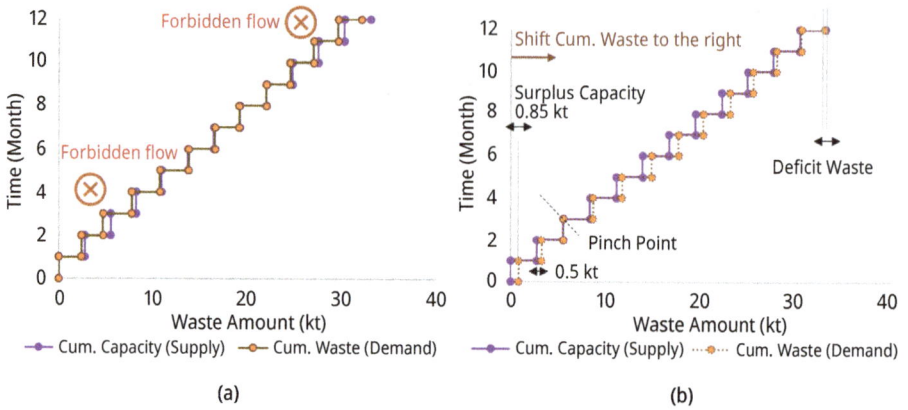

Fig. 11.3: The Solid Waste Management Pinch Analysis using time as quality indicator. The integrated (Clusters A, B, and C) plots with (a) original curves and (b) the shifted demand curves.

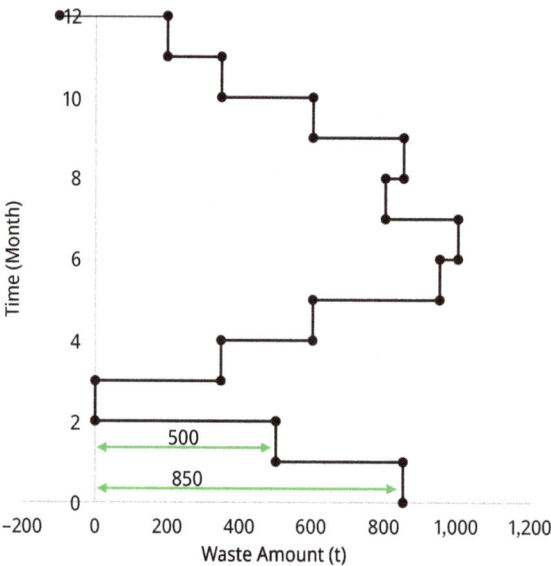

Fig. 11.4: The Grand Composite Curve as a complementary representation of Fig. 11.3b, showing the surplus capacity at each time interval.

and 11.3b. There is a surplus capacity of 850 t in January. Additional waste (outside Clusters A, B, and C) could be treated in Clusters A and C. In February 500 t of waste storage from January will be treated. The waste flow decision (treat in which Clusters) is based on the distance between clusters where priority is given to the shortest distance (lowest GHG footprint). Within its Cluster 350 t out of 500 t from Cluster A is treated, where the remaining 150 t from Cluster B is treated in Clusters A and C; 100 t is transferred to Cluster C as it is nearer to Cluster B, and that is the maximum capacity that could be taken by Cluster C:

$$EF_{min} = \min \sum_{k,t} EI_k + TI_{k,t} - AE_k + EL_k + TL_{k,t} \tag{11.1}$$

where EF_{min} = The minimal total environmental footprint, k is an index of region (e.g. different country or cluster), t is an index of transportation mode (e.g. heavy lorry, light lorry, and train), EI_k = emission releases during incineration and pre-processing process, TI_k = transportation emission to the incineration plant, AE_k = avoided emission from recovered product or utilities, EL_k = emission release from landfill, and TL_k = transportation emission to the landfill.

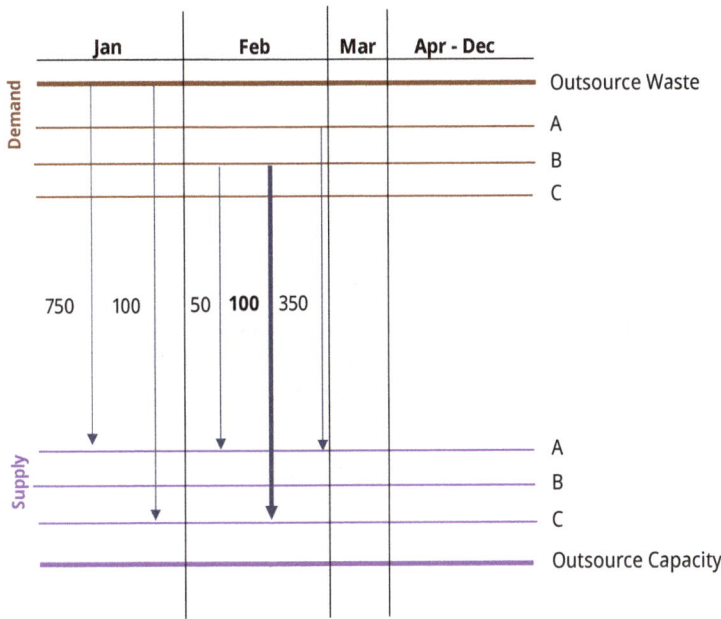

Fig. 11.5: One of the identified designs (January–February) Before the Pinch Point.

By applying eq. (11.1), the GHG emission of the without integration design for January and February is 295 kg GHG/t of waste. With energy recovery 4,500 t of waste is incinerated, and 150 t is landfilled. The integrated design, without waste ending in the landfill, as shown in Fig. 11.5, offers GHG emission of 283 kg GHG/t of waste. It is calculated based on 4,650 t of waste being incinerated with energy recovery, considering the transporting emission from Cluster B to Cluster C (100 t) and Cluster B to Cluster A (50 t). The integrated design offers a potential GHG saving of 12 kg GHG/t of waste, with a surplus capacity in January to treat 850 t of waste from December, within or outside the Clusters.

11.3 Solid waste management targeting: moisture content as quality indicator

This section introduces Solid Waste Pinch Analysis using moisture content/amount of water (Fan et al., 2021b) as a quality indicator, as described in Section 11.3.1. It is developed to facilitate integrated waste treatment planning, matching the waste quality and the tolerance of the waste treatments to the defined waste quality. The proposed targeting method could maximise the efficiency of waste valorisation and minimise the need for drying or fresh water utilisation. The dry waste deficit and the waste surplus could be determined and facilitate further integration (recovery or other treatment approaches). For example, a hot stream from Heat Integration could be used for pre-drying, reducing the need for cold utility while mitigating the dry waste deficit in Solid Waste Integration. The wastewater surplus in Water Integration could also be applied to treat the solid waste surplus with additional tolerance to water. For example, to enhance the anaerobic digestion process using wastewater. The applicability of the method is demonstrated in Sections 11.3.2 and 11.3.3.

11.3.1 Step-by-step guidance

Fig. 11.6 illustrates the overall concept of Grand Total Site Integration (Fan et al., 2021b) consisting of a combination of several Process Integrations, e.g. Solid Waste, Heat, Water, Power, or other Integration introduced in the other chapters. This section describes the step-by-step guidance for Solid Waste Integration (Fig. 11.6b) using moisture content as the quality indicator.

As summarised in Fig. 11.6b, the developed Pinch Analysis for Solid Waste Integration consists of two axes. It comprises two curves: one is for the Cumulative Waste Recovery/Treatment (Sink), and another is for the Cumulative Waste Supply (Source). Different waste recovery processes have different requirements for moisture content specification, and each waste supply has different moisture content. The y-axis represents the amount of water, and the x-axis represents the waste amount. The gradient represents the moisture content, as listed in Step 1 (Fig. 11.7). The succeeding steps of Solid Waste Pinch Analysis using moisture as a quality indicator are summarised in Fig. 11.7. Stacking sequencing (Step 3) of the Cumulative Curves (Step 2) is one of the keys in ensuring the functionality of the shifting (Step 4) in Pinch Analysis. The stacking sequencing of this particular approach (Solid Waste Pinch Analysis using moisture content as a quality indicator) starts with the "high quality" waste to "low quality" waste. It depends on whether high moisture content is viewed as high quality or low quality in the assessed recovery treatments according to the requirement and specification. For example, waste with lower moisture content is preferable for waste to energy and could be viewed as higher quality. In other words, the stacking could start with curves with the lowest gradient to the highest gradient, as shown in Fig. 11.6b, or the highest gradient to the lowest gradient. This is different from the method introduced in Section 11.2 with time as a quality indicator. The cumulative sequencing with time as a quality indicator is fixed (y-axis) as the arrangement is based on the supply and demand availability at a particular time interval. The shifting (Step 4) is complete when none of the Waste Supply (WS) Cumulative Curve crosses over or behind the Waste Recovery (WR) Cumulative Curve, forming a Pinch Point. The identified targets (in Step 5), e.g. the waste surplus or deficit, after the integration at the own site could be extended to the Total Site level. For example, integrating the solid waste from different processes or sectors (urban site and industrial site). The integration of different forms of resources (e.g. waste heat for drying and waste to energy) could be further conducted for Grand Total Site Integration (Step 6). Steps 5 and 6 are further discussed in Section 11.3.3 through a numerical example.

11.3.2 Working session

The assignment for this exercise is to:
1. Perform the solid waste management targeting using moisture content as the quality indicator (Steps 1–5, Fig. 11.7) based on the case study and information listed in Tabs. 11.3 and 11.4 for two different Clusters.
2. Perform the Total Site Solid Waste Integration and identify the supply or deficit potential for further Heat Integration by referring to the conversion efficiency data tabulated in Tabs. 11.3 and 11.4.

11.3.3 Solution

1. Solid waste management targeting using moisture content as a quality indicator has been performed for Clusters A and B. The resulting plots for integration within Clusters A and B are shown in Figs. 11.8 and 11.9, with an increasing gradient stacking sequence. Fig. 11.8a shows the original plot comprises waste amount or Waste Supply (WS) with different moisture content and Waste Recovery (WR) options with different tolerance to water content. The WSCC is moved to the right until one of the point meet and the other points are behind WRCC (Fig. 11.8b) without exceeding the water tolerance of the recovery approaches. As indicated in Fig. 11.8c, 20 t of the waste in Cluster A could be converted to energy with a surplus capacity to recover an additional 6 t of dry waste. Due to the deficit in capacity and the high moisture content 2 t of WS4 is not able to recover without processing. Cluster B (Fig. 11.9c) is identified to have a surplus capacity to recover 3 t of dry waste and a surplus capacity to recover 7 t of wet waste, with 20 t of waste within its cluster converted to energy.

General Framework – Heat, Water and Solid Waste

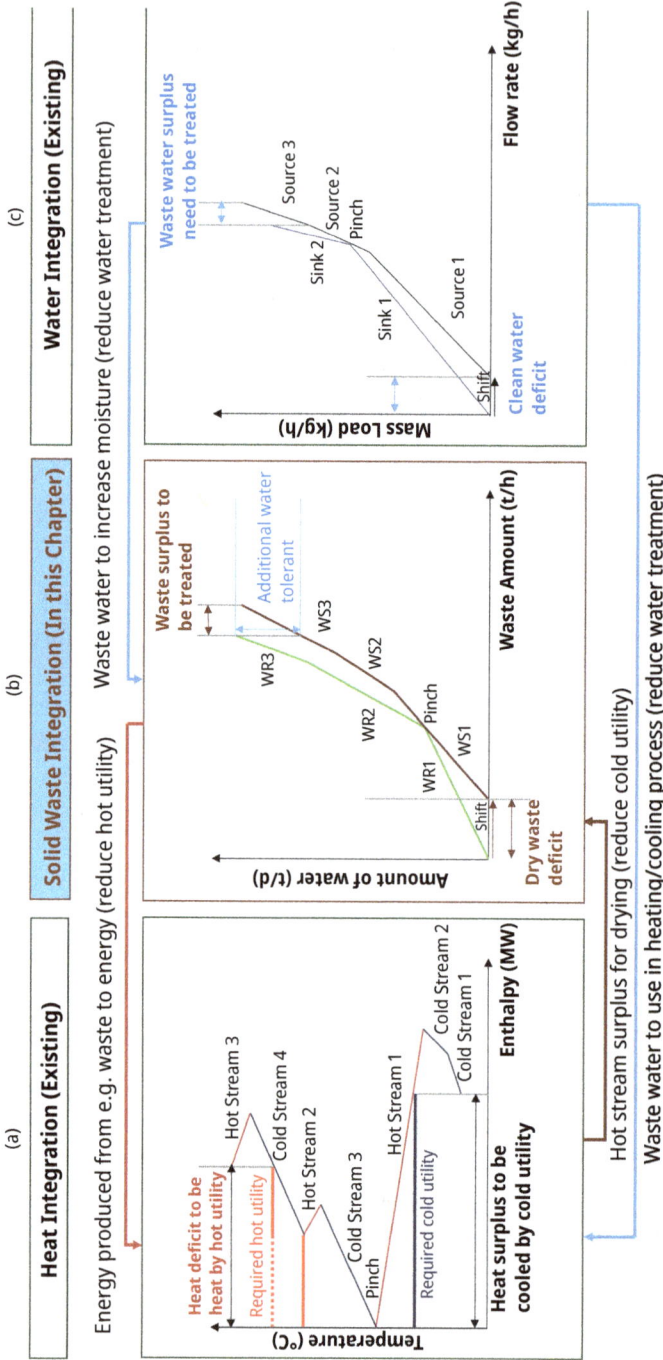

Fig. 11.6: The overall concept of Grand Total Site Integration consisting of (a) Heat Integration (Klemeš, 2022), (b) Solid Waste Integration (novel development), and (c) Water Integration (El-Halwagi et al., 2003) (adapted from Fan et al. (2021b)).

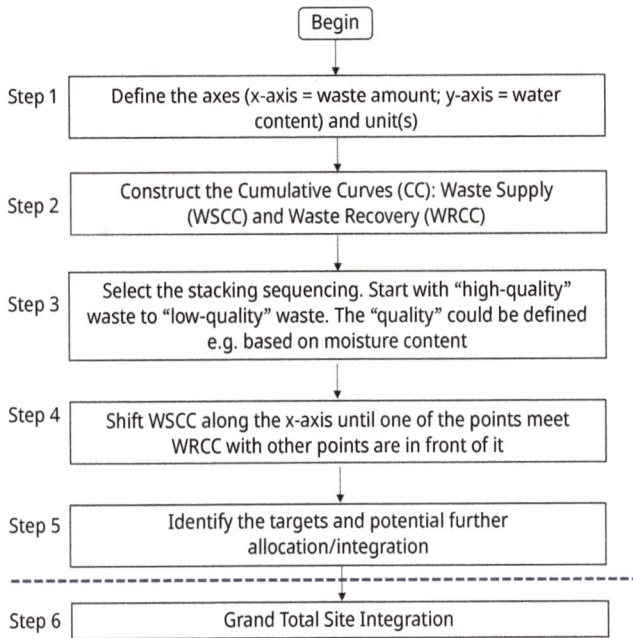

Fig. 11.7: The steps of Solid Waste Pinch Analysis using water/moisture content as the quality indicator (Steps 1–5) follows by further integration (Grand Total Site Integration). Step 6 is not performed in the following working session but has been discussed in Fan et al. (2021b).

Tab. 11.3: Data input of source and sink at Cluster A (Fan et al., 2021b).

Cluster A	Waste amount (t/d)	Water amount (t/d)	Moisture content (%)	LHV (MJ/kg)	Conversion efficiency
Waste Supply 1 (WS1)	12	1.56	13	15	–
Waste Supply 2 (WS2)	3	0.6	20	13	–
Waste Supply 3 (WS3)	4	1	25	10	–
Waste Supply 4 (WS4)	3	2.1	70	7	–
Waste Recovery demand (WR1)	20	2	10	–	2 MWh heat, 2 MWh electricity/t of waste (with LHV of 10 MJ/kg)

Tab. 11.3 (continued)

Cluster A	Waste amount (t/d)	Water amount (t/d)	Moisture content (%)	LHV (MJ/kg)	Conversion efficiency
Waste Recovery demand (WR2)	3	0.9	30	–	4 MWh heat/t of waste (with LHV of 10 MJ/kg), ash (25% of its weight of input)
Waste Recovery demand (WR3)	3	1.5	50	–	5 MWh heat/t of waste (with LHV of 10 MJ/kg), ash (25% of its weight of input)

2. Further Solid Waste Integration could be conducted after the individual integration within its clusters. The resulting plot is shown in Fig. 11.10. As identified in Fig. 11.9c, there is a surplus capacity with a higher tolerance to moisture content (steeper gradient). After integrating within Cluster A, the waste surplus (WS4) can be transferred to Cluster B. Different types of sequencing will propose different targets or allocations. Figs. 11.10(a) and (b) show the other options.

Option 1, in Fig. 11.10(a), is preferable as the dried waste deficit is lower (3.5 t) compared to 5 t in the second option (Fig. 11.10b). The concept is similar to Heat and Water Pinch Analysis, where the aim is to minimise the requirement of resources, e.g. outsource heat utility and clean water. Dried waste offers a wider range of applications, safe waste storage, microbiological inactivation, and lower transporting. Drying is generally regarded as an energy-intensive process. Minimising the dried waste deficit in waste treatment within the same cluster by prioritising the treatment of wet waste as much as possible, whenever the process allowed, could minimise the overall environmental footprints and cost. The total heat that can be recovered in Cluster A (Fig. 11.8) is 67.9 MWh/d, and Cluster B (Fig. 11.10a) is 65.43 MWh/d. They could be used for the Heat Integration process as a Fresh Utility.

In addition, Cluster B (Fig. 11.10a) has a spare capacity of 4.5 t/d, which enables the handling of additional waste from the other Clusters. The identified integrated design (65.43 MWh/d) increased the energy recovered from the solid waste by 11.39 MWh/d compared to the unintegrated design (54.04 MWh/d). Two tonnes per day of waste in Cluster A (WS4) is diverted from the landfill. At the local level, both parties (Clusters A and B) gain from the recovered energy, avoided landfill disposal fees, drying costs or carbon tax, if applied.

Tab. 11.4: Data input of source and sink at Cluster B (Fan et al., 2021b).

Cluster B	Waste amount (t/d)	Water amount (t/d)	Moisture content (%)	LHV (MJ/kg)	Conversion efficiency
Waste Supply 5 (WS5)	3	0.6	20	10	–
Waste Supply 6 (WS6)	5	1.5	30	9	–
Waste Supply 7 (WS7)	5	1.75	35	7	–
Waste Supply 8 (WS8)	7	4.55	65	6	–
Waste Recovery demand (WR4)	15	4.5	30	–	4.5 MWh heat/t of waste (with LHV of 10 MJ/kg), 0.7 t digestate/t waste
Waste Recovery demand (WR5)	10	5	50	–	4.0 MWh heat/t of waste (with LHV of 10 MJ/kg), 0.7 t digestate/t waste
Waste Recovery demand (WR6)	5	3	60	–	600 kg compost/t

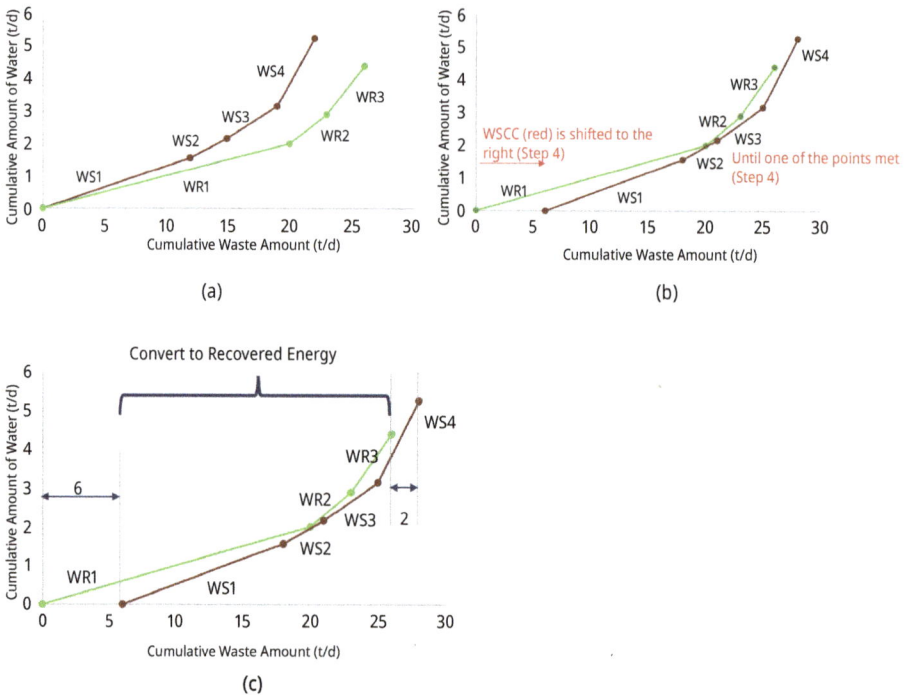

Fig. 11.8: The Solid Waste Management Pinch Analysis using moisture content as quality indicator (a) for Cluster A – before shifting, (b) Cluster A – after shifting, and (c) Cluster A – after shifting with indication (adapted from Fan et al. (2021b)).

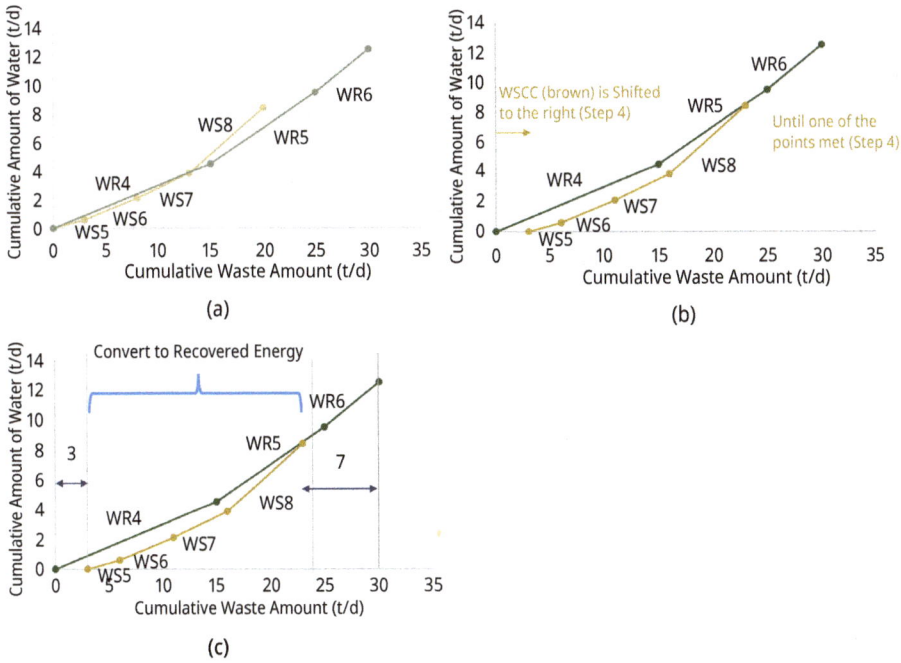

Fig. 11.9: The Solid Waste Management Pinch Analysis using moisture content as a quality indicator:
a) for Cluster B – before shifting, (b) Cluster B – after shifting, and (c) Cluster B – after shifting with
indication (adapted from Fan et al. (2021b)).

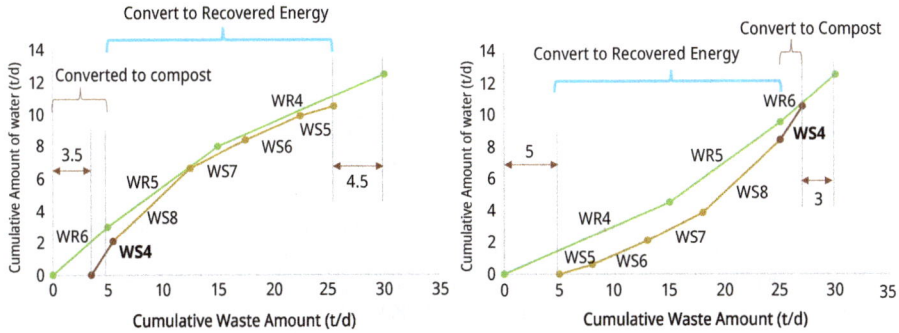

Fig. 11.10: Total Site Integration – Cluster B and WS4 from Cluster A with different stacking sequences
(Step 3): (a) first option and (b) second option.

References

Chee W C, Ho W S, Mah A X Y, Klemeš J J, Fan Y V, Bong C P C, Muis Z. (2022). Maximising the Valorisation of Organic Waste Locally Available via Carbon-to-nitrogen Ratio Supply Composite Curve Shifting, Journal of Cleaner Production, 132389.

El-Halwagi M M, Gabriel F, Harell D. (2003). Rigorous Graphical Targeting for Resource Conservation via Material Recycle/reuse Networks, Industrial & Engineering Chemistry Research, 42(19), 4319–4328.

Fan Y V, Klemeš J J, Chin H H. (2019). Extended Waste Management Pinch Analysis (E-wampa) Minimising Emission of Waste Management: EU-28, Chemical Engineering Transactions, 74, 283–288, doi:10.3303/CET1974048.

Fan Y V, Jiang P, Klemeš J J, Liew P Y, Lee C T. (2021a). Integrated Regional Waste Management to Minimise the Environmental Footprints in Circular Economy Transition, Resources, Conservation and Recycling, 168, 105292.

Fan Y V, Varbanov P S, Klemeš J J, Romanenko S V. (2021b). Urban and Industrial Symbiosis for Circular Economy: Total EcoSite Integration, Journal of Environmental Management, 279, 111829.

Fan Y V. (2022). Chapter 22 – Extension of Pinch Analysis to Sustainable Solid Waste Management, Handbook of Process Integration: Minimisation of Energy and Water Use, Waste and Emissions, 2nd ed. In: Klemeš J.J. (ed.). Woodhead Publishing, Elsevier, Sawston, United Kingdom, doi:10.1016/B978-0-12-823850-9.00003-7.

Gai L, Varbanov P S, Fan Y V, Klemeš J J, Nižetić S. (2022). Total Site Hydrogen Integration with Fresh Hydrogen of Multiple Quality and Waste Hydrogen Recovery in Refineries, International Journal of Hydrogen Energy, 47(24), 12159–12178.

Ho W S, Hashim H, Lim J S, Lee C T, Sam K C, Tan S T. (2017). Waste Management Pinch Analysis (WAMPA): Application of Pinch Analysis for Greenhouse Gas (GHG) Emission Reduction in Municipal Solid Waste Management, Applied Energy, 185, 1481–1489.

Klemeš J J, Varbanov P S, Walmsley T G, Jia X. (2018). New Directions in the Implementation of Pinch Methodology (PM), Renewable and Sustainable Energy Reviews, 98, 439–468.

Klemeš J J. (ed.). (2022). Handbook of Process Integration (PI): Minimisation of Energy and Water Use, Waste and Emissions. Woodhead Publishing, Elsevier Limited, Sawston, United Kingdom, ISBN 9780128238509.

Linnhoff B, Townsend D W, Boland D, Hewitt G F, Thomas B E A, Guy A R, Marsland R H. (1982). A User Guide on Process Integration for the Efficient Use of Energy. IChemE, Rugby, UK, Last Edition 1994.

Marques J P, Matos H A, Oliveira N M, Nunes C P. (2017). State-of-the-art Review of Targeting and Design Methodologies for Hydrogen Network Synthesis, International Journal of Hydrogen Energy, 42(1), 376–404.

Ooi R E, Foo D C, Ng D K, Tan R R. (2013). Planning of Carbon Capture and Storage with Pinch Analysis Techniques, Chemical Engineering Research and Design, 91(12), 2721–2731.

Singhvi A, Madhavan K P, Shenoy U V. (2004). Pinch Analysis for Aggregate Production Planning in Supply Chains, Computers & Chemical Engineering, 28(6–7), 993–999.

Tan R R, Foo D C. (2007). Pinch Analysis Approach to Carbon-constrained Energy Sector Planning, Energy, 32(8), 1422–1429.

Walmsley M R, Walmsley T G, Atkins M J, Kamp P J, Neale J R, Chand A. (2015). Carbon Emissions Pinch Analysis for Emissions Reductions in the New Zealand Transport Sector through to 2050, Energy, 92, 569–576.

Wang Y P, Smith R. (1994). Wastewater Minimisation, Chemical Engineering Science, 49(7), 981–1006.

12 Conclusions and sources of further information

Following the extensive development of the field of Process Integration (PI), the current Third Edition of the book has added novel developments to the initial scope. This chapter summarises the provided knowledge. It has been divided into thematic parts dealing with Heat Integration, Total Site on the heat side, and Mass and Water Integration on the mass and waterside, followed by some generic developments.

Heat Integration was developed earlier and provided the base for the methodology that has been step-by-step extended in other directions such as Mass and Water Integration. However, there have been some further developments such as Hydrogen Network Design and Management, Integrated Waste Management design, Oxygen Pinch, Targeting for carbon emissions and footprints, and Property Pinch. The chapter concludes with an overview of the most prominent modelling tools suitable for applying Process Integration.

12.1 HEN targeting, synthesis, and retrofit

A critical review with an annotated bibliography for the literature on HENS up to the year 2000 was published by Furman and Sahinidis (2002), while Anantharaman (2011) presented an HENS bibliography for the following period up to the year 2010. A more recent review: Heat transfer enhancement, intensification and optimisation in Heat Exchanger Network (HEN) retrofit and operation, was presented by the team from SPIL, the Brno University of Technology, and Xi'an Jiaotong University (Klemeš et al., 2020).

The HEN targeting and synthesis methodologies were developed and pioneered by the Department of Process Integration, UMIST (now the Centre for Process Integration, CEAS, The University of Manchester) in the late 1980s and 1990s. The main books published are as follows: first, the "red book" (Linnhoff et al., 1982, 1994), followed by a "blue book" Smith et al., 1995), updated later in an extended form (Smith, 2005). Another edition of Linnhoff's user guide was published by Kemp (2007). Applications of Heat Integration in the food industry are presented by Klemeš and Perry (2007) and included in an edited book by Klemeš et al. (2008). Some specialised parts providing more detailed instructions later in the retrofit methodology of Network Pinch can be found in (Asante and Zhu, 1997) and (Kemp, 2007). More recent results are presented in the *Handbook of Process Integration* (PI) (Klemeš (ed.), 2013) and most recently in the second updated revision (Klemeš (ed.), 2022).

The use of heat exchanger enhancements provides a cost-effective alternative to the purchase of new heat exchangers and/or investment in piping in the retrofit design of Heat Recovery systems, as discussed by Wang et al. (2012). Key elements are low investment cost and avoiding fundamental structural modifications to the network. Bulatov (2005) has presented a retrofit optimisation framework for compact heat exchangers.

https://doi.org/10.1515/9783110782981-012

Varbanov and Klemeš (2000) used the concept of heat load paths for the debottlenecking of HENS. A review of the developments in retrofit methodologies for HENS is provided by Smith et al. (2010). The recent developments include Wang et al. (2022a) dealing with the pressure drop influence the retrofit and a review by Klemeš et al. (2020).

HEN synthesis and retrofit still develop further, featuring more sophisticated methods. A few typical examples can be given:

– Isafiade et al. (2015) presented a method for HENS problems simultaneously considering multi-period and multi-utility aspects.
– Yong et al. (2015) developed a retrofit procedure based on the Shifted Retrofit Thermodynamic Diagram, an extension of the Grid Diagram, which shows both temperature and heat-capacity flowrates of the process streams. This allows for easily identifying and screening options for improving the HEN heat recovery.
– Heat transfer enhancement can be used in the context of HEN retrofit (Akpomiemie and Smith, 2016). This paper presents a method for cost-effective retrofit based on the area ratio approach.
– A more extended approach to HEN retrofit using heat transfer intensification has been published by Pan et al. (2016), who also account for pressure drop constraints and fouling mitigation.
– Pressure recovery within HENs (Onishi et al., 2013) and the integration of compressors and expanders with HENs (Fu and Gundersen, 2016) have also been considered. These are advanced studies, considering work and heat simultaneously and optimising the systems involving a combination of HEN and various pressure-altering equipment.
– Optimal HEN synthesis with operability and safety considerations (Hafizan et al., 2016)
– Nemet et al. (2018) published a two-step MILP/MINLP approach for the synthesis of large-scale HENs.
– HEN synthesis considering prohibited and restricted matches developed by Wang et al. (2021a).
– HEN synthesis considering detailed thermal-hydraulic performance: Methods and perspectives are published by Li et al. (2022).

12.2 Total Site Integration

Although Total Site Analysis and targeting have been used extensively for a number of years, there remains considerable academic and industrial interest in the area. Consequently, new journal and conference papers continue to be published in a number of research journals, although books are considerably less numerous.

Total Site methodology was first introduced by Dhole and Linnhoff (1993). Their work has been further developed by Raissi (1994) and extended by Klemeš et al. (1997).

Smith (2005), in Chapter 23 of his book, devoted considerable space to the development of Total Sites and Total Site targeting including Total Site Profiles, Total Site Composites, and Site Utility Grand Composite Curve. He particularly pays attention to the use of the Site Utility Grand Composite Curves in relation to the cogeneration shaft work and the methods used to calculate the cogeneration shaft work. Kemp (2007) also refers to Total Site Analysis in his chapter on Utilities, Heat, and Power Systems. He briefly examines the construction of the Site Source and Site Sink Profiles.

A number of authors have extended the Total Site targeting methodology. Perry et al. (2008) made use of Total Site targeting using processes with different ΔT_{min} values to represent heat sources and sinks for the integration of small-scale chemical processes with a hospital complex and domestic buildings and office complexes. The work also examined the possibility of integrating renewable energy (RE) sources to reduce carbon-based emissions; for more details, see Chapter 4. All previous Total Site research up to 2010 was overviewed in the recent book of Klemeš et al. (2010).

Matsuda et al. (2009) applied Total Site Analysis to the Kashima industrial area in Japan, which consists of 31 sites. They extracted data from the heaters and coolers from the Total Site in order to produce the Site Source and Site Sink Profiles.

Hackl et al. (2011) also applied Total Site Analysis to a cluster of five chemical companies producing a variety of products with variable supplies and demands for heating and cooling. They used the Total Site Profiles and Total Site Composite Curves to identify potential Heat Recovery and the identification for the need for a site-wide hot water circuit.

On the subject of cogeneration modelling, the recommended sources include the articles by Mavromatis and Kokossis (1998a,b), who proposed the steam turbine-based model for steam network optimisation and cogeneration targeting; Shang and Kokossis (2004), who on that basis developed a comprehensive model for steam level optimisation, also building upon Total Site (Klemeš et al., 1997) for steam level integration; Varbanov et al. (2004a) on the further comprehensive model development and validation; also Varbanov et al. (2004b) for the improved targeting using the power-to-heat ratio and employing utility system optimisation to target processes for potentially beneficial process-level HEN retrofit.

Wang et al. (2021b) developed an implementation of renewable energy sources – Total Site Heat Integration benefiting from geothermal energy for heating and cooling implementations. Fan et al. (2021) developed an urban and industrial symbiosis for the circular economy, suggesting a novel development of Total EcoSite Integration to facilitate resource management. This is an extension of the Total Sites into a new field, which deserves more extensive research. It aims for symbiosis integration between different sectors considering the interaction between different resources. The surplus and deficit in Heat Integration and Water Integration could combine with Solid Waste Integration for a coherent system. For example, the energy recovered from solid waste could be the hot utility of Heat Integration, and heat surplus could be used for solid waste drying etc.

as well as integrating with water surplus or deficit. In line with this motivation, a new Chapter 12 was introduced to this book addressing Solid Waste Integration.

12.3 Total Site Methodology addressing variable energy supply and demand

In order to allow the reduction of the carbon emission footprint from energy users, Perry et al. (2008) conceptually extended the TSHI by integrating greater community servicing within residential areas, with service and business centres as additional heat sinks and renewable energy as heat sources. The inherent variability in the heat supply and demand increases the difficulties in handling and controlling the system. Varbanov and Klemeš (2011) introduced the Time Slices into the Total Site description, with a heat storage system for accommodating the variations. In that paper, the Total Site Heat Cascade was introduced for visualising the heat flows across processes, the steam system, and the heat storage system.

Liew et al. (2012) introduced a numerical solution for the TSHI system to address variable availabilities. This work presented a Total Site Heat Storage Cascade (TS-HSC) for addressing the heat storage facilities required by the TS system.

Nemet et al. (2012) discussed the approaches needed to maximise the usage of renewable energy sources with a fluctuating supply. They introduced a framework for integrating the production processes with solar energy, allowing a user to determine the amount of potential solar thermal energy that could be used within a process.

Varbanov et al. (2012) revisited the global minimum allowed temperature difference (ΔT_{min}) used in the previous method. This work suggested that the values for ΔT_{min} should be specified individually for each site process. They demonstrated that the assumption of a global ΔT_{min} for the entire Total Site could be oversimplified, leading to potentially inadequate Total Site Heat Recovery targets. The modified targeting procedure allows for obtaining more realistic Total Site Heat Recovery targets.

Liew et al. (2012) recently introduced a numerical method known as the Total Site Problem Table Algorithm (TS-PTA) for targeting the Total Site Utility Requirement, intended to allow automation of obtaining the TS targets. This method is easier to apply and provides faster and more accurate results compared to the graphical approach, which has a tendency to include a graphical error during curve shifting. The Total Site Utility Distribution (TSUD) Table was also introduced to visualise the heat flows between the processes and the utility system. A numerical tool, the Total Site Sensitivity Table (TSST), for exploring site sensitivity was also proposed in the same paper. There are several mathematical models for the plant utility system planning process that incorporates the TSHI theory.

The recent development in this area includes the Total Site Trigeneration targeting and design method by Jamaluddin et al. (2022). Besides trigeneration, the method can handle utility availability variations by designing the systems with appropriate utility storage buffers. A method extension and a case study on geothermal energy integration (Wang et al., 2021) have also been presented.

12.4 Utility system optimisation accounting for cogeneration

Mavromatis and Kokossis (1998a) introduced models for a steam system in the TSHI with two main objectives: the selection of the pressure steam levels and the determination of the operating unit configuration for the steam levels. Top-level analysis (Varbanov et al., 2004b) is another mathematical modelling methodology that can be used for these concepts. This method allows for "scoping", i.e. selecting the site processes for targeting HI improvements. The current steam and power demands can be optimised, and the potential benefit of reducing the steam demand can be assessed. A set of curves for the marginal steam prices can be produced for the system under consideration via top-level analysis.

Chen et al. (2011) proposed a systematic optimisation approach for designing a Steam Distribution Network (SDN) of steam systems to obtain improved energy utilisation within the network. In this model, the operating conditions of the SDN were treated as design variables to be optimised. In another development on the TSHI, Bandyopadhyay et al. (2010) proposed a simplified methodology for targeting cogeneration potential based on the Salisbury (1942) approximation. This method is simple and linear using the rigorous energy balance at the steam header.

Kapil et al. (2012) introduced a new model based on an isentropic expansion. These model results were favourable compared with the results from the detailed isentropic design methods and also included an optimisation study, which systematically determined the levels of the steam mains.

The TSST was proposed by Liew et al. (2012). This tool could be used to systematically determine the minimum and maximum boiler and cooling utilities' capacities under different operating conditions. However, this tool makes a major assumption, which treats the hot and cold utilities as the same types of utilities. This is because the steam generated cannot be used to satisfy the cooling requirement at the same steam level. Furthermore, the tool can be further developed for exploring the potential of cascading the excess energy at a high temperature in order to satisfy the LPS requirement during the process operational change scenario.

However, when implementing HI and TS HI in the industrial environment, some important issues should be considered to deal with the model which is close to reality (Klemeš and Varbanov, 2010). Later, Chew et al. (2013) and followed by more extensive work from the similar team (Chew et al., 2015). Among the numerous works on utility system synthesis, analysis, retrofit, and planning have been a multiperiod formulation by Iyer and Grossmann (1998) and complete utility system synthesis by Varbanov et al. (2005) – slightly extended by a more formal description of the environmental impacts over the Life Cycle by Wang et al. (2022b). Luo et al. (2016) treat the utility system and process heat recovery simultaneously. The planning of utility systems has been addressed by Liew et al. (2013), accounting for the variation and uncertainty in the supply and demand for utilities. They considered the potential storage of utilities, variation scenarios, and sizing of the main facilities and backup generators.

A comprehensive book published by De Gruyter (Varbanov et al., 2020) discusses all essential topics of utility systems and how to make them sustainable – starting from the thermodynamic basics, machine design and simple machine integration in power generation, up to the modelling of utility networks, and advanced Total Site Integration.

12.5 Maximum water recovery targeting and design

Water minimisation is a special case of Mass Integration that involves both mass transfer and non-mass transfer operations. There are two popular categories of approaches for water minimisation. These are the insight-based technique (also known as Water Pinch Analysis) and the mathematical optimisation technique. Professor Robin Smith and his coworkers at The University of Manchester, previously known as the University of Manchester Institute of Science and Technology (UMIST), introduced the mass transfer-based WPA approach in the mid-1990s with the first work published in Wang and Smith (1994). The core developments in Water Integration are thoroughly discussed in Chapters 6 to 9. There have been recently advanced developments in targeting and design of Water Networks with multiple contaminants (Chin et al., 2021c) and the inclusion of Water Mains in the targeting (Chin et al., 2021a).

12.5.1 Recommended books for further reading

In his book, *Pollution Prevention through Process Integration: Systematic Design Tools*, El-Halwagi (1997) introduces the Pinch Analysis concept for Mass Integration, which also covers the water minimisation case. The author proposes the use of a targeting technique involving water elimination, segregation, recycling, interception, and sink/source manipulation.

Another good source of information on WPA with industrial applications is the book by Mann and Liu (1999). It covers water network planning, single and multiple contaminant targeting, and network design for wastewater minimisation through water reuse. The book also examines the design of distributed effluent-treatment systems and reviews methods to target water regeneration, reuse, recycling, and process changes. Detailed applications of the methods are demonstrated via a case study on a petrochemical complex.

The book by Smith (2005) covers water network targeting and design, water regeneration, and methods to handle process changes based on the WPA concept. The methods are, however, limited to systems involving single contaminants. It has also received an updated edition recently (Smith, 2016).

Klemeš et al. (2008) provide a comprehensive review of the legislation and techniques to improve the efficiency of water and energy use as well as wastewater treatment in the food industry. The book is divided into six parts. In Part 1, the book reviews the key drivers to improve water and energy management in food processing. In Part 2, methods to assess water and energy consumption and the strategies to efficiently design the water and energy systems are presented. Part 3 covers good housekeeping procedures, measurement, and process control to minimise water and energy consumption. Part 4 reviews the methods to minimise energy consumption in food processing, retail, and waste treatment. Part 5 describes water reuse and wastewater treatment in the food industry. In the last part, water and energy minimisation in several industrial sectors are presented.

Further information on the optimisation of water and energy use in food processing can be found in the book *Waste Management and Coproduct Recovery in Food Processing*, published by Waldron in two separate volumes. In Volume 1, Waldron (2007) reviews the latest developments in the area of waste management and coproduct recovery in food processing in terms of legislative issues and technology for coproduct separation and recovery. In Volume 2, Waldron (2009) discusses the life cycle analysis and closed-loop production system to minimise environmental impacts in food production. It also reviews methods to exploit coproducts such as food and non-food ingredients.

Another overview of Water Integration is provided by Klemeš et al. (2010). The authors explain the use of the material recovery Pinch Diagram with an application to a fruit juice case study and the use of mathematical optimisation with an application to a brewery case study.

In Foo et al. (2012), a compilation of several works on water minimisation techniques covering WPA, mathematical models, batch wastewater minimisation, and adaptive swarm-based simulated annealing is presented in Section 2 of the book.

Foo (2012) has published a textbook on the latest Process Integration methods for water minimisation, gas recovery, and property integration. The book is divided into two parts. The first part of the book covers the insight-based, Pinch analysis graphical, and algebraic targeting technique for direct reuse/recycle, regeneration, process changes, waste treatment, pretreatment, batch, and inter-plant. This is followed by a description of procedures for network design and evolution. The second part of the book is focused on mathematical optimisation.

The reviews from the time when the second edition of this textbook was published include Klemeš et al. (2010) and, more recently, Klemeš et al. (2020).

The most recent PI handbook: *Handbook of Process Integration (PI): Minimisation of Energy and Water Use, Waste and Emissions*, 2nd updated edition, was just published (Klemeš (ed), 2022). It has been an edited handbook with 37 chapters which was contributed by most of the world's leading Process Integration researchers.

12.5.2 State of-the-art review

The first state-of-the-art critical review on water allocation and water treatment solutions that are based on the WPA and mathematical programming was published by Bagajewicz (2000). The roadmap towards zero liquid discharge and energy-integrated solutions is also presented. One of the first successful implementations of the PI method was developed and presented by Thevendiraraj et al. (2003): Water and Wastewater Minimisation on a Citrus Plant. It demonstrated that PI has substantial real-life potential.

Foo (2009) presented an updated literature review only for the WPA approach for up to the year 2009, covering the targeting and network design stage from two angles, the *fixed load* and *fixed flowrate* problem. It also provides a review of the advanced targeting techniques for multiple freshwater sources, the threshold problem, impure freshwater source(s), water regeneration, wastewater treatment, and total water network (TWN). The author also provided some commentaries on future research directions that may involve other areas of resource conservation problems including the multiple impurity problems, simultaneous targeting and design of water networks, simultaneous heat and water recovery, water network retrofit, and interplant Water Integration.

A very detailed review of water network design methods with literature annotations from journals and conferences up to mid-2009 was presented by Jeżowski (2010). The description of each paper is divided into scope, method, and special features. The paper presents reviews on the total water network (TWN), water-using network (WUN), and wastewater treatment network (WWTN) problems for continuous and batch-wise operations, Simultaneous Heat and Water Integration, and uncertain data and interplant integrations. A total of 264 literature annotations up to 2009 were presented by the author.

A review of the WPA, mathematical programming techniques and combined water-energy minimisation for up to the year 2012 was done by Klemeš (2012). A review of the water footprint and life cycle analysis, as well as a short overview of several case studies, was also included.

The WPA approach has seen further developments in 2013. Parand et al. (2013) recently proposed an improved methodology for Water Cascade Analysis and Material Recovery Pinch Diagram for the threshold problem with external utilities. Shenoy and Shenoy (2013) developed the Unified Targeting Algorithm (UTA) and the Nearest-Neighbours Algorithm (NNA) to target and design cooling water networks.

Liu et al. (2013) presented a new insight-based method to design the distributed effluent treatment systems for systems with multiple contaminants. The new method can reduce unreasoning stream-mixing in the design of distributed wastewater treatment networks. Wastewater degradation caused by unreasoning stream-mixing will increase the total treatment flowrate, which will subsequently increase the treatment cost.

Agana et al. (2013) demonstrated the application of an in-series integrated water management strategy for two large Australian manufacturing companies that produce different products and generate different wastewater types. The integrated strategy to identify water conservation opportunities consists of a water audit, Pinch Analysis, and a membrane process application. This strategy can be used as a guide for other manufacturing sites to develop their water management plans.

Wang et al. (2020) presented a study about Water Footprints, and Virtual Water Flows Embodied in the Power Supply Chain, pointing out the importance and consumption of water for power generation.

Several novel research results were developed by Chin et al. (2021b) dealing with the Extension of Pinch Analysis to target and synthesise water recycling networks with multiple contaminants. As these developments are providing substantial novel opportunities, a new Chapter 11 was introduced to this book.

Due to the growing concern about water scarcity as well as water pollution, there has been renewed interest shown by researchers in solving water minimisation problems. Consequently, more works on water minimisation are expected to be developed and published in the near future by researchers across the globe.

12.6 Analysing the designs of isolated energy systems

An isolated energy system serves the energy demand of a location by providing energy near its point of utilisation and can also provide guidance for highly self-sufficient regions. Isolated energy systems, independent of the national grid, are employed for remote electrification. A very good description has been provided by Bandyopadhyay (2013): Besides electrical systems, similar isolated energy systems can also meet the thermal demand. The advantages of such systems are their low fuel cost, non-polluting capacities, modularity, and ease of extensibility. The disadvantages are associated with them mostly being unpredictable and the low energy densities (kW/m^2) of different renewable resources, and the higher capital investments. It is also important to provide an energy storage system for improving the rate of utilisation of renewable energy sources. The performance depends upon the proper sizing of the overall system. Optimisation of the entire system is needed for sizing, which satisfies certain cost and reliability criteria. Pinch Analysis is applied to the design and optimisation of isolated renewable energy systems.

The cumulative energy generated and required may be plotted on time vs energy axis. These curves are equivalent to the Composite Curves of Pinch Analysis. Typical Composite Curves, Supply Composite Curves, and Demand Composite Curves were plotted by Bandyopadhyay (2011). The energy is equivalent to the heat transfer, and the time is equivalent to the temperatures of the Composite Curves of Pinch Analysis and even more to Total Site problems. Energy generated in renewable generators can be supplied to the energy required by the load. Energy generated during or before the demand can be supplied; energy generated after the demand cannot be supplied to the demand. The Energy Supply Composite Curve has to lie above the Energy Demand Composite Curve. Fulfilling this condition, the energy Supply Composite Curve may be shifted upwards until it touches and lies completely above the Demand Composite Curve. Another work in a similar direction has been presented by Ho et al. (2013).

12.7 PI contribution to supply chain development

The introduction of supply chain design or supply chain management aims at minimising the total supply chain cost for meeting a present and known demand (Shapiro, 2001).

However, the conventional approach to supply chain modelling is outdated, mainly due to the lack of consideration of carbon emissions. According to EIA (2005), the transportation and industrial sectors appear to be the sectors with the largest CO_2 emissions. The techniques derived from Pinch Analysis, which provides a clear insight into the PI concept, also bring a significant contribution to the development of a supply chain design (Lam et al., 2011). The supply chain design using PI provides an analysis pathway. The combination of techniques allows for a supply chain design with an offset between cost and environmental impact for a compromised volume of carbon emission.

Lam et al. (2013) provided an overview of conventional supply chain applications, followed by the advantages that PI has brought to the supply chain regarding the aspects of social, economic, environmental impact, and resource utilisation.

A supply chain system engages production facilities, warehouses, distributors, suppliers, retailers, and customers. One of them usually acts as a "bottleneck" or Pinch that imposes limitations on the system's capacity. The PI of the supply network provides an alternative graphical solution of quality vs quantity. The graphical plot through the Pinch methodology allows for the identification of the Pinch Location. The "debottlenecking" process removes the system's restriction, which may lead to an improvement in the supply chain system's performance.

12.8 Hydrogen networks design and management

The further evolution of Pinch Technology extended Mass Integration to hydrogen management systems. Environmental considerations and resource availability constraints, such as low aromatics gasoline and low sulphur diesel, shift to heavier crude oil, and more upgrading, lead to increased demand for hydrogen in refineries and, at the same time, reduce the sources of hydrogen traditionally available to refineries. This can result in a deficit in the hydrogen balance of a refinery and in need for investment in additional hydrogen production facilities or importing it from outside suppliers. If the available hydrogen resources can be used more efficiently, the requirements for additional production capacity or imports can be decreased.

In one of the early works, Alves (1999) proposed a Pinch Approach to targeting the minimum hydrogen utility. This technology was based on an analogy with the process of Heat Recovery. Similar to the distribution of energy resources in a plant which can be analysed and designed by using Pinch Technology, the distribution of hydrogen resources in a refinery has a number of potential sources, each capable of producing a different amount of hydrogen and a number of hydrogen sinks with different requirements. However, the engineer has more flexibility in deciding the hydrogen loads of individual units by varying throughputs of units and operating many processes over a range of conditions. This leads to considerable potential for optimisation of refinery performance. Using Hydrogen Pinch Analysis, the engineer can make the best use of hydrogen resources in order to meet new demands and improve profitability.

A very good example is the work by Liu et al. (2013), which examines the targeting of hydrogen networks with purification and a recent publication from Gai et al. (2022) presenting Total Site Hydrogen Integration with fresh hydrogen of multiple quality levels and waste hydrogen recovery in refineries. A more detailed description of Hydrogen Pinch can be found elsewhere (Klemeš et al., 2010) and recently by Klemeš (ed) (2022).

12.9 Oxygen Pinch Analysis

Another extension of the Pinch concept was Oxygen Pinch Analysis (Zhelev and Ntlhakana, 1999). Their idea of Oxygen Pinch Analysis is to analyse the system in order to come up with targets prior to designing a system with minimum oxygen consumption by the microorganisms for waste degradation. The next step is to design a flowsheet and operating conditions providing the target. In most cases, oxygen is supplied to the microorganisms through agitation. Aeration requires energy, so eventually, the analysis based on Oxygen Pinch principles leads to the original application associated with energy conservation.

The Oxygen Pinch concept focuses on Mass Transfer Pinch analysis but exceeds the classical Pinch targets expectations. Using the chemical oxygen demand as a common ground for a range of organic contaminants, it sets quantitative targets (oxygen solubility, residence time, oxidation energy load) as well as additional qualitative targets, namely the growth rate directly addressing age and health of microorganisms (Zhelev and Bhaw, 2000).

The main idea of the method is the oxidation/aeration energy minimisation required for the biodegradation of organic waste through aerobic digestion. The targeting procedure is based on a composite plot of the chemical oxygen demand (COD) against the dilution rate, which builds on the Monod model of biodegradation (Monod, 1949).

The reason why the researchers (Zhelev and Bhaw, 2000) applied COD (organic load or chemical oxygen demand) vs the dilution rate (D) is to reverse the graph and build the limiting concentration profile from process streams, not from the dissolved oxygen stream, which might cause problems in data collection and analysis. The analysis of information is followed by matching the oxygen supply line to the Composite Curve (touching it at the point of Pinch) and provides the following predesign information:

1. Microorganisms growth rate;
2. Oxygen solubility;
3. Residence time;
4. Oxidation energy load.

This method brings us back to the energy-saving domain in Pinch Technology development, which later was developed in the environmental area: Water Pinch. The Oxygen Pinch approach gets back to energy but is accompanied by extra information concerning environmental issues. An important contribution of this method is its ability to target in parallel with the concentration and the total energy required, a quality characteristic: the microorganisms' health assessed through their reproduction rate. The design guidelines follow the Water Pinch analogy. This method provides an improved tool for the design of cost-efficient biotreatment processes.

12.10 Pressure drop considerations and heat transfer enhancement in Process Integration

Various factors such as flowrate, composition, temperature, and phase can affect heat capacity C_p. Pressure is another very important factor to be taken into account. Back in 1990, Polley et al. (1990) extended the targeting procedure by considering pressure drops. They used the relationship between the pressure drop ΔP, heat transfer coefficient h, and the heat transfer area A:

$$\Delta P = KAh^m. \tag{12.1}$$

Instead of specifying the heat transfer coefficients for streams, the allowable pressure drop was specified for each stream. The heat transfer coefficients for the streams are calculated iteratively to minimise the total area. The target area procedure is modified based on the fixed pressure drops rather than fixed-film coefficients.

Ciric and Floudas (1989) suggested a mathematical programming-based two-stage approach. It included a match selection and an optimisation stage. The match selection stage used an MILP transhipment model to select process stream matches and match assignments. The optimisation stage used NLP formulation to optimise the match order and flow configuration of matches.

Nie and Zhu (1999) developed a strategy for considering the pressure drop for HEN retrofit. They assume that additional area should be concentrated on a small number of units to minimise the piping and civil work. The optimisation procedure consists of two stages. The first stage is the Screening for a small number of units which require additional area, and the second stage is the Consideration of serial or parallel shell arrangements for those units.

The topology change options are first determined by applying the Network Pinch Method (Asante and Zhu, 1997). Then their two-stage optimisation is used for area distribution and shell arrangement with pressure drop constraints. Area distribution and shell arrangement are the two most common options which have a main effect on pressure drop.

Václavek et al. (2003) classified in more detail when pressure plays an important role in Energy Process Integration. Rather than individual process streams available for Heat Recovery, whole combinations of process streams (tracks) are considered. They formulated some heuristic rules.

Aspelund et al. (2007) described a new methodology to account for pressure drops for process synthesis, which extends traditional Pinch Analysis with exergy calculations: Extended Pinch Analysis and Design (ExPAnD). The authors focus on the thermomechanical exergy, which is the sum of pressure and temperature-based exergy.

Compared with traditional Pinch Analysis, the problem that Aspelund et al. (2007) consider (sub-ambient process-based) is much more complex, with a large number of alternatives for the manipulation and integration of streams. Aspelund et al. (2007) also provide a number of heuristics (general and for specific streams) that complement the ExPAnD methodology. In a further development, Aspelund and Gundersen (2007), using the concept of "Attainable Region", suggest a graphical representation of all possible Composite Curves for a pressurised cold stream below ambient to include the cooling effect when the stream is expanded to its target pressure. The Attainable Region is an addition to the ExPAnD methodology, a new tool for Process Synthesis extending Pinch Analysis by explicitly accounting for pressure and including exergy calculations. The methodology shows great potential for minimising total shaft work in sub-ambient processes.

To account for debottlenecking retrofit (which should be distinguished from the energy saving retrofit), Panjeshahi and Tahouni (2008) suggested a pressure drop optimisation method. Its main stages are as follows.
(i) Simulation of the existing process operating at the desired increased throughput. The additional utility is used to restore the required temperatures in the process.
(ii) Area efficiency specification in the existing network after increasing throughput using an area-energy plot. A new virtual area is introduced, which is named the pseudo network.

Zhu et al. (2000) suggested a heat transfer enhancement procedure for HEN retrofit. The methodology has two stages: the targeting stage and selection. The approach developed provides a quick way to determine the application of enhancement in the conceptual design.

The method is based on the results of Network Pinch Analysis. The limitation of the method is that heat transfer enhancement is only used to take the place of additional area.

Smith et al. (2013) provide an introduction to the enhancement methods of commercial shell-and-tube heat exchangers and their engineering application. The key aspects of enhancing shell-and-tube heat exchangers in Heat Exchanger Network (HEN) retrofit are presented and illustrated with examples involving both energy saving and fouling considerations. This analysis has been based on a series of works by Pan. Pan et al. (2011a) studied the improvement of energy recovery in HENs with intensified tube-side heat transfer. Pan et al. (2011b) introduced an optimisation method for retrofitting HENs with intensified heat transfer. Pan et al. (2012) presented an MILP-based (Mixed Integer Linear Programming) iterative method for the retrofit of HENs with intensified heat transfer, and later Pan et al. (2013) developed an optimisation procedure for the retrofit of large-scale HENs with intensified heat transfer.

Heat transfer can be intensified in shell-and-tube heat exchangers for improved heat transfer performance. Such intensification has been widely studied from the point of view of individual heat exchangers. The use of heat transfer enhancement in Process Integration has many benefits. Firstly, enhanced heat exchangers require less heat transfer area for a given heat duty because of higher heat transfer coefficients. Secondly, the heat transfer capacity for the given heat exchanger can be increased without changing the physical size of the exchanger. Thirdly, the use of enhancement can reduce pumping requirements in some cases, as enhanced heat exchangers can achieve higher overall heat transfer coefficients with lower velocities, which may lead to lower friction losses. Using enhancement techniques has practical advantages in HEN retrofit, it can avoid modification of the network structure. The implementation of enhancement is a relatively simple task that can be easily achieved within a normal maintenance period, allowing production losses to be kept to a minimum level and no civil engineering required.

12.11 Power (electricity) and Hybrid Pinch

Wan Alwi et al. (2013) stated that a Hybrid Power System (HPS) generates electrical power by a combination of several renewable energy and fossil fuel generators. The Power Pinch Analysis concept has been recently implemented for the optimal design of an HPS. This work introduces a new graphical tool known as the Outsourced and Storage Electricity Curves (OSEC) to visualise the required minimum outsourced electricity and the current storage capacity at each time interval during HPS startup and continuous operation. Heuristics for load shifting for the integrated HPS system that can lead to further reductions of the maximum storage capacity and the Maximum Power Demand (MPD) have also been introduced in this work. The routine load-shifting strategies in energy management cannot be used without knowledge of how the Integrated HPS components interact with one another. Application of that new approach to case studies demonstrated that the OSEC could provide vital insights for designers to perform the correct load shifting. The results show that up to 50% reduction in the maximum storage capacity and the MPD can be achieved.

Further development of Hybrid Power Systems (HPS) by Mohammad Rozali et al. (2015) consists of different renewable generators, which produce electricity from renewable energy (RE) sources required by the load. An optimal sizing method is a key factor in achieving the technical and economic feasibility of the HPS. The power Pinch Analysis (PoPA) method has been applied to set the guidelines for proper HP's sizing. Different scenarios for RE generators allow the designers to choose the best alternative for their systems.

Those new enhancements of Process Integration have been receiving considerable attention.

12.12 Computational and modelling tools suitable for applying PI

Process Integration, modelling, and intensification problems are complex in terms of scale and relationships. Solving is very much aided by relevant computer software. There are a large number of efficient tools available, each with its particular advantages. Here the main tools used in the research and applied calculations are discussed. In Klemeš et al. (2010), "Sustainability in the Process Industry – Integration and Optimisation", a comprehensive list of software tools is provided. However, from these, only some are directly suitable for applying PI.

12.12.1 Heat and power PI applications

There are dedicated tools for Heat Integration. The earlier (legacy) tool is named SPRINT. This is a software package for designing Heat Recovery systems for individual processes (SPRINT, 2012). This software provides energy targets and optimises the choice of utilities for an individual process; it also performs Heat Exchanger Network (HEN) design automatically for the selected utilities. Both new design and retrofit are carried out automatically, though the designer maintains control over network complexity. SPRINT can also be used for HEN operational optimisation. The successor of SPRINT is HEAT-int (2013). This is a product of Process Integration Ltd. The program is used to improve the energy performance of individual processes on a site. It is the next-generation development of the SPRINT software (by the same team of developers) to a commercial standard and offers more user-friendly interface features. HEAT-int has been more recently superseded by i-Heat (PIL Software, 2017), which has an overall software update and new features.

Another very reputable software package for Process Integration modelling is SuperTarget (KBC, 2013). SuperTarget is mainly used to improve Heat Integration in new design and retrofit projects by reducing operating costs and optimally targeting capital investment. It is also a tool for day-to-day application by novices or occasional users, and it makes Pinch Analysis a routine part of process design.

STAR is a software package for the optimisation of site utility and cogeneration systems (STAR, 2012). It analyses the interactions between site processes and the utility system, locally fired heaters, and cooling systems. The analysis can be used to reduce energy costs and to plan infrastructure adjustments. STAR can also be used to investigate flue gas emissions, which must often be reduced to meet tighter environmental regulations. The STAR package incorporates several tools, as described next. One of the advanced procedures possible with STAR is top-level analysis, which is used for identifying potential processes for Heat Integration retrofit based on limited information from the utility system only, saving investment and time for industrial companies.

Like HEAT-int, SITE-int (2013) is a product of Process Integration Ltd. It is the successor of STAR. SITE-int is a state-of-the-art software package for the design, optimisation, and integration of site utility systems in process industries. Its main features are similar to those of STAR, but updated and commercial support is included. Both have been superseded by i-Steam (PIL Software, 2017).

One particularly interesting tool is Aspen Energy Analyzer (Aspentech, 2013). It is intended to provide an environment for optimal Heat Exchanger Network design and retrofit. In this sense, it is a direct competitor to HEAT-int and SPRINT. It also provides integration and interaction with other Aspentech tools such as Aspen Plus and Aspen HYSYS.

It is also possible to perform the targeting parts and other calculations using spreadsheets. An example is the spreadsheet tool developed as an addendum to the PI book by Kemp (2007).

12.12.2 Water Pinch software

There is much commercial software available that can optimise water networks in the industry either based on the WPA or mathematical modelling. The software tools can help users with limited knowledge of the technology to perform water minimisation in their facilities. There follows a brief review of available software.

WaterTarget (2013) by KBC Energy Services consists of two stand-alone programs, i.e. WaterTracker and WaterPinch. WaterTracker can be used to develop the site water balances and reduce time in data collection. It helps the user to identify where to measure next, how to resolve data conflicts and can reconcile data points. Water-Pinch allows data to be imported from WaterTracker. Flowsheets can be developed, and constraints for water reuse can be specified. Distances, pump, and treatment unit capacity can also be included in the software. The software also provides operating cost (water purchase, treatment, pumping, and discharge) analysis, optimal network connection analysis, and sensitivity analysis.

ASPEN Water (2013) has been developed by Aspen Technology Inc., USA. A user can develop a water flowsheet with the respective unit operations by using this software. Software tools can be linked to ASPEN Properties to model the water chemistry of the plant. It can also perform data reconciliation to provide estimates of flow and compositions. This software uses optimisation techniques to minimise water usage and minimise water network costs. The software tool allows users to determine the water cost, connection cost, fixed and variable regeneration cost, and discharge cost in its model. ASPEN

Water can also perform contaminant sensitivity analysis, process modifications, regeneration, wastewater treatment selection, and optimisation.

Water (2013) has been developed by the Centre for Process Integration, The University of Manchester. It can minimise freshwater consumption and wastewater generation through maximum reuse and regeneration. The software can cater for systems with multiple contaminants and also multiple freshwater sources. It can automatically design water reuse and effluent treatment networks. The software also provides a trade-off analysis between freshwater, effluent treatment and pipework/sewer cost.

Optimal Water (2013) (previously known as Water MATRIX$^{©}$) is the software developed by Process Systems Engineering Centre (PROSPECT), Universiti Teknologi Malaysia. The software can perform minimum water targeting for the single contaminant system. It uses the Water Cascade Analysis method developed by Manan et al. (2004) and the Balanced Composite Curves as well as Water Surplus Diagram method by Hallale (2002). This software can automatically generate water network designs based on the Source and Sink Mapping Diagram (Polley and Polley, 1998). The software is also capable of performing simultaneous water and energy minimisation and minimum regeneration targeting.

OptWatNet (Teles et al., 2007) has been developed by DMS/Instituto Nacional de Engenharia Tecnologia e Inovação, Portugal. It can design water-using networks with multiple water sources and contaminants. The software uses LP/NLP models and solves the water minimisation problem by using GAMS. The user interface is developed by using Microsoft Visual Studio 2005.

Water Design (2013) has been developed by Virginia Polytechnic Institute and State University. It employs the Water Composite Curves method to target the maximum water recovery and can generate the optimum water network design. The software tool adopts the method developed by Wang and Smith (1994) that is limited to single contaminant systems.

12.13 Challenges and recent developments in Pinch-based PI

Bandyopadhyay (2006) presented another extension to the Source Composite Curve for waste reduction and Atkins et al. (2010) presented a Carbon Emissions Pinch Analysis (CEPA) for emissions reduction in the electricity sector. A graphical targeting tool for the planning of carbon capture and storage issues was analysed by Ooi et al. (2012).

Gerber et al. (2013) defined optimal configurations of geothermal systems using process design and PI methods. The problem of soft data in PI was studied and industrially applied by Walmsley et al. (2013), an important issue which deserves extended attention in PI problems and data reconciliation (Manenti et al., 2011).

PI issues not covered in this overview but still important are targeting and design of batch processes; see, e.g. Foo et al. (2012) and cost-effective Heat Exchanger Network design for non-continuous processes by Morrison et al. (2012).

Chin et al. (2021a) developed a novel direction for the Total Site Material Recycling Network Design and Headers Targeting Framework with Minimal Cross-Plant Source Transfer, which seems very promising,

The recent overviews of PI, in most cases related to sustainability, include Klemeš et al. (2010), *Sustainability in the Process Industry – Integration and Optimisation*; El-Halwagi (2012), *Sustainable Design through PI – Fundamentals and Applications to Industrial Pollution Prevention, Resource Conservation, and Profitability Enhancement*; Foo (2012), PI for Resource Conservation; and Foo (2013), *A Generalised Guideline for Process Changes for Resource Conservation Networks*. Recently an edited Handbook on PI was published to which most leading researchers contributed (Klemeš (ed.), 2013).

The other recent overview publications are Klemeš and Kravanja (2013), *Forty Years of Heat Integration: Pinch Analysis (PA) and Mathematical Programming (MP)*; and Klemeš et al. (2013), *Recent Developments in Process Integration*; and also Klemeš and Varbanov (2013), *Process Intensification and Integration: an assessment* (Klemeš and Varbanov, 2013).

12.14 PRES Conferences on Process Integration, Modelling, and Optimisation for Energy Saving and Pollution Reduction

This list can never be fully comprehensive, and very substantial research results are being published virtually every week. The research and new results in all directions of Process Integration continue, and its growth has even been accelerating. This provides very strong evidence that Process Integration methodology, both based on Pinch and complemented by mathematical programming, has still not reached its saturation point, and many extensions are still to be envisaged.

To closely follow the most recent developments, the PRES Conferences (2013) are probably the best option. The 16th Conference on Process Integration, Modelling and Optimisation for Energy Saving and Pollution Reduction – PRES'13, took place in Rhodes, Greece; the 17th Conference PRES'14 (Varbanov et al., 2014) in Prague, the Czech Republic, and the 18th PRES'15 (Varbanov et al., 2015) in Malaysia, Sarawak, Borneo.

The 19th PRES'16 (Varbanov et al., 2016) was in Prague, and the jubilee 20th PRES'17 (Varbanov et al., 2017) was again in Asia in the Chinese city of Tianjin. An overview was also presented an overview of important PI-related research works for the recent period (Klemeš, Varbanov and Lam, 2017). 21st PRES'18 was the last joint venue with CHISA. PRES has become very popular and well-attended, and for co-organising, the number of delegates over 1,000 has become too extensive.

22nd PRES'19 was on the beautiful Greek island of Crete, and everything looked rosy and promising.

However, the worldwide and threatening epidemic of COVID-19 also influenced conferences heavily. Unlike some other conferences, which did not been organised, PRES developed a very efficient hybrid way (Tao et al., 2021) based on a virtual platform. The 23rd was the first hybrid well-organised by Xi'an Jiaotong University in China, and 24th PRES'21 by SPIL, Brno University of Technology, Czech Republic. The COVID-19 pandemic in Europe eased a little, and the Silver jubilee 25th PRES'22 returned to pre-pandemic levels (Fig. 12.1) at another beautiful island, Brać near Split in Croatia (Klemeš et al., 2022). 26th PRES'23 is scheduled in Greece again in the historical city of Thessaloniki.

PRES'22 — Presenters by Countries

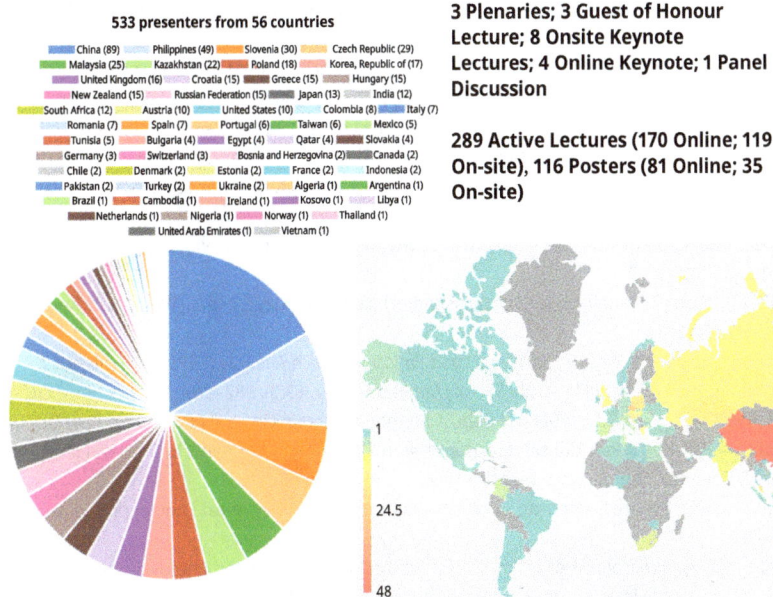

533 presenters from 56 countries

China (89) Philippines (49) Slovenia (30) Czech Republic (29)
Malaysia (25) Kazakhstan (22) Poland (18) Korea, Republic of (17)
United Kingdom (16) Croatia (15) Greece (15) Hungary (15)
New Zealand (15) Russian Federation (15) Japan (13) India (12)
South Africa (12) Austria (10) United States (10) Colombia (8) Italy (7)
Romania (7) Spain (7) Portugal (6) Taiwan (6) Mexico (5)
Tunisia (5) Bulgaria (4) Egypt (4) Qatar (4) Slovakia (4)
Germany (3) Switzerland (3) Bosnia and Herzegovina (2) Canada (2)
Chile (2) Denmark (2) Estonia (2) France (2) Indonesia (2)
Pakistan (2) Turkey (2) Ukraine (2) Algeria (1) Argentina (1)
Brazil (1) Cambodia (1) Ireland (1) Kosovo (1) Libya (1)
Netherlands (1) Nigeria (1) Norway (1) Thailand (1)
United Arab Emirates (1) Vietnam (1)

3 Plenaries; 3 Guest of Honour Lecture; 8 Onsite Keynote Lectures; 4 Online Keynote; 1 Panel Discussion

289 Active Lectures (170 Online; 119 On-site), 116 Posters (81 Online; 35 On-site)

Fig. 12.1: The scope and outreach of the PRES'22 International Conference.

Valuable contributions have been regularly selected from PRES conferences and published in Chemical Engineering Transactions (2013–2022) and also as special issues of the Journal of Cleaner Production, Energy, Renewable and Sustainable Energy Reviews, Applied Thermal Engineering, Cleaner Technologies and Environmental Policy, Theoretical Foundation of Chemical Engineering, and Resources, Conservation and Recycling, Computers and Chemical Engineering, Energies, Thermal Science and Engineering Progress.

The most recent development witnessed Virtual Special Issues in new emerging journals as well as Cleaner Chemical Engineering, Cleaner Energy Systems and Cleaner Engineering and Technology.

References

Agana B A, Reeve D, Orbell J D. (2013). An Approach to Industrial Water Conservation – A Case Study Involving Two Large Manufacturing Companies Based in Australia, Journal of Environmental Management, 114, 445–460.

Akpomiemie M O, Smith R. (2016). Retrofit of Heat Exchanger Networks with Heat Transfer Enhancement Based on an Area Ratio Approach, Applied Energy, 165, 22–35.

Alves J. (1999). Analysis and Design of Refinery Hydrogen Systems, PhD Thesis, UMIST, UK.

Anantharaman R. (2011). Energy efficiency in process plants with emphasis on heat exchanger networks – Optimisation, thermodynamics and insight, PhD Thesis, Norwegian University of Science and Technology, Trondheim, Norway.

Asante N D K, Zhu X X. (1997). An Automated and Interactive Approach for Heat Exchanger Network Retrofit, Chemical Engineering Research and Design, 75(A), 349–360.

Aspelund A, Berstad D O, Gundersen T. (2007). An Extended Pinch Analysis and Design Procedure Utilizing Pressure Based Exergy for Subambient Cooling, Applied Thermal Engineering, 27(16), 2633–2649.

ASPEN Properties (2013). Aspen Technology, Inc. www.aspentech.com/products/aspen-properties.aspx, accessed 06/07/2013.

AspenTech (2013). Aspen Energy Analyzer – AspenTech., www.aspentech.com/products/aspen-hx-net.aspx, accessed 21/08/2017.

ASPEN Water (2013). Aspen Technology, Inc., www.aspentech.com/brochures/Aspen_Water_Brochure.pdf, accessed on 06/07/2013.

Atkins M J, Morrison A S, Walmsley M R. (2010). Carbon Emissions Pinch Analysis (CEPA) for Emissions Reduction in the New Zealand Electricity Sector, Applied Energy, 87(3), 982–987.

Bagajewicz M J, Rivas M, Savelski M J. (2000). A Robust Method to Obtain Optimal and Sub-optimal Design and Retrofit Solutions of Water Utilisation Systems with Multiple Contaminants in Process Plants, Computers & Chemical Engineering, 24, 1461–1466.

Bandyopadhyay S. (2006). Source Composite Curve for Waste Reduction, Chemical Engineering Journal, 125(2), 99–110.

Bandyopadhyay S. (2013). Applications of Pinch Analysis in the Design of Isolated Energy Systems. In Klemeš. (ed.), Process Integration Handbook. Woodhead Publishing, Cambridge, UK.

Bandyopadhyay S. (2011). Design of Renewable Energy Systems Incorporating Uncertainties through Pinch Analysis, Computer Aided Chemical Engineering, Elsevier, 29, 1994–1998.

Bandyopadhyay S, Varghese J, Bansal V. (2010). Targeting for Cogeneration Potential through Total Site Integration, Applied Thermal Engineering, 30(1), 6–14.

Bulatov I. (2005). Retrofit Optimisation Framework for Compact Heat Exchangers, Heat Transfer Engineering, 26, 4–14.

Chemical Engineering Transactions, www.aidic.it/cet, accessed 12/08/2022.

Chen C L, Lin C Y. (2011). Design and Optimization of Steam Distribution Systems for Steam Power Plants, Industrial & Engineering Chemistry Research, 50(13), 8097–8109.

Chew K H, Klemeš J J, Wan Alwi S R, Manan Z A. (2013). Industrial Implementation Issues of Total Site Heat Integration, Applied Thermal Engineering, 61(1), 17–25, doi:10.1016/j.applthermaleng.2013.03.014.

Chew K H, Klemeš J J, Wan Alwi S R, Manan Z A. (2015). Process Modifications to Maximise Energy Savings in Total Site Heat Integration, Applied Thermal Engineering, 78, 731–739.

Chin H H, Jia X, Varbanov P S, Klemeš J J, Liu Z-Y. (2021a). Internal and Total Site Water Network Design with Water Mains Using Pinch-Based and Optimization Approaches, ACS Sustainable Chemistry & Engineering, 9, 6639–6658, doi:10.1021/acssuschemeng.1c00183.

Chin H H, Liew P Y, Varbanov P S, Klemeš J J. (2021b). Extension of Pinch Analysis to Targeting and Synthesis of Water Recycling Networks with Multiple Contaminants, Chemical Engineering Science, 248(Part B), 117223, doi:10.1016/j.ces.2021.117223.

Chin H H, Varbanov P S, Klemeš J J, Alwi S R W. (2021c). Total Site Material Recycling Network Design and Headers Targeting Framework with Minimal Cross-Plant Source Transfer, Computers & Chemical Engineering, 151, 107364, doi:doi: j.compchemeng.2021.107364.

Ciric A R, Floudas C A. (1989). A Retrofit Approach for Heat Exchanger Networks, Computers & Chemical Engineering, 13(6), 703–715.

Dhole V R, Linnhoff B. (1993). Total Site Targets for Fuel, Co-generation, Emissions, and Cooling, Computers & Chemical Engineering, 17, S101–S109.

El-Halwagi M M. (1997). Pollution Prevention through Process Integration: Systematic Design Tools. Academic Press, San Diego, USA.

El-Halwagi M M (2012), Sustainable Design through Process Integration – Fundamentals and Applications to Industrial Pollution Prevention, Resource Conservation, and Profitability Enhancement, Amsterdam, The Netherlands: Elsevier, www.knovel.com/web/portal/browse/display?_EXT_KNOVEL_DISPLAY_bookid$=$5170\&VerticalID$=$0, accessed 06/07/2016.

Fan Y V, Varbanov P S, Klemeš J J, Romanenko S V. (2021). Urban and Industrial Symbiosis for Circular Economy: Total EcoSite Integration, Journal of Environmental Management, 279, 111829.

Foo D C Y. (2009). State-of-the-art Review of Pinch Analysis Techniques for Water Network Synthesis, Industrial & Engineering Chemistry Research, 48(11), 5125–5159.

Foo D C Y. (2012). Process Integration for Resource Conservation. CRC Press, Boca Raton, Florida, USA.

Foo D C Y. (2013). A Generalised Guideline for Process Changes for Resource Conservation Networks, Clean Technologies and Environmental Policy, 15(1), 45–53.

Foo D C Y, El-Halwagi M M, Tan R R. (2012). Advances in Process Systems Engineering. Vol. 3. Recent Advances in Sustainable Process Design and Optimization. World Scientific, Singapore.

Fu C, Gundersen T. (2016). Correct Integration of Compressors and Expanders in above Ambient Heat Exchanger Networks, Energy, 116, 1282–1293.

Furman K C, Sahinidis N V. (2002). A Critical Review and Annotated Bibliography for Heat Exchanger Network Synthesis in the Twentieth Century, Industrial & Engineering Chemistry Research, 41, 2335–2370.

Gai L, Varbanov P S, Fan Y V, Klemeš J J, Nižetić S. (2022). Total Site Hydrogen Integration with Fresh Hydrogen of Multiple Quality and Waste Hydrogen Recovery in Refineries, International Journal of Hydrogen Energy, 47(24), 12159–12178, doi:10.1016/j.ijhydene.2021.06.154.

Gerber L, Fazlollahi S, Maréchal F. (2013). A Systematic Methodology for the Environomic Design and Synthesis of Energy Systems Combining Process Integration, Life Cycle Assessment and Industrial Ecology, Computers & Chemical Engineering, 59, 2–16.

Hackl R, Andersson E, Harvey S. (2011). Targeting for Energy Efficiency and Improved Energy Collaboration between Different Companies Using Total Site Analysis (TSA, Energy, 36, 4609–4615.

Hafizan A M, Wan Alwi S R, Manan Z A, Klemeš J J. (2016). Optimal Heat Exchanger Network Synthesis with Operability and Safety Considerations, Clean Technologies and Environmental Policy, doi:10.1007/s10098-016-1222-z.

HEAT-int (2013). HEAT-int |process Integration Limited, www.processint.com/chemical-industrial-software/heat-int, accessed 02/08/2017.

Ho W S, Hashim H, Lim J S, Klemeš J J. (2013). Combined Design and Load Shifting for Distributed Energy System, Clean Technologies and Environmental Policy, 15(3), 433–444.

Isafiade A, Bogataj M, Fraser D, Kravanja Z. (2015). Optimal Synthesis of Heat Exchanger Networks for Multi-period Operations Involving Single and Multiple Utilities, Chemical Engineering Science, 127, 175–188.

Iyer R R, Grossmann I E. (1998). Synthesis and Operational Planning of Utility Systems for Multiperiod Operation, Computers & Chemical Engineering, 22(7/8), 979–993.

Jamaluddin K, Wan Alwi S R, Abd Manan Z, Hamzah K, Klemeš J J. (2022). Design of Total Site-Integrated TrigenerationSystem Using Trigeneration Cascade Analysis considering Transmission Losses and Sensitivity Analysis, Energy, 252, 123958, doi:10.1016/j.energy.2022.123958.

Ježowski J. (2010). Review of Water Network Design Methods with Literature Annotations, Industrial & Engineering Chemistry Research, 49(10), 4475–4516.

Kapil A, Bulatov I, Smith R, Kim J K. (2012). Site-wide Low-grade Heat Recovery with a New Cogeneration Targeting Method, Chemical Engineering Research and Design, 90(5), 677–689.

KBC (2018). SuperTarget™ – KBC Advanced Technologies Plc, www.kbcat.com/energy-utilities-software /supertarget, accessed 02/08/2013.

Kemp I C. (2007). Pinch Analysis and Process Integration. A User Guide on Process Integration for Efficient Use of Energy. Elsevier, (authors of the first edition: Linnhoff, B., Townsend, D.W., Boland, D., Hewitt, G.F., Thomas, B.E.A., Guy, A.R. and Marsland, R.H, Amsterdam, The Netherlands, (1982, latest edition 1994), A User Guide on Process Integration for the Efficient Use of Energy, Rugby, UK: IChemE).

Klemeš J J. (ed.). (2022). Handbook of Process Integration (PI): Minimisation of Energy and Water Use, Waste and Emissions, 2nd updated. Woodhead/Elsevier, Cambridge, UK, ISBN: 9780128238509.

Klemeš J J, Wang Q W, Varbanov P S, Zeng M, Chin H H, Lal N S, Li N Q, Wang B, Wang X C, Walmsley T G. (2020). Heat Transfer Enhancement, Intensification and Optimisation in Heat Exchanger Network Retrofit and Operation, Renewable and Sustainable, Energy Reviews, 120, 109644.

Klemeš J J, Varbanov P S, Fan Y V, Seferlis P, Wang X C, Wang B. (2022) Silver Jubilee of PRES Conferences: Contributions to Process Integration Towards Sustainability; Chemical Engineering Transactions, doi: 10.3303/CET2294001

Klemeš J J, Varbanov P S, Fan V Y, Lam H L. (2017). Twenty Years of PRES: Past, Present and Future – Process Integration Towards Sustainability, Chemical Engineering Transaction, 61, 1–24, doi:10.3303/ CET1761001.

Klemeš J J, Varbanov P V, Walmsley T G, Jia X X. (2018). New Directions in the Implementation of Pinch Methodology (PM), Renewable and Sustainable, Energy Reviews, 98, 439–468.

Klemeš J J, Varbanov P S, Kravanja Z. (2013). Recent Developments in Process Integration, Chemical Engineering Research and Design, 91(10), 2037–2053.

Klemeš J J, Kravanja Z. (2013). Forty Years of Heat Integration: Pinch Analysis (PA) and Mathematical Programming (MP, Current Opinion in Chemical Engineering, 2(4), 461–474.

Klemeš J J, Varbanov P S. (2013). Process Intensification and Integration: An Assessment, Clean Technologies and Environmental Policy, 15(3), 417–422.

Klemeš J J. (2012). Industrial Water Recycle/reuse, Current Opinion in Chemical Engineering, 1, 238–245.

Klemeš J, Dhole V R, Raissi K, Perry S J, Puigjaner L. (1997). Targeting and Design Methodology for Reduction of Fuels, Power and CO_2 on Total Sites, Applied Thermal Engineering, 7, 993–1003.

Klemeš J, Friedler F, Bulatov I, Varbanov P. (2010). Sustainability in the Process Industry: Integration and Optimisation. McGraw Hill Companies Inc, New York, USA.

Klemeš J, Perry S. (2007). Process Optimisation to Minimise Energy Use in Food Processing. In Waldron K. (ed.), Handbook of Waste Management and Coproduct Recovery in Food Processing, Vol. 1. Woodhead, Cambridge, UK, pp 59–89.

Klemeš J, Smith R, Kim J-K. (eds.). (2008). Handbook of Water and Energy Management in Food Processing. Woodhead, Cambridge, UK.

Lam H L, Klemeš J J, Kravanja Z, Varbanov P. (2011). Software Tools Overview: Process Integration, Modelling and Optimisation for Energy Saving and Pollution Reduction, Asia-Pacific Journal of Chemical Engineering, 6(5), 696–712.

Lam H L, Klemeš J J, Varbanov P S, Kravanja Z. (2013). P-graph Synthesis of Open Structure Biomass Networks, Industrial & Engineering Chemistry Research, 52(1), 172–180.

Li N, Klemeš J J, Sunden B, Wu Z, Wang Q, Zeng M. (2022). Heat Exchanger Network Synthesis considering Detailed Thermal-hydraulic Performance: Methods and Perspectives, Renewable and Sustainable, Energy Reviews, 168, 112810, doi:10.1016/j.rser.2022.112810.

Liew P Y, Wan Alwi S R, Varbanov P S, Manan Z A, Klemeš J J. (2012). A Numerical Technique for Total Site Sensitivity Analysis, Applied Thermal Engineering, 40, 397–408.

Liew P Y, Wan Alwi S R, Varbanov P S, Abdul Manan Z, Klemeš J J. (2013). Centralised Utility System Planning for a Total Site Heat Integration Network, Computers & Chemical Engineering, 57(104–111).

Linnhoff B, Townsend D W, Boland D, Hewitt G F, Thomas B E A, Guy A R, Marsland R H. (1982). A User Guide to Process Integration for the Efficient Use of Energy. IChemE, Rugby, UK, latest 1994.

Liu Z, Shi J, Liu Z. (2013). Design of Distributed Wastewater Treatment Systems with Multiple Contaminants, Chemical Engineering Journal, 228, 381–391.

Liu G, Li H, Feng X, Deng C, Chu K H. (2013). A Conceptual Method for Targeting the Maximum Purification Feed Flow Rate of Hydrogen Network, Chemical Engineering Science, 88, 33–47.

Luo X, Huang X, El-Halwagi M M, Ponce-Ortega J M, Chen Y. (2016). Simultaneous Synthesis of Utility System and Heat Exchanger Network Incorporating Steam Condensate and Boiler Feedwater, Energy, 113, 875–893.

Manenti F, Grottoli M G, Pierucci S. (2011). Online Data Reconciliation with Poor-redundancy Systems, Industrial & Engineering Chemistry Research, 50(24), 14105–14114.

Mann J G, Liu Y A. (1999). Industrial Water Reuse and Wastewater Minimization. McGraw Hill, New York, USA.

Manan Z A, Tan Y L, Foo D C Y. (2004). Targeting the Minimum Water Flow Rate Using Water Cascade Analysis Technique, AIChE Journal, 50(12), 3169–3183.

Matsuda K, Hirochi Y, Tatsumi H, Shire T. (2009). Applying Heat Integration Total Site Based Pinch Technology to a Large Industrial Area in Japan to Further Improve Performance of Highly Efficient Process Plants, Energy, 34(10), 1687–1692.

Mavromatis S P, Kokossis A C. (1998a). Conceptual Optimisation of Utility Networks for Operational Variations – I. Targets and Level Optimisation, Chemical Engineering Science, 53(8), 1585–1608.

Mavromatis S P, Kokossis A C. (1998b). Conceptual Optimisation of Utility Networks for Operational Variations – II. Network Development and Optimisation, Chemical Engineering Science, 53(8), 1609–1630.

Mohammad Rozali N E, Wan Alwi S R, Manan Z A, Klemeš J J, Hassan M Y. (2015). A Process Integration Approach for Design of Hybrid Power Systems with Energy Storage, Clean Technologies and Environmental Policy, 17(7), 2055–2072.

Monod J. (1949). The Growth of Bacterial Cultures, Annual Review of Microbiology, 3(1), 371–394.

Morrison A S, Atkins M J, Walmsley M R W. (2012). Ensuring Cost-effective Heat Exchanger Network Design for Non-continuous Processes, Chemical Engineering Transaction, 29, 295–300.

Nemet A, Klemeš J J, Varbanov P S, Kravanja Z. (2012). Methodology for Maximising the Use of Renewables with Variable Availability, Energy, 44(1), 29–37.

Nemet A, Isafiade A, Klemeš J J, Kravanja Z. (2018). Two-step MILP/MINLP Approach for the Synthesis of Large-scale HENs, Chemical Engineering Science, doi:10.1016/j.ces.2018.06.036.

Nie X R, Zhu X X. (1999). Heat Exchanger Network Retrofit considering Pressure Drop and Heat-transfer Enhancement, AIChE Journal, 45(6), 1239–1254.

Onishi V C, Ravagnani M A S S, Caballero J A. (2013). Simultaneous Synthesis of Heat Exchanger Networks with Pressure Recovery: Optimal Integration between Heat and Work, AIChE Journal, 60(3), 893–908.

Ooi R E H, Foo D C Y, Ng D K S, Tan R R. (2012). Graphical Targeting Tool for the Planning of Carbon Capture and Storage, Chemical Engineering Transaction, 29, 415–420.

Optimal Water. (2013). Process Systems Engineering Centre (PROSPECT), Universiti Teknologi Malaysia, www.cheme.utm.my/prospect, accessed on 7 July 2013.

Pan M, Bulatov I, Smith R, Kim J-K. (2011a). Improving Energy Recovery in Heat Exchanger Network with Intensified Tube-side Heat Transfer, Chemical Engineering Transaction, 25, 375–380, doi:10.3303/CET1125063.

Pan M, Bulatov I, Smith R, Kim J-K. (2011b). Novel Optimisation Method for Retrofitting Heat Exchanger Networks with Intensified Heat Transfer, Computer Aided Chemical Engineering, 29, 1864–1868, doi:10.1016/B978-0-444-54298-4.50151-3.

Pan M, Bulatov I, Smith R, Kim J-K. (2012). Novel MILP-based Iterative Method for the Retrofit of Heat Exchanger Networks with Intensified Heat Transfer, Computers & Chemical Engineering, 42, 263–276, doi:10.1016/j.compchemeng.2012.02.002.

Pan M, Bulatov I, Smith R. (2016). Improving Heat Recovery in Retrofitting Heat Exchanger Networks with Heat Transfer Intensification, Pressure Drop Constraint and Fouling Mitigation, Applied Energy, 161, 611–626.

Pan M, Bulatov I, Smith R, Kim J-K. (2013). Optimisation for the Retrofit of Large Scale Heat Exchanger Networks with Different Intensified Heat Transfer Techniques, Applied Thermal Engineering, 53(2), 373–386.

Panjeshahi M H, Tahouni N. (2008). Pressure Drop Optimisation in Debottlenecking of Heat Exchanger Networks, Energy, 33(6), 942–951.

Parand R, Yao H M, Tadé M O, Pareek V. (2013). Targeting Water Utilities for the Threshold Problem without Waste Discharge, Chemical Engineering Research and Design, doi:10.1016/j.cherd.2013.05.004.

Perry S, Klemeš J, Bulatov I. (2008). Integrating Waste and Renewable Energy to Reduce the Carbon Footprint of Locally Integrated Energy Sectors, Energy, 33(10), 1489–1497.

Polley G T, Panjeh Shahi M H, Jegede F O. (1990). Pressure Drop Considerations in the Retrofit of Heat Exchanger Networks. Transactions of the Institute of Chemical Engineers, 68.

Polley G T, Polley H L. (2000). Design Better Water Networks, Chemical Engineering Progress, 96(2), 47–52.

PRES Conference, www.confrencepres.com, accessed 12 August 2013.

Raissi K. (1994). Total Site Integration, Doctoral dissertation, The University of Manchester, UK.

Salisbury J K. (1942). The Steam-turbine Regenerative Cycle – An Analytical Approach, Transaction of the ASME, 64(4), 231–245.

Shapiro J F. (2001). Modeling and IT Perspectives on Supply Chain Integration, Information Systems Frontiers, 3(4), 455–464.

Shenoy A U, Shenoy U V. (2013). Targeting and Design of CWNs (Cooling Water Networks, Energy, 55, 1033–1043.

SITE-int (2013). HEAT-int |process Integration Limited, www.processint.com/chemical-industrial-software/site-int, accessed 02/08/2013.

PIL Software (2017). Software, www.processint.com/chemical-industrial-software, accessed 22/06/2017.

Smith R. (1995). Chemical process design. McGraw-Hill, New York, United States, ISBN 978-0-07-059220-9.

Smith R. (2005). Chemical Process Design and Integration. John Wiley & Sons, Chichester, UK.

Smith R. (2016). Chemical Process Design and Integration, 2nd ed. John Wiley & Sons, Chichester, UK.

Smith R, Jobson M, Chen L. (2010). Recent Developments in the Retrofit of Heat Exchanger Networks, Applied Thermal Engineering, 30, 2281–2289.

Smith R, Pan M, Bulatov I. (2013). Heat Exchanger Networks with Heat Transfer Enhancement. In Klemeš J J. (ed.), Process Integration Handbook. Woodhead Publishing, Cambridge, UK, pp 966–1035.

Shang Z, Kokossis A. (2004). A Transhipment Model for the Optimisation of Steam Levels of Total Site Utility System for Multiperiod Operation, Computers & Chemical Engineering, 28(9), 1673–1688.

STAR (2022). Process Integration Software (Centre for Process Integration, School of Chemical Engineering and Analytical Science, The University of Manchester, UK), www.ceas.manchester.ac.uk/research/centres/centreforprocessintegration/software/packages/star, accessed 26/10/2009.

SPRINT (2022). Process Integration Software (Centre for Process Integration, School of Chemical Engineering and Analytical Science, The University of Manchester, UK), www.ceas.manchester.ac.uk/research/centres/centreforprocessintegration/software/packages/sprint, accessed 26/10/2009

Tao Y, Steckel D, Klemeš J J, You F. (2021). Trend Towards Virtual and Hybrid Conferences May Be an Effective Climate Change Mitigation Strategy, Nature Communications, 12(1), 7324, doi:10.1038/s41467-021-27251-2.

Teles J P, Castro P M, Novais A Q. (2007). OptWatNet – A Software for the Optimal Design of Water-using Networks with Multi-contaminants, Computer Aided Chemical Engineering, 24, 497–502.

Thevendiraraj S, Klemeš J J, Paz D, Aso G, Cardenas G J. (2003). Water and Wastewater Minimisation on a Citrus Plant, International Journal of Sustainable Resource Management and Environmental Efficiency, 37(3), 227–250.

Václavek V, Novotná A, Dedková J. (2003). Pressure as a Further Parameter of Composite Curves in Energy Process Integration, Applied Thermal Engineering, 23(14), 1785–1795.

Varbanov P S, Doyle S, Smith R. (2004a). Modelling and Optimisation of Utility Systems, Chemical Engineering Research and Design, 82(A5), 561–578.

Varbanov P, Perry S, Makwana Y, Zhu X X, Smith R. (2004b). Top-level Analysis of Site Utility Systems, Chemical Engineering Research Design, 82(A6), 784–795.

Varbanov P, Perry S, Klemeš J, Smith R. (2005). Synthesis of Industrial Utility Systems: Cost-effective De-carbonisation, Applied Thermal Engineering, 25(7), 985–1001.

Varbanov P S, Fodor Z, Klemeš J J. (2012). Total Site Targeting with Process Specific Minimum Temperature Difference (Δt_{min}, Energy, 44(1), 20–28, doi:10.1016/j.energy.2011.12.025.

Varbanov P S, Klemeš J. (2000). Rules for Paths Construction for HENs Debottlenecking, Applied Thermal Engineering, 20, 1409–1420.

Varbanov P, Klemeš J. (2011). Integration and Management of Renewables into Total Sites with Variable Supply and Demand, Computers & Chemical Engineering, 35(9), 1815–1826.

Varbanov P S, Klemeš J J, Liew P Y, Yong J Y. (eds.). (2014). 17th International Conference on Process Integration, Modelling and Optimisation for Energy Saving and Pollution Reduction – PRES 2014, AIDIC/CSChI, CET vol. 39, Milan, Italy/Prague, Czech Republic.

Varbanov P S, Klemeš J J, Wan Alwi S R, Yong J Y, Liu X. (2015). 18th International Conference on Process Integration, Modelling and Optimisation for Energy Saving and Pollution Reduction – PRES 2016, AIDIC/CSChI, CET vol. 39, Milan, Italy/Prague, Czech Republic, First-edition, AIDIC Servizi S.r.l.

Varbanov P S, Liew P-Y, Yong J-Y, Klemeš J J, Lam H L. (2016). 19th International Conference on Process Integration, Modelling and Optimisation for Energy Saving and Pollution Reduction – PRES 2016, AIDIC/CSChI, CET vol. 39, Milan, Italy/Prague, Czech Republic, First-edition, AIDIC Servizi S.r.l.

Varbanov P S, Škorpík J, Pospíšil J, Klemeš J J. (2020). Sustainable Utility Systems: Modelling and Optimisation, Sustainable Utility Systems. De Gruyter, Berlin, Germany, ISBN 978-3-11-063009-1, doi:10.1515/9783110630091.

Varbanov P S, Su R, Lam H L, Liu X, Klemeš J J. (2017) 20th International Conference on Process Integration, Modelling and Optimisation for Energy Saving and Pollution Reduction – PRES 2016, AIDIC/CSChI, CET vol. 61, Milan, Italy, First-edition, AIDIC Servizi S.r.l.

Waldron K. (ed.). (2007). Waste Management and Coproduct Recovery in Food Processing. Woodhead Publishing Limited, Cambridge, UK, pp 90–118.

Waldron K. (ed.). (2009). Waste Management and Coproduct Recovery in Food Processing. Vol. 2. Woodhead Publishing Limited, Cambridge, UK, Vol. 391, 6, pp 134–168.

Walmsley T G, Walmsley M R, Atkins M J, Neale J R. (2013). Improving Energy Recovery in Milk Powder Production through Soft Data Optimisation, Applied Thermal Engineering, 61(1), 80–87.

Wan Alwi S R, Tin O S, Rozali N E M, Manan Z A, Klemeš J J. (2013). New Graphical Tools for Process Changes via Load Shifting for Hybrid Power Systems Based on Power Pinch Analysis, Clean Technologies and Environmental Policy, 15(3), 1–14.

Wang B, Klemeš J J, Varbanov P S, Zeng M, Liang Y. (2021a). Heat Exchanger Network Synthesis considering Prohibited and Restricted Matches, Energy, 225, 120214.

Wang B, Klemeš J J, Varbanov P S, Shahzad K, Kabli M R. (2021b). Total Site Heat Integration Benefiting from Geothermal Energy for Heating and Cooling Implementations, Journal of Environmental Management, 290, 112596, doi:10.1016/j.jenvman.2021.

Wang B, Arsenyeva O, Klemeš J J, Varbanov P S. (2022a). A Novel Pressure Drop Grid Diagram for Energy Saving in Heat Exchanger Network Retrofit, Chemical Engineering Transaction, doi:10.3303/CET2294020.

Wang Q, Han X, Zhao L, Ye Z. (2022b). Sustainable Retrofit of Industrial Utility System Using Life Cycle Assessment and Two-Stage Stochastic Programming, ACS Sustainable Chemistry & Engineering, 10, 13887–13900, doi:10.1021/acssuschemeng.2c05004.

Wang L, Fan Y V, Varbanov P S, Wan Alwi S R, Klemeš J J. (2020). Water Footprints and Virtual Water Flows Embodied in the Power Supply Chain, Water, 12(11), 3006.

Wang Y, Pan M, Bulatov I, Smith R, Kim J-K. (2012). Application of Intensified Heat Transfer for the Retrofit of Heat Exchanger Networks, Applied Energy, 89, 45–59.

Water Design (2013). Virginia Polytechnic Institute and State University Virginia. www.design.che.vt.edu/waterdesign/waterdesign.html, accessed 7/7/2013.

Water (2013). Centre for Process Integration, The University of Manchester, www.ceas.manchester.ac.uk/media/eps/schoolofchemicalengineeringandanalyticalscience/content/researchall/centres/processintegration/WATER.pdf, accessed 7/7/2013.

Wang Y P, Smith R. (1994). Wastewater Minimisation, Chemical Engineering Science, 49(7), 981–1006.

WaterTarget (2013). KBC Energy Services, www.kbcat.com/Products-and-Services/Software/Energy-Optimisation-Software-/WaterTarget/, accessed 7/7/2013.

Yong J Y, Varbanov P S, Klemeš J J. (2015). Heat Exchanger Network Retrofit Supported by Extended Grid Diagram and Heat Path Development, Applied Thermal Engineering, 89, 1033–1045, doi:10.1016/j.applthermaleng.2015.04.025.

Zhelev T K, Ntlhakana J L. (1999). Energy-environment Closed-loop through Oxygen Pinch, Computers & Chemical Engineering, 23, S79–S83.

Zhelev T K, Bhaw N. (2000). Combined Water–oxygen Pinch Analysis for Better Wastewater Treatment Management, Waste Management, 20(8), 665–670.

Zhu X X, Zanfir M, Klemes J. (2000). Heat Transfer Enhancement for Heat Exchanger Network Retrofit, Heat Transfer Engineering, 21(2), 7–18.

Index

https://doi.org/10.1515/9783110782981-013

www.ingramcontent.com/pod-product-compliance
Lightning Source LLC
Chambersburg PA
CBHW080705220326
41598CB00033B/5313

*9 783110 782837 *